GAME THEORY, EXPERIENCE, RATIONALITY
FOUNDATIONS OF SOCIAL SCIENCES, ECONOMICS AND ETHICS
IN HONOR OF JOHN C. HARSANYI

VIENNA CIRCLE INSTITUTE YEARBOOK [1997]

5

VIENNA CIRCLE INSTITUTE YEARBOOK [1997]
5

Institut 'Wiener Kreis'
SOCIETY FOR THE ADVANCEMENT OF THE SCIENTIFIC WORLD CONCEPTION

Editorial Address:
Institut 'Wiener Kreis',
Museumstraße 5/2/19, A-1070 Wien, Austria,
Tel./Fax: +431/5361005 (international) or
01/5261005 (national)
email: i_v_c@ping.at
homepage: http://hhobel.phl.univie.ac.at/wk

The titles published in this series are listed at the end of this volume.

GAME THEORY, EXPERIENCE, RATIONALITY

FOUNDATIONS OF SOCIAL SCIENCES, ECONOMICS AND ETHICS

In Honor of John C. Harsanyi

Edited by

WERNER LEINFELLNER

University of Nebraska and University of Graz

and

ECKEHART KÖHLER

Institute Vienna Circle, Vienna and University of Vienna

KLUWER ACADEMIC PUBLISHERS

DORDRECHT / BOSTON / LONDON

A C.I.P. Catalogue record for this book is available from the Library of Congress.

ISBN 978-90-481-4992-6

Published by Kluwer Academic Publishers,
P.O. Box 17, 3300 AA Dordrecht, The Netherlands.

Sold and distributed in the U.S.A. and Canada
by Kluwer Academic Publishers,
101 Philip Drive, Norwell, MA 02061, U.S.A.

In all other countries, sold and distributed
by Kluwer Academic Publishers,
P.O. Box 322, 3300 AH Dordrecht, The Netherlands.

Gedruckt mit Förderung des Österreichischen Bundesministeriums für
Wissenschaft, Verkehr und Kunst
Printed with financial support of the Austrian Ministry for Science, Transport, Casinos
Austria and Bank Austria

02-0799-200 ts

Printed on acid-free paper

Printed in the Netherlands.

Table of Contents

VI. Ethics and Game Theory

VII. Applications of Game Theory

Report – Documentation

Review Essay

Reviews

Activities of the Institute Vienna Circle

EDITORIAL

About 50 years ago, Oskar Morgenstern's article, "Logistik und die Sozialwissenschaften", was published in the journal *Zeitschrift für Nationalökonomie*. Here he directed attention to the rich potential of the "new logic" (Karl Menger and Kurt Gödel) or "logistics" (Russell and Whitehead) as a pure formal discipline. For economics Morgenstern explicitly endorsed the theory of types (Bertrand Russell), axiomatics (David Hilbert) as well as the use of an exact scientific language, of a so-called "scientific logic" in Rudolf Carnap's sense. He concluded his article by referring to the relevance of these methods for the social sciences as well, including theoretical economics and political economics. To this end, he summed up the main ideas of Karl Menger's book *Morality, Decision and Social Organization. Toward a Logic of Ethics* (1934, transl. 1974).

It should be noted that John von Neumann attended a Vienna Circle congress (Königsberg 1930) and Menger's famous "Mathematical Colloquium" in the twenties and thirties.

It would be a misleading exaggeration to claim that the Vienna Circle was the main source of ideas for today's decision and game theories. It was, however, one of the principal contributors, inter alia, through Menger's work. I would only like to recall the intellectual setting on which the modern theory of science in the natural and social sciences is based. What I am addressing here also has more to do with a general approach, namely, the attempt to formulate solutions in clear language with precise formal methods and by recourse to empirical methods. Still today this approach is relevant in that it provides principles of research. In the final analysis, this approach is also characteristic of Harsanyi's work on economics, and has a bearing on social theory as well as ethics.

Moreover, issues such as 'experience' and 'rationality', 'chance' and 'determinism' and their relation were already very seriously discussed issues for Logical Empiricism.

The evolutionary view of science can be traced from the work of Mach and Boltzmann to today's evolutionary game theory. Since the seventies, the dynamic perspective has spread from biology to the rest of science (e.g., mathematization of Darwinian evolutionary theory). What is at stake here is apparently an all-embracing game of nature *and* society!

I would like to express my thanks to the following participants of our "game" who helped make the symposium "Game Theory, Experience, Rationality", held on June 12 to 15, 1996 in Vienna, possible, which herewith is documented in this *Yearbook*:

- the *Bundesmisterium für Wissenschaft, Verkehr und Kunst* (Federal Ministry for Science, Transportation and Art) for funding this conference

- the *Zentrum für Internationale und Interdisziplinäre Studien der Universität Wien* (Center for international and interdisciplinary Studies of the University of Vienna) (ZiiS)
- the *Österreichische Gesellschaft für Operations Research* (Austrian Society for Operations Research) (ÖGOR)

Thanks also go to both our private sponsors

- the BANK AUSTRIA, the generous host of our conference
- the CASINOS AUSTRIA for helping in financing this event and for organizing a concluding reception at the Casino Baden.

and, last but not least, the reliable team of the Institute Vienna Circle (IVC).

Vienna, August 1997 Friedrich Stadler

WERNER LEINFELLNER

INTRODUCTION

When von Neumann's and Morgenstern's *Theory of Games and Economic Behavior* appeared in 1944, one thought that a complete theory of strategic social behavior had appeared out of nowhere. However, game theory has, to this very day, remained a fast-growing assemblage of models which have gradually been united in a new social theory – a theory that is far from being completed even after recent advances in game theory, as evidenced by the work of the three Nobel Prize winners, John F. Nash, John C. Harsanyi, and Reinhard Selten. Two of them, Harsanyi and Selten, have contributed important articles to the present volume. This book leaves no doubt that the game-theoretical models are on the right track to becoming a respectable new theory, just like the great theories of the twentieth century originated from formerly separate models which merged in the course of decades. For social scientists, the age of great discoveries is not over. The recent advances of today's game theory surpass by far the results of traditional game theory. For example, modern game theory has a new empirical and social foundation, namely, societal experiences; this has changed its methods, its "rationality." Morgenstern (I worked together with him for four years) dreamed of an encompassing theory of social behavior. With the inclusion of the concept of evolution in mathematical form, this dream will become true. Perhaps the new foundation will even lead to a new name, "conflict theory" instead of "game theory."

As the title of this book indicates, and as the contributions to it show in detail, the new role of societal experience, of game dynamics, and of the new cognitive and prescriptive methods clashes with traditional game theory – and this has led to a crisis. The book explains why traditional game theory got into this trouble, and how this crisis has been solved by giving the rationality of practical reasoning a new grounding. This development will radically change our understanding of modern societies and their evolutionary dynamic of conflict solving.

The crisis began with the widening of the gap between many of the abstract, traditional game-theoretical models and their empirical interpretations: Most of the abstract models did not help us to solve our empirical societal conflicts.

The bridging of this gap between experience and traditional idealistic rationality in today's game theory, then, will open the door to a unified new theory. It seemed that game theory was on the best way to this goal until a fight broke loose over which "rationality" would be the best, not only theoretically but also empirically.

The contributions of this book bridge this gap. They greatly enhance our understanding of how decision making and conflict solutions function in our (democratic) societies. Most of the articles critically investigate the usefulness of the new methods when we want to compute optimal solutions of societal conflicts, and they do so from a realistic societal standpoint. They stress the necessi-

W. Leinfellner and E. Köhler (eds.), Game Theory, Experience, Rationality, 1–8.

ty of confirming empirically the theoretically possible optimal solutions in our society. According to Harsanyi, societal conflicts are conflicts between individual interests and the common interests of groups or society as a whole. They include, firstly and contrary to traditional game theory, risky conflicts vis-à-vis randomness in our society and in our environment which affect our solutions of societal conflicts. Secondly, the development of a unified game theory has brought about a search for new, empirically applicable, dynamic methods of mathematical reconstruction, a common rational method of how to solve, predict, and empirically test the theoretically obtained optimal solutions. This search triggered the quest for a common underlying rationality that would guide our attempts to solve serial societal conflicts. But this was not to be: The models of today's game theory do not use a unique and common method but divergent methods and many "rationalities." In 1982, Harsanyi listed seven kinds; later, their number was reduced to two main types: the deductive theoretical one, used by game theorists until the seventies, and a new, practical, inductive, or Bayesian "dynamic" rationality. Traditional game theorists prefer the classical, linear, deductive, or pseudodeductive method, the same as used by classical scientists in the last century. Classical rationality permits the theoretical computation of optimal outcomes (solutions of societal conflicts) of competitive and cooperative societal games; but it failed in many cases where the theoretical solutions were put to empirical tests by being realized. When they were applied to solve empirical societal conflicts, classical game-theoretical models clashed with reality and produced well-known paradoxes: Arrow's, Allais', Tversky's, and the Bergen paradox, and many others. An unbridgeable gap appeared between the theoretical methods, that is, between the rationality of traditional game theory and its empirical interpretation and application.

Therefore, many game theorists nowadays view most of the traditional game-theoretical models with suspicion. And this simply, because most of them are merely mathematical, highly idealized models which provide us with theoretically possible solutions but are very difficult to interpret empirically. An exception are the applications of competitive and cooperative games in the microeconomic theory of the market.

After the eighties, a renaissance in game-theoretical research set in. Firstly: New applications in evolutionary biology stimulated the game-theoretical investigations of serial conflicts; this increased the chances for the unification of all game-theoretical models in one dynamic and evolutionary theory. Secondly: Harsanyi's research changed one tenet of game theory by asking how societal conflicts are solved in our *present* democratic societies. Conflict solving by models and testing their solutions by empirical realizations became a business for all: economists, social scientists, politicians, managers, and humanists alike. Finally: The call for the empirical confirmation of theoretically computed solutions of societal conflicts pushed forward research in this field.

That the 1994 Nobel Prize for economics was awarded to three game theorists, Nash, Harsanyi, and Selten, marked another decisive step; it may be seen as encouraging for the unification of the different models and methods. Nash's formulation of optimal solutions of societal conflicts by means of the equilibrium concept which bears his name proved to be a successful theoretical solution for all noncooperative conflicts not only in today's game theory but also in economics and in evolutionary game theory. But finding the empirically feasible equilibrium solution among the often numerous theoretically possible equilibria could not be achieved only theoretically. This once again shows that only empirical tests can decide which solution really works in a democratic society.

Three decades after game theory had been invented, Harsanyi's empirical breakthrough, expressed clearly in his Nobel Prize address of 1994, showed that modern game theory uses new methods or "rationalities" that are different from older methods or the old "rationality," just like today's scientific methods in general are different from those of the classical, linear sciences. The methodological differences between the "good old" classical, linear, and deductive methods and the new empirical, Bayesian methods mark a break in terms of how we view the rationality of making decisions and solving societal conflicts within the rulelike framework of our democracies. In short: Harsanyi and many other scholars in this volume have initiated a renaissance of induction by giving it a new probabilistic-stochastic formulation. This new mathematical formulation is a rational method in its own right, Popper notwithstanding. The new inductive, probabilistic, stochastic, and nonlinear methods successfully integrate the impact of randomness on solutions of societal conflicts; they permit empirical applications far better than previous methods, as is shown in the articles in this volume. The new inductive methods, very different from Mill's, define a rediscovered form of human rationality used by humans when they solve their complicated societal conflicts. These methods are rational, contrary to what Popper has maintained, since they can be defined mathematically, can be used by everybody, and can be checked empirically.

Here an empirical, mathematically reformulated Bayesian trial-and-error method, often labelled "irrational" in classical literature, can be employed to test which one of the theoretically or mathematically possible Nash-equilibrium solutions works empirically. Harsanyi's pioneering work on a Bayesian foundation of the new game theory resulted in his definition of a new inductive rationality, the rationality of practical reasoning. It is, in addition to Simon's and Selten's "bounded rationality," the first detailed demonstration of how future game-theoretical models can theoretically and practically solve our very complex societal conflicts. Empirists prefer this new "dynamic" rationality, which is "dynamic" since it permits learning. It can be tested by our social, empirical experiences and may be accepted afterwards by democratic consent.

A new question emerges: What role do the different rational methods of conflict solution play? This question opens a pandora box of answers and new questions: Do we really play games in our societies? Or do we try to solve serial empirical conflicts between individual interests and the common interest of our societies? Does it make any sense at all to speak about playing games in the present crisis of our democratic societies? What impact do the multiple democratic rules have on our solutions of societal conflicts? According to Selten, most games of traditional game theory use only abstract, mathematical, and linear models of how we may compute in theory, but not in practice, outcomes or solutions. The next question is: Were most of the traditional game theorists just fascinated by the new mathematical methods and not interested in the practical, empirical, psychological, and social roots of game theory? Like so many abstract concepts, the traditional game-theoretical models run into almost insurmountable difficulties when they are being applied, as the many paradoxes have shown. All this leads to a paradoxical situation: Most of the traditional game-theoretical models rest on deterministic, one-one causal interactions between the individual agents while neglecting the role of statistical causality and multiple causes and the impact of successful or catastrophic random events on our society. This has changed with statistical decision theory, with utility theory under uncertainty and risk, and with the advent of evolutionary game-theoretical models.

It seems that a paradigm change, from "solving games" to "solving societal conflicts," is going to transform traditional game theory into a more practical, social discipline. The big advantage of interpreting games as serial conflicts is, according to Rapaport, Leinfellner, and Schuster, that their models become practical, prescriptive instructions for future use, provided they can be corroborated by their past empirical success and their democratic acceptance. To solve a problem is, according to H. Simon, a theoretical effort; to solve a societal conflict means the successful empirical realization of its solution and the use of social rules. According to Harsanyi's social ethics, empirical societal solutions need the additional, democratically accepted background rules which we impose on the few special, game-theoretical rules. This amounts to embedding the new game theory into our democratic societies, as Harsanyi, Sen, Rawls, Holland, McClelland, and Leinfellner have proposed. Today not only the special game-theoretical rules but also the societal, juridical, cultural, and socio-ethical empirical rules for the realizations of theoretically possible solutions play an important role.

In the last ten years, social scientists and game theorists have begun to doubt whether there exists a unique traditional, deductive, or pseudodeductive rational method in game theory, decision making, and traditional utility theory. Allais' paradox and most of the paradoxes which have haunted and shaken traditional utility and game theory, postmodern constructivism, and even many cognitive sciences, suffer from a reality syndrome. This syndrome reveals the uneasiness, for example, vis-à-vis too many traditional game-theoretical models which look weird and paradoxical when we try to interpret them empirically. But modern game theory is on its best way to solving these paradoxes by following a maxim: It is not the abstract theory alone which counts but also its empirical realization when we apply theoretically possible solutions of societal conflicts.

One lesson of Harsanyi's work is: If the new statistical and stochastic models of today's game theory are in fact reliable cognitive reconstructions of how we mentally solve our societal conflicts, then these models prove that we use, even in everyday solutions of societal conflicts, a rationality of practical reasoning as discussed in this book. When we ponder over optimal solutions of societal conflicts, then we use probabilistic deliberations, lotteries, expected evaluations, and statistical and stochastic cognitive inferences. We did this long before game theory offered us scientific prescriptive models of how to do it. When we solve our societal conflicts, our way of thinking is basically nonlinear, nondeductive, or partially deductive.

This book tries to introduce the reader to the consequences of this ongoing paradigm change and the new concept of rationality, on which certainly not only the new foundation of all social sciences will depend but also our understanding of our practical way of thinking and reasoning when we solve societal conflicts.

The seven chapters of this volume contain a new and fascinating analysis of the paradigm change in the rationality or in the cognitive methods, used today in the new game theory. The first chapter deals with rationality and the foundations of the social sciences; Chapter II with cooperation and rationality; Chapter III with rationality and economics; Chapter IV with Bayesian theory and rationality; Chapter V with evolutionary game theory; Chapter VI with ethics and game theory; and, finally, Chapter VII with the new applications of game theory.

Game theorists, economists, mathematicians, social scientists, social philosophers, statisticians, and biologists, the Nobelist Selten among them, have honored in their contributions the work of John C. Harsanyi, who received the Nobel Prize for welfare economics and game theory in 1994.

For example, Harsanyi's rationality of practical reasoning is based on a new probabilistic version of Bayesian rationality (Harsanyi 1976: 38, 46) which differs from the received deductively oriented rationality. But the latter remains a useful first approach and an overall criterion for the removal of obvious inconsistencies and open contradictions from the theoretical part of our models. Models which are totally free from contradictions exist only in a platonistic armchair philosophy. As Gödel has shown, there is no guarantee that we will not find a hidden contradiction in any richer cognitive model or theory.

Harsanyi's postulates of practical reasoning are intellectually very compelling; no argument has yet been proposed that would cast serious doubt on his new criteria for a new mode of rational decision making. One of the new criteria is: repeated and dynamic learning from new empirical experiences and past realizations of solutions of societal conflicts (Harsanyi 1969: 517, fn. 9).

The problems and consequences of the new rationality of practical reasoning are discussed especially in chapters I-V. These chapters offer us an astonishing insight into the new foundations of the social sciences, one of the main topics of this book. If the decision maker follows the Bayesian rationality when he/she has to solve a societal conflict, then – according to Harsanyi – this will be equivalent to an optimization of his/her and, at the same time, of the common expected utility. Whatever the merits of Bayesian rationality may be, it has revolutionized traditional game theory. It enables us to include new psychological, social, and, according to R. Damasio, even brain physiological, multicausal factors into the calculation of expected solutions.

At first glance, it seems difficult to imagine that Bayesian rational behavior can explain solutions of games where the outcomes seem to depend on so many partial causes. Here we are often caught in the idealistic trap of traditional game theory: that we know everything with certainty, for example, which specific solution our adversary or partner will choose from all possible ones. For an empirist this is something we can't know for sure, namely, what is going on in our partner's mind. This problem, however, has been settled ingeniously in Harsanyi's Nobel Prize lecture. Only Bayesian conditional probabilities in lotteries associated with the possible outcomes can handle incomplete information. By repeating the possible conflict and the possible solutions, in mente and/or in practice, we may learn to change incomplete information into perfect information, to use Harsanyi's term. "Perfect" here means that incomplete information has become, with highest probability, "sufficient" to solve a conflict. Harsanyi's general theory of practical reasoning unifies four, until today separate, game-theoretical models, often called "theories," into a truly grand theory: (i) the nonlinear utility models under risk and uncertainty; (ii) statistical decision models, formerly called "games against nature"; (iii) the new probabilistic-stochastic game theory of noncooperative and cooperative games and evolutionary game theory (Weibull, Leinfellner); (iv) finally, Harsanyi's own theory of solving societal-ethical conflicts between individual, egoistic welfare and common welfare under the constraints of democratically accepted juridical and ethical rules.

Throughout this book, it will be argued that the new probabilistic and stochastic foundations of economic theory, utility theory, decision theory, and especially evolutionary game theory have made important contributions to exploring the new rationality of practical reasoning and its role in solving societal conflicts. But one should not forget that there exist today two different ways of solving societal conflicts. One is based on our natural, "inborn" rationality of practical reasoning which needs no game-theoretical models in order to function. This is the natural way of how we evaluate possible solutions, make inferences in mente, and solve our societal conflicts in practice. We learn from past experiences and use successful methods again and again to solve new societal conflicts or to invent new solutions. The other track begins when we use cognitive, scientific models which we regard as reconstructions of our natural, mental, and practical way of solving societal conflicts. It is similar to the situation where humans used the syllogism to make deductive conclusions, long before the Persian logicians and Aristotle reconstructed them, formalized them, and made them public. Afterwards, this newly formulated deductive system of rules served as a prescriptive pattern of how to make sound deductive conclusions. It has been taught in schools and survives up until today, not unlike a grammar, for the purpose of deducing logically correct conclusions. Maybe the same will happen to the models of the unified game or conflict theory. Actually, many prescriptive game-theoretical models help us already to effectively solve our societal conflicts and to implement the solutions in our societies. We have to come to terms with the mathematically reformulated inductive Bayesian methods, with handling probabilities, evaluating risks, and making inferences. According to Savage, thinking in terms of probabilities and evaluating risk is an inborn ability of humans, used long before they could handle numbers. It seems strange that we cannot predict the choices of single individuals but only solutions for a group, that is, average solutions. It is much easier to reach a conclusion in a syllogism than to optimally solve societal conflicts. But the disadvantage of deductive predictions is that they have to use a perfect, deductive logic which is valid in all possible societies or in all possible worlds. But there is no all-embracing rationality for everybody and for all societies. Even modern physicists have given up the idea that there exists a theory of everything. The advantage of inductive Bayesian methods is that they permit a mathematical, albeit nonlogical, formulation and explanation of everyday conflict solutions, even when we don't know much or when randomness prevails. Likewise, quantum logic deviates from classical predicate logic, which is an offspring of the old syllogistic. Classical probability axioms are not valid in nonlinear utility theory under uncertainty and risk because empirically the independence axiom of classical probability does not hold. Everywhere in modern science, we can find new methods where the empirical applications determine new methods or rationalities. It is the empirical experience in parts of our world and in our society which dictates the use of appropriate new cognitive methods, rules, and that mode of rationality which we need to get reliable scientific knowledge.

To sum it up: By a solution of a societal conflict we do not mean the abstract solutions of the many linear mathematical models of classical game theory which can be redeemed only in fictitious societies. On the contrary: Theoretically optimal solutions of societal conflicts which use the new rationality of practical reasoning must be empirically testable in (democratic) societies.

Chapter IV on Bayesian theory and rationality deals with mathematical details and the applicability of Bayesian probabilistic and stochastic methods; they alone take into account the impact of random events on decision processes and conflict solutions. The empirical examples show drastically the limits and the breakdown of classical rationality in favor of the new rationality of practical reasoning.

The articles of chapter V on evolutionary game theory and game dynamic turn to a central issue of the theory of societal and cultural evolution, an issue that evolutionists have been avoiding for too long: How can the traditional qualitative theory of evolution be quantified? Evolutionary game theory offers a new original method. Serial realizations of consecutive solutions of societal conflicts, even across generations, change our societies evolutionarily. Under the selective pressure of randomness, new solutions of societal conflicts arise. Here our traditional concepts of a one-dimensional, linear, quasi-deductive rationality breaks down completely in favor of the new probabilistic and stochastic, dynamic methods of the rationality of practical reasoning. This evolutionary process permits Bayesian learning in serial conflicts across generations. Presently, evolutionary game theory is a fast-moving new branch of game theory; it explores how we solve serial or consecutive societal conflicts which may change our societal evolution. Evolutionary solutions aim at a relative individual and social stability, at relative or evolutionarily stable equilibria – but, interestingly, not at the maximization of expected utility. Evolutionary game theory takes up ideas from evolutionary biology and Bayesian rationality. It highlights the role of randomness (creative mutations) and the mechanisms of selection and adaptation, albeit not in a Darwinian competitive but in a cooperative fashion. Dynamic stability, Maynard Smith's evolutionarily stable solution, is a special version of Nash's equilibrium solution. An alternative to the matrix method usually used in game theory computes solutions with ordinary differential equations, called "evolution equations."

The contributions in chapter VI on the ethical foundation of the social sciences all come to the conclusion that we have to give up the idea that there exists a strict boundary between economic, social, political, and cultural decisions in democratic welfare societies and in social ethics. According to Harsanyi, the empirically feasible equilibria between egoistic individual interests and common interests of the society as a whole rest on a Bayesian rationality and on learning processes.

Chapter VII, the last chapter, deals with applications of game theory.

The book is a concise synopsis of recent advances, of new problems, and of the new foundation of today's modernized dynamic game theory and its impact on the foundations of the social sciences. The most important task of all social sciences certainly is to be prepared to solve emerging societal conflicts and to realize the optimal solutions to the benefit of the individual and of society, even under risk and uncertainty. This is the real challenge of our time.

REINHARD SELTEN

GAME THEORY, EXPERIENCE, RATIONALITY*

In this paper I want to deal with game theory, experience and rationality, and I will try to put these into historical perspective.

GAME THEORY

What is game theory? I'm often asked this question by journalists – to say in one minute what is game theory. Well, I can say it is something like this: It is mathematical modelling and analysis of purposeful interaction in conflict and cooperation. Maybe twenty years ago I would have said: Interpersonal interaction, but you cannot say this anymore because it might now be animals or plants which interact. This is not necessarily rational interaction as had been originally assumed for game theory, but it may be thought of as purposeful interaction.

Now let us look back at the game theory of von Neumann and Morgenstern [1944] and at the way they did it. They already had all of the basic modelling approaches which we use now. The most important ones were: extensive games, normal form games and coalition games, i.e., games based on a characteristic function. They viewed these models as being derived from each other. Thus the extensive game was the most detailed description, but this was transformed into the normal form game by going to strategies. The normal form game was thought to be sufficient for all game analyses. (Later it turned out that this is not quite so.) Then the normal form game was further simplified to a coalition game, a characteristic function game. So they did not look at them as different kinds of models but models related to each other by reduction.

What was von Neumann's and Morgenstern's picture of rationality? For individual behavior it was maximization of objectively expected utilities. They did not consider subjective probabilities, nothing of the sort was done

What is Game Theory?
It is mathematical modelling and analysis of purposeful interaction in conflict and cooperation
Game Theory according to von Neumann and Morgenstern 1944:
Models Used
Extensive Game
Normal Form Game
Coalition Game (characteristic function)
The Rationality Concepts they had for Individual Behavior:
- Maximization of objectively Expected Utility
- Maximin approach to Strategic Uncertainty

Group Behavior:
Exhaustion of all Cooperative Possibilites (Coase Theorem)

W. Leinfellner and E. Köhler (eds.), Game Theory, Experience, Rationality, 9–34.

at that time. The Neyman-Pearson view on statistics prevailed and people did not like subjective probabilities at all. Therefore they also took the maximin approach to strategic uncertainty, not a Bayesian approach as we do today.

They also had rationality on the group level. Group behavior was viewed as the idea that all cooperative possibilities will be exhausted, all players in the game would exhaust all the cooperative possibilities. The idea was that, if there are any profitable cooperation possibilities at all, they will be taken. This is what in economics is sometimes called the Coase theorem. We only learned later that this is also a very dubious thing. Cooperation needs stability, you cannot cooperate merely because profitable cooperation is possible. It may be a necessary condition for cooperation, but not a sufficient one.

> **Determinate and Indeterminate Games**
> Determinate: the players' maximal security levels add up to the maximally obtainable payoff sum
> Fundamental Theorem: *2-person zero-sum games are determinate*
> **Solution Concepts**
> *Maximin strategies* for determinate games
> No theory of individual behavior for indeterminate games
> *Stable sets* (von Neumann–Morgenstern solutions) are sets of cooperative outcomes with two stability properties (internal and external)
> **Rationality and Experience**
> **Naïve Rationalism:**
> Apart from unimportant deviations, what is rational is real, and what is real is rational

Von Neumann and Morgenstern also did not yet have the distinction between cooperative and non-cooperative games; they had another distinction between determinate and indeterminate games. Determinate games are games in which the players' maximal security levels add up to the maximally obtainable payoff sum. It was the fundamental theorem of game theory that two-person zero-sum games are determinate. Other games are only determinate exceptionally, they are usually indeterminate. And that was the great distinction: because of the Coase theorem, they thought that, if a game is determinate, then it makes sense that people become non-cooperative in some way, but that if it is indeterminate then they must take the opportunity to cooperate. Therefore you then have to go to characteristic function theory in order to solve the game.

What were von Neumann and Morgenstern's solution concepts? Obviously maximin strategies for determinate games. However, there was no theory of individual behavior for indeterminate games. They did not say what individual behavior would be for these cases, they just defined cooperative behavior with respect to the characteristic function as stable sets, also called the von Neumann-Morgenstern solution. These stable sets are sets of cooperative outcomes with two stability properties, namely internal and external stability. I do not want to explain this in detail, but the concept of the von Neumann-Morgenstern solution is a very beautiful concept. It has all the beauty of a truly circular concept: balanced in itself. But it turned out to be very difficult to apply and it does not always exist.

What was their position with respect to the relation between rationality and experience? I would say they were naïve rationalists. What is a naïve rationalist? A naïve rationalist is anybody who thinks that, apart from unimportant deviations (there are always some crazy people), what is rational is real, and what is real is rational, as Hegel observed – of course, in a very different sense. People cannot but behave rationally, and if you have found the rational solution to something you have also described behavior. Behavior in such situations is hence not really an empirical question, you just deduce it from the principle of rationality. This is the position which I call naïve rationalism. It has prevailed in economics for a long time and it is still a very strongly held position in economics-but it is weakening. I think people are getting more defensive and trying to be a little less naïve about rationalism.

Cooperative Theory
What is Rationality in Coalition Games?
Some of the many different answers:
Stable Set, von Neumann and Morgenstern 1944.
Shapley Value, Shapley 1953.
Core, Gillies 1953, Edgeworth 1881.
Bargaining Set, Aumann, and Maschler 1964.
Nucleolus, Schmeidler, 1969.
Stable Demand Vectors, Albers 1975.

Thus von Neumann-Morgenstern. What came in the next two decades? The main interest of game theorists was concentrated on cooperative solution concepts, because as long as you took the point of view of von Neumann-Morgenstern, maximin theory was closed, there was not much to do there any more. Of course there were mathematicians who proved maximin theorems in Banach spaces and so on, but actually maximin was not such a hot issue. Thus a number of cooperative theories for characteristic functions were developed and originally one hoped that game theory would overcome the traditional confusion of oligopoly theory.

But this introduced new confusions, on a higher level, namely the confusions among cooperative solution concepts. I have listed here only the most influential ones. I would say, maybe the most influential one was the stable set. Then I will mention the Shapley value, and also the core, attributed to Gillies (1953). But at this time, it was also pointed out that at least the core in bilateral monopoly had already been considered by Edgeworth under the name of contract curves – this is what Shubik found out. But *within* game theory it was introduced by Gillies. Then came the bargaining set by Aumann and Maschler (1964); the nucleolus, which is a point in the bargaining set in 1969; and the stable demand vectors worked out by Albers; and similar concepts like aspirations by Bennet and so on.

These are the developments that I want to mention. I have left out some important things. For example, I left out what John Harsanyi did in cooperative game theory and will make up for this a little later. I left out what I did in cooperative game theory at this time, and I left out dozens of other theories also. But it was very exciting to think about these things, and it is *still* worth thinking about. At the moment there is not that much interest left in these things, but they are still interesting subjects.

EXPERIENCE I – EXPERIMENTAL GAMES

What also came about was the experimental approach, also connected with these cooperative theories. The first experiment on cooperative game theory was by Kalish, Milnor, Nash and Nering (1954). If you have this bewildering multiplicity of cooperative concepts you may think that people act rationally, but we have not yet found out *what* the rationality of people is, because we are now confused about that. But let's look at them and see what they do.

What Kalish, Milnor, Nash and Nering found was that even rationality is in question. Maschler in 1978 wrote about experiments which he already did much earlier. A list of his results circulated around 1961 already. (This is one of the stories that people do not publish something for a long time because they delay and delay even if it is important.) Then came my own experiments with Schuster in 1968; then I wrote a paper on equal-share analysis in 1972 where I presented an empirical theory, summarizing what I found in the experiments available at this time. At this time I could only find 207 games played. Now there are many more available.

The Experimental Approach
Kalish, Milnor, Nash and Nering 1954.
Maschler 1978 (1961).
Selten and Schuster 1968.
Selten 1972 – *Equal Share Analysis*
Kahan and Rapoport 1984
Selten 1987–*Equal Division Payoff Bounds*
Uhlich 1989 – *Proportional Payoff Bounds*
Emergence of *descriptive theory* aimed at *boundedly rational* behavior observed in laboratory experiments

Kahan and Rapoport also did important experimental work. They primarily supported the bargaining set, but I think it is more important to look at their data. How you *should* look at data in general is an important methodological question which has also a philosophical dimension, but I do not want to go into this here. In this respect I differ from Kahan and Rapoport.

Then I developed my equal-division payoff bounds theory, which fits the data much better than the bargaining set, at least for three-person games. It does not extend beyond three-person games because if you do empirical theories you have to remain close to the data. Most games which have been played are three-person games; not enough four-person and five-person games have been played experimentally. The parameter space is enormous; one is hesitant to generalize beyond three persons. So this is virtually a three-person theory, and it is only for zero-normalized games, for games which have zero payoffs for one-person coalitions. I think everything changes if you have positive payoffs for one-person coalitions. Uhlich then wrote about his proportional payoff bounds and he already had access to a database of 3000 games.

In this way a descriptive theory aimed at boundedly rational behavior observed in laboratory experiments emerged. Theorizing was no longer based on the idea of full rationality, but rather on that of bounded rationality. In order to give you an idea of how this looks, I show you a comparison for a very simple case.

Normative and descriptive theory

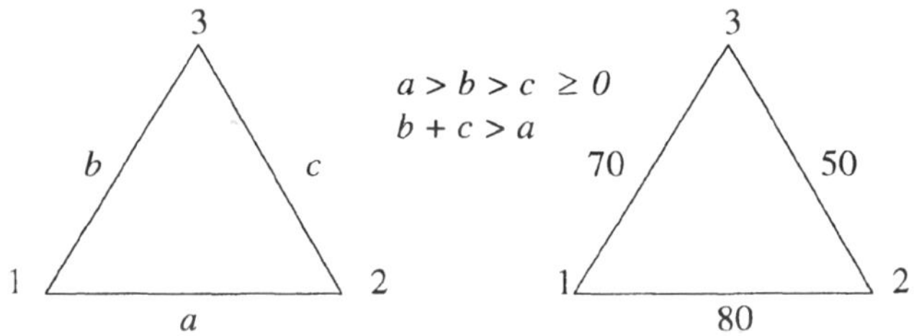

Example: Fully asymmetric 3-person quota games without the grand coalition and with zero payoffs for 1-person coalitions

The simple case which I want to consider is the fully asymmetric three-person quota game without a grand coalition and with zero payoffs for one-person coalitions. Now this game can be seen in the triangle of the accompanying figure. You can look at the three players 1, 2 and 3. Here 1 and 2 together get *a*, 1 and 3 together get *b* and 3 and 2 together get *c*; and *a* is greater than *b*, *b* is greater than *c* – this is just a question of numbering the players. These values are all greater than or equal to zero, and then *b* plus *c* is greater than *a*. This is a kind of triangular equation. An example is given here, where they get 80, 70 and 50, respectively.

For the class of games considered here several important solution concepts reduce to what might be called "quota theory". The quotas are three numbers q_1, q_2, and q_3, defined as payoffs for the players 1, 2, and 3 with the property that the quotas of two players sum up to the value of their coalition. Thus the sum of q_1 and q_2 is a, the sum of q_1 and q_3 is b, and the sum of q_2 and q_3 is c.

If the inequalities of the accompanying figure are satisfied, the quotas are positive numbers. I now want to sketch the intuitive background of the quotas. If we speak of a "quota agreement" we mean a coalition of two players i and j in which both players receive their quotas. Quota agreements have a special stability property. In order to see this suppose that players 1 and 2 consider to form a coalition with payoffs x for player 1 and y for player 2. Assume that x is greater than player 1's quota q_1. Then y is smaller than player 2's quota q_2. This means that a quota agreement with player 3 is more advantageous for player 2 than the coalition with player 1. Player 1, however, would get less than x in a quota agreement with player 3. In this sense player 2 can offer more to player 3 than player 1. If x is smaller than player 1's quota, a similar imbalance prevails: player 1 can offer more to player 3 than player 2. Only quota agreements avoid this kind of imbalance.

Quotas	**Equal division payoff bounds**
$q_1+q_2=a$ $q_1+q_3=b$ $q_2+q_3=c$	$1 > 2 > 3$ (order of strength)
Example: $q_1=50$ $q_2=30$ $q_3=20$	Lower bounds s_1,s_2,s_3 for payoffs in 2-person coalitions Stronger member claims equal share $s_1=a/2, \quad s_2=c/2$
(Circular Definition)	Upper bounds for 1 and 2 in (1,2) $h_1=a-s_2, \quad h_2=a-s_1$ 3's competitive bound $s_3=\max[0,\min(b-h_1,c-h_2)]$ Example: $s_1=40$, $s_2=25$, $s_3=10$ (Algorithmic definition)

With equal-division payoff bounds, the reasoning is completely different. You see here, in order to find the quotas, you must solve a system of simultaneous equations. Our subjects in the laboratory usually solve this simultaneous equation system. That is something they rarely do. Ordinarily they don't go into this circularity. They do everything in steps. The first step that you see them make is that they see that player 1 is stronger than player 2 and player 2 is stronger than player 3. This is easy to see because 1 has better opportunities than 2, and 2 has better opportunities than 3. Then the idea is: if I am the stronger one in the coalition and I am player 1 and I form a coalition with 2, I have to get at least 40, i.e. I have to get at least half of the total payoffs because I am stronger.

Then player 2 will say: well, if I form this coalition with player 3, I am the stronger one; I will have to get at least 25. This gives you already two lower bounds for the payoffs, and then the question is what player 3 should get as his lower bound. He looks at upper bounds for the players 1 and 2. The most 1 can get in the coalition (1,2) is h_1, i.e. $a - s_2$, the most 2 can get is h_2 , i.e. $a - s_1$. Then 3 has to be willing to give both of them that which they can get, at most, in order to compete. In the worst case he has to be willing to do this, hopefully not more. So what he has to be willing to concede, is the minimum of what is left over for him in the coalition (1,3) and what is left over for him in the coalition (2,3). Only in this way he can compete with the most attractive coalition possibility (1,2). In our numerial example the bounds are 40, 25, 10. It has been shown that the equal-division payoff bounds have a much greater predictive success than the point-predictions of the quotas. Even if you take an interval (or a region) around the quota, it is better to have the bounds. The bounds give a much better predictive result.

NON-COOPERATIVE GAMES

Now I turn to the rise of non-cooperative theory. Non-cooperative theory actually began very early on, but it did not take off for a long time. Nash introduced the game-theoretical notion of equilibrium and mixed strategies already in 1951. He then also introduced the distinction between non-cooperative and cooperative theory and he proved an existence theorem for equilibria in finite games. It was, of course, very important that an equilibrium always exists for finite games. He also began to work on what is called the Nash program. The Nash program is the idea that cooperation should be modelled non-cooperatively. You describe the situation within which cooperation occurs, view it as a non-cooperative extensive game, and then you see how cooperation emerges in this extensive game as a non-cooperative equilibrium point.

The Nash-program was not really recognized at first as such. Actually the name "Nash program" is due to Lloyd Shapley. We discussed this question in our game theory workshop of 1965, which was one of the first important meetings on a small scale, the first workshop in a series of work-

The rise of non-cooperative theory
Nash 1951:
The game theoretical notion of *equilibrium*
The distinction between *non-cooperative* and *cooperative* theory
Existence theorem for finite games
The Nash Program
Non-cooperative modelling of cooperation
Nash 1950,1953 - cooperative Nash solution
Aumann 1959 - *supergames*
Harsanyi 1974 - *model generating* modified stable sets
Selten 1981 - *proposal* model
Rubinstein 1982 - *alternating bid* bargaining
Incomplete Information
Harsanyi 1967/68 - Reduction to complete information by distributions over combinations of player types.

shops. We were only seventeen people at this time and we talked about this problem of modelling cooperation non-cooperatively. Shapley said one could call it the Nash-program. Nash initiated this direction of research with his Nash-bargaining theory in the papers of '50 and '53.

Then something very important occurred early on: Aumann's theory about *supergames* where he essentially connected the core with supergames, i.e. with repeated games. That was a very natural and simple way of modelling cooperation non-cooperatively. Later Harsanyi presented a bargaining model for coalition games which generated a somewhat modified version of the stable sets introduced by von Neumann and Morgenstern. I developed a very simple model of coalition bargaining, the proposal model, which has a connection to the stable demand vectors of Albers. I would also like to mention Rubinstein's famous analysis of his non-cooperative alternating-bid two person bargaining model which provides an alternative interpretation of Nash's cooperative theory of two-person bargaining with fixed threats. Many people have contributed to the non-cooperative modelling of cooperation. I cannot even mention them here. The problem remains an important one, since we cannot understand the strategic aspects of cooperation without modelling cooperation non-cooperatively.

A very important innovation in non-cooperative game theory was Harsanyi's concept of incomplete information. In games with incomplete information players do not have exact knowledge about the payoffs of their opponents or maybe not even about their own payoffs. Harsanyi modelled such situations by the introduction of player types. A player may have several types which differ with respect to payoffs and subjective probabilities for opponent type constellations. Harsanyi's path breaking work on games with incomplete information was published as early as '67, '68. He came with an early version of his paper to the '65 conference. The participants were fascinated by Harsanyi's ingenious reduction of incomplete information to complete information. There was a heated discussion about what is now sometimes called the Harsanyi-doctrine: The idea that there is a common joint prior probability distribution over the types of all players behind everybody's beliefs. This means that types of the same player or of different players do not differ with respect to their basic picture of the world but only by some additional factual information. The Harsanyi-doctrine is often assumed in applications, but the theory does not depend on this. One can also treat the case without a common prior.

Now I want to turn to rationality requirements on equilibria. Equilibrium is usually defended as a necessary condition for the rational solution of a game. A non-cooperative solution for a game can be looked upon as a general recommendation which specifies a strategy for every player. If the solution is an equilibrium, then nobody can gain by playing another strategy if he expects that the others follow the recommendation. Suppose that something is proposed as a solution which is not an equilibrium. Then it is profitable for at least one player to deviate from the proposed solution, if she expects that the others stick to it. This means that such a proposal is a self-destroying prophecy! If everybody believed that the recommendation were followed, then it would be better to deviate from it. What a theory of an acceptable solution *cannot be* if it is a *rational* theory is a self-destroying prophecy.

Rationality requirements on equilibria
Equilibrium is a necessary rationality condition
Non-equilibria are self-destroying prophecies
What are sufficient conditions?
Refinement
Counterfactually necessary conditions in unreached parts of the game
Selten 1965, 1975 - *perfect*
Myerson 1978 – *proper*
Kreps & Wilson 1982 – *sequential*
Kalai & Samet 1984 – *persistent*
Kohlberg & Mertens 1986 – *stable*

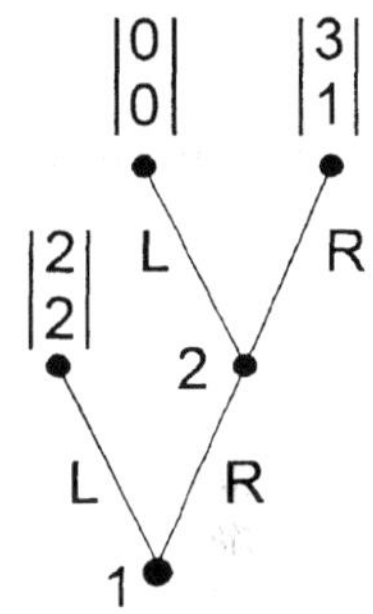

equilibria:
(L,L) and (R,R)

But is this really the only condition? Are there other necessary conditions? What are sufficient conditions for rationality? In fact, there are other necessary conditions and that led to the refinement literature. There are counterfactually necessary conditions in unreached parts of games. Let us now look at the simple example of an extensive 2-person game with perfect information in the figure above. First player 1 has to choose between left (L) and right (R). If he chooses L, the game ends with payoffs of 2 for each of both players. If player 1 selects R, then player 2 has to choose between left (L) and right (R). The game ends after player 2's choice. Both players get zero payoffs if player 2 chooses L. If he chooses R, then player 1 receives 3 and player 2 receives 1.

Each of both players has two pure strategies, L and R. There are four pairs of pure strategies (L,L), (L,R), (R,L), and (R,R); here the first letter stands for player 1's strategy and the second one indicates player 2's strategy. A natural way of looking for a solution of a perfect information game is backward analysis. One begins near the end and works one's way down to the beginning. For our example this means, one first looks at player 2. If he has to make a choice, then R is better for him than L. Therefore player 1 expects player 2 to choose R, if he himself selects R. This means that player 1 expects payoffs 2 for L and

3 for R. Consequently he chooses R. This reasoning leads to the equilibrium (R,R).

(R,R) is the natural solution, but not the only equilibrium of the game. (L,L) is an equilibrium, too. If player 1 thinks that player 2 selects L after R, then he expects payoffs 2 for L and zero for R. Therefore player 1 cannot gain if he alone deviates from (L,L). Player 2 cannot gain either by deviating alone from (L,L). If player 1 selects L then player 2's payoff is 2, regardless of what player 2's choice would be after player 1 selected R.

I discovered this problem in the context of a theoretical investigation of a dynamic oligopoly model with demand inertia (1965). In order to solve it I introduced the idea of *subgame perfectness*. This concept requires equilibrium not only in the game as a whole, but also in all its subgames, regardless of whether they are reached by equilibrium play or not. In our example a subgame begins at the decision point of player 2. The only equilibrium of this subgame requires player 2 to choose R. Therefore the game has only one subgame perfect equilibrium, namely (R,R).

The idea of *subgame perfectness* was later embedded into a more general concept of *perfectness* by myself. This opened up a whole literature on *refinements*. Many game theorists thought that my perfect equilibrium is not perfect enough and each of them found something which imposes even stronger conditions on equilibrium. Among the more important refinement concepts are Myerson's *proper* equilibrium, the *consistent* equilibrium of Kalai and Samet, and the mathematically very sophisticated *stable* equilibrium of Kohlberg and Mertens. I look at perfectness as a necessary rationality condition, not a sufficient one. The additional conditions imposed by these stronger concepts look suspicious to me, because it seems to me that they are not the right ones. I have more sympathy with the concept of *sequential equilibrium* introduced by Kreps and Wilson which imposes slightly weaker conditions than perfectness.

But refinement is not everything. There is, for example, a problem of equilibrium selection in the unanimity game shown in the figure below. Player 1 has two pure strategies T (top) and B (bottom). Player 2 also has two pure strategies L (left) and R (right). Each field shows player 1's payoff for the corresponding strategy pair in the upper left corner and

Equilibrium selection
The search for a unique rational recommendation:
If an equilibrium is played for rational reasons, it must be the *unique rational* solution

(Harsanyi and Selten 1988)

	L	R
T	5 / 5	0 / 0
B	0 / 0	3 / 9

equilibria:
(T, L) and (B,R)

player 2's payoff in the lower right corner. The game has two equilibria in pure strategies, (T,L) and (B,R). (There is also a third mixed equilibrium not important here.) Both equilibria (T,L) and (B,R) satisfy all conditions imposed by any refinement concept. Refinement does not tell us how to choose one of them, regardless of which concept is used.

A rational theory which provides a recommendation how to play the game must select a unique equilibrium. If an equilibrium is played for rational reasons it must be the unique rational solution of the game. What should the players choose on the basis of rationality (T,L) or (B,R)? The answer is very clear. A very convincing axiomatic system shows that it should be (B,R). More about this can be found in the book by Harsanyi and myself. I agree that in this book there are many things which are debatable. But for some games the rational non-cooperative solution is not debatable, e.g. for the unanimity game discussed above. But for lack of time I cannot tell you why.

EXPERIENCE II – EVOLUTIONARY GAMES

In the '70s something happened which surprised many game theorists, and this was the emergence of biological game theory. This was very strange because we thought that game theory is connected to rationality, i.e. to rational behavior of individual agents. Now certainly animals are not rational in the sense that they make decisions based on careful deliberation and sophisticated computation. Therefore it was a surprise that 1973 a path-breaking article by Maynard Smith and Price initiated a biological literature which proposed game theoretic explanations of behavioral and morphological features of animals and plants as well. That is what this is all about. Not just animals, also plants – why plants have big flowers, or small flowers etc. – questions like these gave rise to game theoretical models. The idea is that natural selection tends to maximize fitness, where fitness roughly speaking amounts to the expected number of offspring of members of a species in social interaction among themselves. This leads to game equilibrium, to non-cooperative game equilibrium and Nash equilibrium. This was first introduced by Maynard Smith and Price, and they introduced the idea of an evolutionarily stable strategy, which is a symmetric equilibrium strategy with certain additional stability requirements. They typically treated only symmetric games.

The additional requirement says: every alternative best reply does worse against itself than the equilibrium strategy does. So if you match a best reply to the equilibrium strategy, which is different from the equilibrium strategy *against itself*, then the equilibrium strategy matched against the alternative best reply is more successful. This prevents invasion of mutants which have alternative best replies. Now a problem arose about the genetic foundation of all this. It works alright if you have a non-sexually reproducing population. It works also alright if you have only one locus, i.e. if you have phenotypes which are determined by one gene locus only. But what about multi-locus systems? There the problem arises already without any game interactions. Moran (1964) showed examples of decreasing mean fitness in a standard two-locus model. Here you have a completely standard two-locus model in which fitness is decreasing instead of increasing. There is no selective pressure towards fitness maximization. The reason for this is that the crossing-over mechanism prevents the transfer of fitness-enhancing features to offspring.

> **Biological game theory**
> Game theoretic explanation of animals and plants
> Natural selection tends to maximize fitness.
> **Fitness:** Expected number of offspring; in social interaction between members of the same species, this leads to game equilibrium
> Maynard Smith and Price 1973
> Overview: Hammerstein and Selten 1994
> **Evolutionarily stable strategy**
> Symmetric equilibrium strategy with an additional stability property: *Every alternative best reply does worse against itself than the equilibrium strategy does*

It could be that there are some genotypes which are more fit than others (have more offspring), but that the genotypic distribution of offspring of these fitter types is nevertheless the same as that of not-so-fit types in the kind of dynamic equilibrium reached. This caused great disappointment with Darwinian theory among some mathematical population geneticists. I think most biologists did not really recognize what a crisis Darwinism had entered. Karlin and his school of population geneticists in Stanford took the problem seriously. However the highly mathematical work of the population geneticists was hardly noticed by the empirical workers in the field; they continued to rely on Darwinian natural selection in their explanations of natural phenomena.

> **Population-genetic foundations**
> **Problem:** *for sexually reproducing populations, evolution within a fixed gene pool does not necessarily maximize fitness*
> Moran 1964: Examples of decreasing mean fitness in a standard 2-locus model
> Crossing over prevents the transfer of fitness-enhancing features to offspring; the difficulty arises even without game interaction

Luckily the difficulty raised by the possibility of decreasing fitness in standard two-locus models can be overcome by an ingenious idea of Eshel and Feldman (1984). They introduced a distinction between *short run evolution* by changing gene frequencies within a fixed gene pool and *long run evolution* by the emergence of succesful mutants. Short run evolution may fail to increase fitness, but in a long run evolutionary equilibrium fitness is maximized.

When Hammerstein and myself wrote our survey paper on game theory and evolutionary biology (1994) we found it difficult just to report what Eshel and Feldman had done and therefore tried to improve a little on their work. (We partially failed to do this since there is a gap in our proof of theorem 2. This has been pointed out by Franz Weissing. However, the proof of theorem 1 is right. In this talk, I only explain theorem 1.)

Hammerstein sometimes refers to the theory of short run and long run evolution as the "streetcar theory". The reason for this name will soon become clear.

The idea of the streetcar theory is that, for certain periods of time, this short-run evolution is quite rapid. Evolution proceeds gradually until an equilibrium is reached and then, once equilibrium is reached, a mutant may get on and the mutant will upset the equilibrium. We picture this by a part of a map of a city

The Streetcar Theory
Hammerstein and Selten 1994
based on Eshel and Feldman 1984
Short-run evolution
Adaptation of gene frequencies without mutation
Long-run evolution
Invasion of mutants

- Stop – short-run equilibrium
- Last Stop – long-run equilibrium

At a streetcar stop not yet in long-run equilibrium, a mutant gets on and upsets the short-run equilibrium
At the last stop, no mutant can successfully invade

Theorem: *In game situations, a dynamic long-run equilibrium must be a game equilibrium*

where a streetcar goes: It starts at some point, goes around town and then reaches the last stop. Now the stops of the streetcar are points where the process is in short-run equilibrium. Then a mutant gets on and the equilibrium is upset, moving the streetcar to a new stop. Again a mutant may get on and some other genes may also get off, so there are always people getting into the streetcar and getting out of it, and you get to the next stop and so on, until you come to a last stop and the last stop is such that no mutant can sucessfully invade any longer.

What we can prove is that in the standard model, in game situations a dynamic long-run equilibrium must be a game equilibrium, must be a Nash equilibrium. It is not guaranteed that such a process reaches an equilibrium, but if it reaches a last stop it must be a game equilibrium.

RATIONALITY

I now want to turn to the subject of methodological dualism, to some methodological questions. In my view a sharp distinction must be made between descriptive and normative game theory. Biological game theory is very *descriptive*, it really intends to *describe* the behavior and the morphological structure of animals and plants. Nobody would ever accept a paper which is not directed towards a phenomenon known in biology. An empirical phenomenon must exist and be discovered first before you write a paper. I will return to this later.

A sharp distinction has to be made in my view between *descriptive* and *normative* game theory. Both are important; I am not saying that either of them is unimportant. Normative game theory strives to mold a balanced mathematical structure of ideal rationality out of conflicting inherent tendencies of the human mind. This is my view of what normative game theory is about. Vague ideas exist about what rationality should ideally be, but these ideas conflict with each other. There is much discussion about this in game theory. But there also exists in humans a capacity to balance these out and therefore the discussion will – eventually – bring forth a balanced mathematical structure of ideal rationality.

To take an example, in my view Bayesianism is a balanced structure of decision making under uncertainty. It is, of course, not undisputed; there are other ideas about how this should be structured. But at the moment it seems to be the most successful in this normative discus-

Methodological dualism
A sharp distinction must be made between **normative** and **descriptive** game theory
Normative game theory
Normative game theory strives to mold balanced mathematical structure of *ideal rationality* out of conflicting inherent tendencies of the human mind.
The problem is *philosophical*, not empirical – empirical arguments are *irrelevant*.
Descriptive game theory
The aim of *descriptive* game theory is the explanation of observed behavior of man, animals or plants.
The problem is *empirical* – only empirical arguments count.
Need for the distinction
Experimental evidence on human game playing refutes *naïve rationalism*.
Therefore, a descriptive game theory about human players must be developed.

sion. The problem of normative game theory is philosophical, not empirical. Only if you are a naïve rationalist can you think it is empirical. Empirical arguments are really irrelevant in this discussion. It does not matter what people really do, it is what *human thinking strives toward*, not what behavior is going toward. Maybe people find it too difficult, too demanding, to follow complete rationality – just as they also find it too demanding to follow complete morality. But it is an important question. Mankind has to know itself, it strives to understand itself, and normative game theory is involved with this very important activity.

What is *descriptive* game theory? This is something very different. The aim of descriptive theory is the explanation of observed *behavior* of men, animals or plants, and has nothing to do with normative game theory. The problem here is empirical and *only empirical* arguments count, nothing else. The need for this distinction arises because there is experimental evidence from human game players which refutes naïve rationalism. Naïve rationalism could have been right, but it is not, it is refuted by experimental evidence. So we have to make this distinction and we also have to develop a descriptive game theory about human players. This is what I now turn to.

Example: A sells firm to B
Value for A: v — v random, uniform
Value for B: 1.5v — over $0 \leq v \leq 100$
A knows v, B knows distribution only
B names price x – A accepts or rejects
Always feedback information on v to B
A (simulated by computer) sells if $v \leq x$
Conditional averages
for A: ½x, for B: ¾x, loss for B: ¼x
Optimum $x = 0$
Experiments by Ball et al.
20 repetitions, 37 subjects
Average bids stay near the interval $50 \leq x \leq 60$
No tendency towards optimum
Explanation by learning direction theory
Ex-post rationality:
Tendency to move in the direction of what would have been better last time, if behavior is changed
After $x > v$ downward tendency
After $x < v$ upward tendency
Unpublished experiments by Selten, Abbink, and Cox confirm this explanation

An example I will show you now is a very simple decision situation in which we observe a phenomenon called *winner's curse*. The phenomenon often occurs in auctions in which objects are sold whose value is not known to the bidders. The participant who wins the auction by making the highest bid is likely to be one of those who most overestimate the value of the object and therefore incurs a loss by bidding too high. This may go so far that those who succeed to buy in such auctions lose money, not only on single occasions but also in the average. One can observe this in a very simple experimental decision situation which I am going to explain to you now.

A is willing to sell a firm to B. The value of this firm for A is v. The value v is a random variable uniformly distributed over the interval $0 \leq v \leq 100$. Whatever the value of this firm is for A, the potential buyer B is the better manager and for B the value is 1.5 times v or in other words 50% more. A knows v exactly but B knows only the distribution of v.

B has to place a bid x and A accepts in the case $x \geq v$ and rejects the bid otherwise. In the experiments A is simulated by the computer. All subjects are in the role of B and they know that the computer behaves like this.

Subjects often think that they can bid something near to 50 or even more, since the expected value of v is 50 for A and therefore the expected value for B is 75. Actually this reasoning is wrong. B, if he bids x, receives the object only in the case $0 \leq v \leq x$. The conditional expectation of v for this case is x/2. Therefore the conditional expectation of the value for B is 50% higher, i.e. .75x. This means that B has a conditionally expected loss of .25x, since he has to pay x but receives only .75x on the average. It follows that any positive bid x is unprofitable. $x = 0$ is the only optimal bid.

In an experiment by Ball, Bazerman, and Caroll (1991), 37 subjects were exposed to the situation 20 times. After each repetition they received feedback on the value of v, regardless whether the object was sold or not. The average hovered between 50 and 60 or near to this interval for all 20 periods. There was no tendency towards the optimum at $x = 0$.

The fallacious argument based on absolute instead of conditional expectations explains why subjects initially tend to bid in this way, but it does not explain why they fail to learn to come nearer to the optimum. Why don't they learn? They make losses on the average, so they should learn to lower their bids. But they don't learn to.

In an unpublished experiment by Klaus Abbink, Ricarda Cox, and myself we repeated similar decision situations for 100 trials, and the subjects *still* didn't learn the optimum. Actually, there *was* a small lowering. We had an average bid of 38 by trial 10, but by trial 90 the average was still 33. Our subjects were more sophisticated than Ball's; there were many among them who already had heard the course on game theory, and some of them actually computed the optimal solution and played it. But most of them did not. The question arises what is happening here.

There is an explanation which I want to give. I first must say that the behavior of the subjects varies significantly, but one very common type of behavior can be explained by what we call learning-direction theory. This theory is based on a principle of ex-post rationality. One looks at the past in order to see what would have been better than what has been done and then moves in this direction. In normative decision and game theory rationality is ex ante in the sense that expected utility is maximized on the basis of expectations about the future. Ex-post rationality is not concerned with the anticipation of the future but rather with the analysis of the past. Reasoning on the past directly influences behavior.

Learning direction theory does not assert that behavior always is guided by ex-post rationality. It only predicts that, in the case of a change, behavior will have a tendency to move in the direction of what would have been better last time. What does this mean for our example? Suppose that last time the bid x was greater than the value v. Then the object could have been bought for v. It would have been best to bid v. Therefore in the case of a change of x we expect a tendency to lower the bid. Now suppose that the last bid x was smaller than the last value v. Then a profit of v/2 could have been made by bidding v. In this case we would expect a tendency to increase the bid, if it is changed at all.

If a subject bids 50, then values above the bid are equally probable as values below the bid. This means that with the same probability an upward or a downward influence on the next bid will be experienced. The lower the bid is, the higher is the probability of an upward influence and the higher the bid is, the higher is the probability of a downward influence. In this way one can understand why average bids stay in the middle range in the course of many repetitions.

The unpublished experiments by Klaus Abbink, Ricarda Cox and myself confirm learning direction theory as the explanation of the phenomenon that most subjects do not learn to come nearer to optimal bidding, even in 100 trials.

Learning direction theory has been applied successfully to the explanation of observed behavior in about 10 experimental studies. It goes back to a paper by Stoecker and myself (1986). The studies in which it has been applied are in a number of quite different areas, e.g. ultimatum games with incomplete information (Mitzkewitz and Nagel 1993), auctions (Selten and Buchta 1994), 2-person alternating bid bargaining under incomplete information (Kuon 1993), coordination games (Berninghaus and Ehrhart 1996), and centipede games (Nagel and Tang 1997).

Admittedly, learning direction theory makes only weak predictions. It does not claim to be the only influence on behavior. But it seems to be an important influence on behavior, wherever it is applicable.

I would like to emphasize that learning direction theory is not just a theory of learning by trial and error. It is not a theory of reinforcement learning. It does not matter how much the learner got in the last trial. It does matter how much more he *could have gotten* compared with what he did get. It is not so important whether he got much or got little, but his reasoning about what he could have gotten is important, and he will have a tendency to go in the direction indicated by this in the case of a change of behavior.

The subject matter of learning in decision and game situations is very important for descriptive game theory. However, this is not the only research area in this field. Another important question concerns the structure of bounded rationality in strategic situations. I now want to show you an example which suggests that the structure of bounded rationality is very different from that of full rationality. The example is a study by Mitzkewitz, Uhlich and myself (1997) which we colloquially call "measure-for-measure" even if this term does not

Measure-for-measure
Selten, Mitzkevitz, and Uhlich 1997
20-times repeated asymmetric quantity variation duopoly with numerically specified linear costs and demand
3 rounds game playing experience
3 rounds strategy programming with computer tournaments
Motivation by grades – 24 participants
The typical approach to the strategic problem
1. What is my cooperative goal?
Ideal point based on fairness criteria
2. How do I reach cooperation?
Measure-for-measure policy
The typical strategy structure
Initial phase • main phase • final phase
Fixed concessions • ***measure-*** • non-cooperative
up to 4 periods • ***-for-measure*** • up to 4 periods
No prediction – no optimization
Active exertion of influence rather than passive adaptation
Typicity and success

appear in the title of our paper.

The study experimentally investigates behavior in a twenty times repeated asymmetric quantity duopoly with numerically specified linear costs and demand. One player had higher fixed costs and lower marginal costs and the other one had lower fixed costs and higher marginal costs. The experiment was run in the framework of a student seminar with 24 participants.

The students first played the twenty times repeated game against anonymous changing opponents via computer terminals. After three rounds of game playing experience they had to write computer programs specifying strategies for both roles. These computer programs then were matched with each other in a computer tournament. Each program played against all others in both roles. Then every participant got the results of all plays in which his or her program was involved and they could revise their programs. The revised programs were matched in a second computer tournament and the participants again had the opportunity to revise their strategies in the light of the results. Finally a third computer tournament followed.

The students received no monetary payoffs. However, their grades for the seminar mainly depended on their success in the last tournament. They were highly motivated and spent much time and effort on their strategy programs.

The evaluation of the final programs and the documentations the students wrote on them revealed a typical approach to the strategic problem. Already in the three game playing rounds the participants had reached the conclusion that they must cooperate in order to be successful. Actually the notion of subgame perfectness excludes rational cooperation in this game. Most participants were aware of this but nevertheless not impressed by the argument. Therefore the first question they typically ask themselves is the following one: What is my cooperative goal?

The answer to this question is what we call an *ideal point* (or two of them, one for each player role). An ideal point is a pair of supply quantities, one for each player. The ideal points are mainly based on fairness criteria, such as equal profits, or equal surplus over Cournot profits, or profites proportional to Cournot profits.

After the ideal point has been chosen, a second question arises: How do I reach cooperation at my ideal point? Typically a *measure-for-measure policy* is used for this purpose. This is a mode of behavior which can be described as follows: If you increase your supply and thereby go farther away from your quantity in my ideal point, then I increase my supply to a similar extent; if you decrease your supply and thereby move nearer to your quantity in my ideal point then I decrease my supply to a similar extent. However, I do this only within the range limited by my own ideal quantity below and by my equilibrium quantity above.

What is meant by the phrase "to the same extent" differs from participant to participant. It may be the same absolute amount or the same percentage or something more complicated. However, the basic pattern is always the same. A change of the opponent's behavior in the direction of more aggressiveness is punished by a comparable increase of one's own aggressiveness and increase of the opponent's cooperativeness is rewarded by a comparable increase of one's own cooperativeness. This is a kind of reciprocity: As you do unto me, so I do unto you.

Reciprocity is a strong force in human behavior. We know this now from experimental games. I will not go further into this here, because research on reciprocity is more related to motivation than to rationality. However, the reciprocity of a measure-for-measure policy is probably not primarily due to an immediate emotional satisfaction obtained by responding in kind but rather to the idea that reward and punishment will guide the other player toward one's own cooperative goal. In this sense the reciprocity of a measure-for-measure policy seems to be *instrumental* rather than directly motivated.

The typical strategy program has a phase structure. In an *initial* phase of up to 4 periods there is no reaction to the opponent's behavior. Quantities begin high and are gradually decreased, in order to signal a willingness to cooperate. Then a *main phase* follows in which a measure-for-measure policy is used. Finally in an *end phase* of up to 4 periods behavior becomes non-cooperative. In most cases the equilibrium quantity is played in the end phase.

The breakdown of cooperation in the end phase supports the idea that measure-for-measure policies are instrumental. There seems to be no good reason why directly motivated reward and punishment should not go on until the end of the game.

The typical approach to the strategic problem has two striking features expressed by the catchwords "*no prediction*" and "*no optimization*". The term "no prediction" stands for the complete absence of any attempt to make quantitative predictions about the opponent's behavior. "No optimization" means that the quantity chosen is not the outcome of any optimization procedure. This is in sharp contrast to traditional oligopoly theories and game theoretic treatments of the oligopoly problem. These approaches usually involve some kind of profit maximization against quantitative expectations about the behavior of the other oligopolists. It is surprising that more than 150 years after Cournot's pioneering work on oligopoly an experimental study led to a duopoly theory of a completely different nature which was not foreshadowed by the vast literature on the duopoly problem.

Optimization based on quantitative expectations takes the behavior of the other player as given and passively adapts to it. The typical strategy does not take this approach. There is instead an attempt to actively assert influence. The use of a measure-for-measure policy in order to shape the other player's behavior by means of reward and punishment is very different from what is prescribed by normative game theory. Nevertheless this active approach is not non-sensical. It has its own kind of rationality.

Normative game theory represents our picture of ideal full rationality in interactive decision situations. We do not yet have a comprehensive theory of boundedly rational strategic reasoning. The results of our experiment suggest that the structure of the bounded strategic rationality of experienced human players is quite different from that of ideal strategic rationality. Maybe "no prediction" and "no optimization" are features of natural human strategic reasoning in a great variety of complex game situations.

I now am at the end of my talk. I hope that I succeeded in conveying an impression of the problem of rationality in game theory and its relationship to experience and the different facets of this.

SELECTED DISCUSSION

Robin Pope: Would you elaborate more on your notion of rationality as *balance*? Often rationality is defined in terms of optimizing. When you conceive it in terms of *balance*, is this not an equilibrium, a position that doesn't change? How does your sharp distinction between *descriptive* and *rational* relate to bounded rationality? When Simon argues that we can be successful when we are bounded, we might regard this descriptively. I felt that in some of your earlier work, such as your position in Norway, you were also focussing on how successful people do things, which didn't seem to sharply distinguish between the rational and the empirical.

Reinhard Selten: I remember a panel discussion at Oslo[1] among Allais, Harsanyi and myself. Here I had already taken the dualist position of a sharp distinction between normative and descriptive theories. When I say *balanced*, I don't mean this in a precise mathematical way. Obviously there are inherent tendencies of human thinking about rationality. We all have the idea that rationality is something to strive for, at least to understand what it would mean, even if we cannot follow it. Ideas about this conflict, otherwise we would not have had so much discussion about solution concepts and so on in game theory. Obviously these conflicting tendencies have to reach a balance. One has to weigh them, one against the other, one has to reach a balanced synthesis of all the inherent tendencies of the human mind. This is what I think of as full rationality.

Full rationality defined as optimizing is itself actually the outcome of a long discussion, as we have seen. In von Neumann and Morgenstern, it was not yet optimizing; they took a maximin view instead. We now think, in a Bayesian spirit, that rationality has to do with optimization, which is already a balance which has been reached by such a discussion. It does not precede thinking about rationality. *Bounded rationality* is something different: it is not something we strive for, which should be invented in the armchair, the outcome of long discussions. Bounded rationality is something which has to be observed in actual behavior. Bounded rationality is not a substitute for optimization. It has a different structure, it does not merely say that I am bounded in my memory capacity or my data processing capacity, and now I try to optimize within these bounds. If I were to do this, I would again transcend my bounds. We actually find such problems much more difficult than the original [fully rational] optimization problems. The fact is, it is possible to find behavior which is boundedly rational and not optimizing, but sound, as in Measure-for-measure.

Werner Leinfellner: Concerning your sharp distinction between normative and descriptive, the French group, especially Munier, have a third solution: the prescriptive. If you work with a computer, you don't work with the whole program, you just work with a configuration of some programs which you need and the rest you store somewhere. One solution would be to have a *prescriptive* mode which is admittedly *descriptive* from time to time; but at other times we use a description in a Bayesian way, and it takes a prescriptive turn.

I want to add: *everything goes* or in other words: things are much easier in theories where "Small is Beautiful". We typically use small matrices in our game theoretical examples. But once you consider someone's payoff in a concrete situation, computation quickly becomes impossible. With lotteries of three alternatives, perhaps they work; but with huge lotteries, it is impossible. Do you stand by your sharp distinction, or do you think that there is something in between, a mixture of descriptive and normative, for special cases?

Reinhard Selten: I don't exclude the possibility that there is a middle ground between descriptive and normative. But this middle ground can be developed only if the distinction is well understood. The distinction is very important. A type of prescriptive position has been taken by Howard Raiffa in his book on negotiation. He says, we have to look at others as behaving in a descriptive fashion – they are all motivated psychologically, they behave as people behave; but *we* should behave *better*. We should maximize our utilities and behave better.

I think there is something wrong with this approach. One thing that is wrong is expressed by what Marshal Radetzky said: you should never think the enemy is more stupid than you yourself.

My second comment is that, if you want to be prescriptive, even with one-person situations (if you want to be Bayesians), you have to be able to produce consistent probability distributions and utility assignments, and here things are already very bad. People just are not able to produce consistent probability distributions and consistent preferences. Of course you could try to help them to provide such judgments by certain methods. Then it turns out that what you get out of such measurements is very sensitive to the method of measurement. If you use one method you get concave-convex utilities in money, if you use the other, you get convex-concave. This has been shown in the experimental literature: In human behavior, there simply exists nothing like von Neumann-Morgenstern utility. If you force people to construct it, then the outcome depends quite arbitrarily on the way you force them to construct it. Therefore I think even prescriptive theories like Raiffa's which assume others to behave boundedly rationally, but requiring utility functions and subjective probability distributions, are like building your house on sand. You just don't have the inputs! Already just providing the inputs overtaxes people, puts problems at them which they cannot solve. Therefore, prescriptive theory should also have a structure other than Bayesianism. What are my ideas of what prescriptive theories should do? That we explore descriptively what people do, and then we try to teach them how to do the same thing, but much more systematically and consistently. That I could call the semi-normative approach. That may be feasible, but the time is not mature for this. We first have to find out what the behavior is before we can begin to do something like this.

Question by unidentified participant: Could one explanation for the inability of empirical subjects to show consistent preference orderings lie in the fact that decision making is a resource-using process? Rationality itself uses scarce resources. You have to judge how to use scarce resources before making rational decisions. Isn't there a kind of gap in rational decision making?

Reinhard Selten: Of course this fundamental difficulty exists. But normative rationality theory, as it is now, explicitly abstracts away from these bounds. It is a theory not for man but for gods. It is, as I sometimes say, rational theology. But theology is also important. Even somebody who doesn't believe in God might wonder how God should be, were he to exist. God is an important idea of mankind, quite apart from whether he exists or not. This is what normative game theory is about – of course it's not about God, but about a subject of similar nature, i.e., rationality. It has nothing to do with the limitations of human thinking. Of course, if you take human limits seriously, you see that it's not possible. It has often been said that people must optimize taking decision costs into account, but this doesn't work. You are completely right, I agree with you.

Follow-up Question: But wouldn't that mean, assuming rational decision making needs scarce resources, that probably every decision made by a subject able to make rational decisions at all will be rational.

Reinhard Selten: On the contrary, every decision by a human will only be boundedly rational, because the computations to make rational decisions cannot be made. Bounded rationality will necessarily have a different structure from what full rationality is. It will not proceed in the same way. It does not proceed from a general principle, but it proceeds from the situation which it looks at in a very simple way, then if there is still time and resources, it sees that it is more complicated, so gradually a model of the situation is built up and analyzed at the same time. The process stops if resources get too scarce. Then you make the decision which you have reached – or another one, because you don't necessarily make the decision which you've reached by rational deliberation, there are not only cognitive limits to rationality, but also emotional ones. I did not go into this here. All these things require another lecture.

Robin Pope: We would have no decisions to make – we would have made the whole string of them at birth – if it were not for having decision costs. Therefore part of the heart of decision theory relates to there being costs to making decisions. If we don't rationally realize that there are costs, then we start doing computations that we can't finish, and we make worse decisions than if we realized this. It's rational to take our limitations into account. The real god in us would be the one who perfectly takes into account our limitations.

Reinhard Selten: This cannot be done the way decision theory is today. It is in its structure normative. Progress in making actual decisions is strongly inhibited by naïve rationalism. Naïve rationalism is the roadblock preventing this.

NOTES

* Inaugural address delivered at the Symposium "Game Theory, Experience, Rationality", Institute Vienna Circle 1996 (partly transcribed by Eckehart Köhler).

1. "First International Conference on the Foundations and Applications of Utility, Risk and Decision Theories", Oslo 1982; proceedings edited by Sbigum & Wensdop (1983).

SELECTED LITERATURE

W. Albers, "Zwei Lösungskonzepte für kooperative Mehrpersonenspiele", in: *OR-Verfahren*, XXI: 1975, pp. 1-13.

R.J. Aumann, "Acceptable Points in General Cooperative *n*-Person Games", *Contribution to the Theory of Games*, in: *Annals of Mathematics Studies*, 40, 1959, pp. 287-324.

R.J. Aumann/M. Maschler, "The Bargaining Set for Cooperative Games", in: Dresher, Shapley & Tucker (eds.): *Advances in Game Theory*, Annals of Mathematical Studies, 52, Princeton University Press 1964.

S.B. Ball/M.H. Bazerman/J.S. Caroll, "An Evaluation of Learning in the Bilateral Winner's Curse", in: *Organizational Behaviour and Human Decision Process*, 48, 1991, pp. 1-22.

S.K. Berninghaus/K.-M. Ehrhart, *Stochastic Learning in Tacit Coordination Games*, Discussion Paper, Game Theory and Experimental Economics, No. 2, Institut für Statistik und Mathematische Wirtschaftsforschung, Universität Karlsruhe, 1996.

K. Binmore, "Social Contract I: Harsanyi and Rawls", in: *Economic Journal*, 99, pp. 84-102.

K. Binmore, "Social Contract IV: Convention and Evolution", in R. Selten (ed.): *Rational Interaction: Essays in Honor of John Harsanyi*, Springer, Berlin 1990.

K. Binmore/A. Martin/A. Rubinstein, "Non-Cooperative Models of Bargaining", in: *Handbook of Game Theory with Economic Applications*, Amsterdam 1992.

R.H. Coase, "The Problem of Social Costs", in: *Journal of Law and Economics*, 3, 1960, pp. 1-44.

A.A. Cournot, *Recherches sur les principes mathématiques de la théorie des richesses*, Paris 1838; transl. by N.T. Bacon (1899): *Researches into the Mathematical Principles of the Theory of Wealth*, Macmillan, New York; transl. by W.G. Waffenschmidt (1924): *Untersuchungen über die mathematischen Grundlagen der Theorie des Reichtums*, Fischer, Jena.

F.Y. Edgeworth, *Mathematical Psychics*, Kegan Paul, London 1881.

I. Eshel/M. Feldman, "Initial Increase of New Mutants and Some Continuity Properties of ESS in Two Locus Systems", in: *American Naturalist*, 124 (5), 1984, pp. 631-640.

D.B. Gillies, "Some Theorems on *n*-Person Games", Ph.D. dissertation, Princeton University 1953.

P. Hammerstein/R. Selten, "Game Theory and Evolutionary Biology" in: R. Aumann/S. Hart (eds.): *Handbook of Game Theory with Economic Applications* II, Amsterdam 1994.

J.C. Harsanyi, "Approaches to the Bargaining Problem before and after the Theory of Games", in: *Econometrica*, XXIV, 1956, pp. 144-157.

J.C. Harsanyi, "Bargaining and Conflict Situations in the Light of a New Approach to Game Theory", in: *The American Economic Review*, LV, 1965, pp. 447-457.

J.C. Harsanyi, "Games with Incomplete Information Played by 'Bayesian' Players", in: *Management Science*, 14, 1967-68, pp. 159–182, 320–334, 486-502.

J.C. Harsanyi, "An Equilibrium-Point Interpretation of Stable Sets and a Proposed Alternative Definition", in: *Management Science*, 20, 1974, pp. 1472-1495.

J.C. Harsanyi, "The Tracing Procedure", in: *International Journal of Game Theory*, 4, 1975, pp. 61-94.

J.C. Harsanyi, *Rational Behavior and Bargaining Equilibrium in Games and Social Situations*, Cambridge University Press 1977.

J.C. Harsanyi/R. Selten, "A Generalized Nash Solution for Two-Person Bargaining Games with Incomplete Information", in: *Management Science*, 18, 1972, pp. 80-106.

J.C. Harsanyi/R. Selten, *A General Theory of Equilibrium Selection in Games*, MIT Press, Cambridge MA 1988.

J. Kahan/A. Rapoport, "A Test of the Bargaining Set and Kernel Models in Three-Person Games", in Rapoport, A. (ed.): *Game Theory as a Theory of Conflict Resolution*, Reidel, Dordrecht 1974.

E. Kalai/D. Samet, "Persistent Equilibria in Strategic Games", in: *International Journal of Game Theory*, 13, 1984, pp. 129-144.

G.K. Kalisch/J.W. Milnor/J.F. Nash/E.D. Nering, "Some Experimental *n*-Person Games", in: Thrall, R.M., C.H. Coombs and R.L. Davis (eds.), *Decision Processes*, Wiley, New York 1954.

E. Kohlberg/J.-F. Mertens, "On the Strategic Stability of Equilibria", in: *Econometrica* 54, 1986, pp. 1003-1038.

D. Kreps/R. Wilson, "Sequential Equilibrium", in: *Econometrica*, 50, 1982, pp. 863-894.

D. Kreps/R. Wilson, "Sequential Equilibria", in: *Econometrica*, 54, 1986, pp. 1003-1038.

B. Kuon, *Two Person Bargaining with Incomplete Information*, Lecture Notes in Economics and Mathematical Systems, No. 412, Berlin, 1993.

D. Luce/H. Raiffa, *Games and Decisions*, New York, Wiley 1957.

J. Maynard Smith, *Evolutionary Game Theory*, Cambridge University Press, Cambridge 1982.

J. Maynard Smith, *Evolution and the Theory of Games*, Cambridge University Press, Cambridge 1982.

J. Maynard Smith/G.R. Price, "The Logic of Animal Conflict", in: *Nature*, 246, 1973, pp. 15-18.

M. Maschler, "Playing an *n*-Person Game. An Experiment", in: Sauermann (ed.), *Coalition Performing Behavior*, Tübingen 1978.

M. Mitzkewitz/R. Nagel, "Experimental Results on Ultimatum Games with Incomplete Information", in: *International Journal of Game Theory*, 22, 1993, pp. 171-198.

M. Mitzkewitz/G.R. Uhlich, "Duopoly Strategies Programmed by Experienced Players", in: *Econometrica*, 65, No 3, pp. 517-555.

P.A.P. Moran, "On the Nonexistence of Adaptive Topographies", in: *American Human Geneticist*, 27, 1964, pp. 343-383.

R.B. Myerson, "Refinements of the Nash Equilibrium Concept", in: *International Journal of Game Theory*, 7, 1978, pp. 73-80.

R. Nagel/F.F. Tang, *Experimental Results on the Centipede Game in Normal Form: An Investigation of Learning*, Discussion Paper No. 149, Center for Rationality and Interactive Decision Theory, The Hebrew University of Jerusalem, 1997.

J.F. Nash, "Equilibrium Points in n-Person Games", in: *Proceedings of the National Academy of Sciences USA*, 36, 1950, pp. 48-49.

J.F. Nash, "The Bargaining Problem", in: *Econometrica*, 18, 1950, pp. 155-162.

J.F. Nash, "Non-Cooperative Games", in: *Annals of Mathematics*, 54, 1951, pp. 286-295.

J.F. Nash, "Two-Person Cooperative Games", in: *Econometrica*, 21, 1953, pp. 128-140.

H. Raiffa, *The Art and Science of Negotiation*, Harvard University Press, Cambridge MA 1982.

A. Rubinstein, "Perfect Equilibrium in a Bargaining Model", in: *Econometrica*, 50, 1982, pp. 97-109.

A. Rubenstein/M.A. Osborne, *Bargaining and Markets*, San Diego 1990.

B. Sbigum/F. Wensdop, *Foundations of Utility and Risk Theory with Applications*, Reidel, Dordrecht 1983.

R. Selten, "Spieltheoretische Behandlung eines Oligopolmodells mit Nachfrageträgheit", in: *Zeitschrift für die gesamte Staatswissenschaft*, 121, 1975, pp. 01-324.

R. Selten, "Re-examination of the Perfectness Concept for Equilibrium Points in Extensive Games", in: *International Journal of Game Theory*, 4, 1975, pp. 25-55.

R. Selten, "A Note on Evolutionarily Stable Strategies in Asymmetrical Animal Conflicts", in: *Journal of Theoretical Biology*, 84, 1980, pp. 3-101.

R. Selten, "Equal-Division Payoff Bounds", in: R. Selten, *Models of Strategic Rationality*, Kluwer, Dordrecht 1987.

R. Selten/J. Buchta, *Experimental Sealed Bid First Price Auctions with Directly Observed Bid Functions*, Universität Bonn: Sonderforschungsbereich 303, Discussion Paper: B 270.

R. Selten/U. Leopold, "Equilibrium Point Selection in a Bargaining Situation with Opportunity Costs", in: *Économie Appliquée*, 1982, pp. 611-648.

R. Selten/M. Mitzkevitz/G.R. Uhlich, "Duopoly Strategies Programmed by Experienced Players", in: *Econometrica*, 65, No. 3, 1997, pp. 517-555.

R. Selten/K.G. Schuster, "Psychological Variables and Coalition Forming Behavior", in Borch, K. and D. Mossin (eds.), *Risk and Uncertainty*, London 1968.

R. Selten/R. Stoecker, "End Behaviour in Sequences of Finite Prisoner's Dilemma", in: *Journal of Economic Behaviour and Organization*, 7, 1986, pp. 47-70.

L.S. Shapley, "A Value for *n*-Person Games", in: Kuhn and Tucker, *Contributions to the Theory of Games*, II, Princeton University Press 1953.

D. Schmeidler, "The Nucleolus of a Characteristic Function Game", in: *SIAM Journal of Applied Mathematics*, 17, 1975, pp. 1163-1170

G.R. Uhlich, *Descriptive Theories of Bargaining. Lecture Notes in Economics and Mathematical Systems*, Springer Verlag, Berlin 1989.

E. Van Damme, "Extensive Form Games", in: J.Eatwell/M. Milgate/P. Newman (eds.), *Game Theory*. The New Palgrave I: A Dictionary of Economics, Macmillan, London 1987.

J. von Neumann/O. Morgenstern, *Theory of Games and Economic Behavior*, Princeton University Press 1944/1953.

Institut für Gesellschafts- und Wirtschaftswissenschaften
Adenauerallee 24-42
D-53113 Bonn
Germany

ANATOL RAPOPORT

COUNTERPRODUCTIVE RATIONALITY

Let us begin by proposing a sort of common denominator of the various conceptions of rationality, some feature that underlies practically all its definitions.

I submit it is a mark of rationality for a person to take into account the consequences of his or her actions.

In the past few decades a formidable strictly rigorous discipline has been developed in which this minimal conception of "rationality" is accepted as the foundation, namely decision theory. Actually decision theory bifurcates into two branches, a normative one and a descriptive one. The former purports to say how people who deserve to be called "perfectly rational" make decisions; or, in the light of positive connotations of "rationality", how rational people *ought* to make decisions. The latter purports to describe, and sometimes, in the light of a successful theory, to predict, how real people actually make decisions. Here we will be concerned only with the normative theory.

Normative decision theory prescribes courses of action which in some sense, lead to the realization of goals associated with the actor's value system. This system is given (or assumed) *a priori*. Therefore normative decision theory can be regarded as relativistic and, for that reason "value free", since its prescriptions are derived with reference to an arbitrarily given (or chosen) value system. This makes it attractive to circles where science is regarded as a value-free activity or, at any rate, one that should be such.

Techniques developed in normative decision theory serve admirably in determining maxima or minima of a single variable. Unlike mathematical models, however, real life situations present themselves as complexes of desiderata or their opposites. Attempting to make decision problems tractable, decision theorists are strongly tempted to consolidate these components into a single component, thus reducing the problem of optimization to that of finding a maximum or a minimum of a single variable or else to treat the components of the situation one at a time. It is just this temptation that leads to frequent counterproductivity of the paradigm, which in the last few decades emerged as the so called paradigm of rational choice.

Let us begin by examining the use of this paradigm in a concrete situation. The following excerpt is taken from a paper presented at a conference on American national security in 1965. The authors are concerned with the economics of deterrence. Specifically they seek an index, which would reasonably reflect the effectiveness of a given deterrence posture. They take into account not merely getting "more bang for a buck" (as crude estimates of military effectiveness were expressed in those days) but more refined indices of efficiency. That is, they undertake a more sophisticated cost-benefit analysis within the limits of a given budget. Three types of index are examined.

W. Leinfellner and E. Köhler (eds.), Game Theory, Experience, Rationality, 35–48.

> (a) The number and weight of offensive weapons (for a given budget). Criteria of this general type are used in a surprisingly large number of cases by military correspondents, columnists, an other "experts", who should know better ... Little thought is required to dismiss such crude devices. A missile that can carry a small bomb and deliver it within 10 miles of the target presents nothing like the deterrence threat of a missile that can carry a large bomb and deliver it within 2 miles.
>
> (b) The number and value of enemy targets that can be destroyed (for a given budget). This is a criterion that makes a little more sense. It takes into account not only the numbers of our offense bombers and missiles but also their operational effectiveness ... It still is, of course, an ambiguous criterion and requires more precise definition. For example, what target systems - population, industry or military bases - should be used to keep score...?
>
> (c) The number and value of enemy targets that can be destroyed (for a given budget) after an enemy first strike. This is much closer to what we want. It requires us, in allocating our given budget, to reduce the vulnerability of our force whenever money spent on reducing vulnerability ... will increase our "strike second" capability more than the same money spent on additional bombers or missiles and the personnel to cooperate them. (Hitch and McKean, 1965, p.126.)

The language reflects a specific conceptualization of the posed problem. The sources of the conceptualization are entirely apparent. They come partly from business activity, partly from competitive sports, as in reference to the problem of "keeping score" in the conduct of a nuclear war. The optimization imperative and the total acceptance of relativity of values are seen even more clearly in the following more general formulation of the problem.

> Ideally we should choose the course of action, which, with available resources maximizes something like "satisfaction" of an individual, the profits of a firm, the "military worth" of the military establishment, or the welfare of a group. If possible, we should like to ascertain the total amount of these magnitudes under each of various alternative policies. Then we would pick the policy that promised to yield the most satisfaction, the most profits, the most military worth, or the most wellbeing, depending on the identity of the person or organization, whose choice we were advising.

This approach perhaps exemplifies most clearly the ultimate goals of a normative decision theory, namely in terms of effecting a "rational decision", that is, rational from the point of view of a given actor in a particular situation. If one is satisfied that this is the sort of thing that "rational decision making" entails, there is nothing further to say. If, however, one feels that there is something wrong with the consequences of conceiving rationality in this way, then it behooves us to re-examine the basis of the attempted definition.

The first thing that comes to mind is the horizon of possible consequences that are taken into account. Is it rational for a person to set no limits on the amount of food he or she consumes, if consumption gives immediate satisfaction? Yes, if the horizon of perceived consequences is in the immediate vicinity; no, if long term effects are taken into account.

Aside from the time factor, there is also the range of possible consequences to consider. Thus, in setting up criteria of effectiveness in allocating available resources to nuclear war heads and their delivery systems, the authors of the article cited make no mention of what the reactions of the Russians might be to the sort of weapons systems the Americans plan to develop (and vice versa).

More generally, consequences of one's actions practically always depend on events over which the decision maker has no control. To take a trivial example, the consequences of my decision to take an umbrella with me when I go to work in the morning depends on weather conditions beyond my control. I choose between two alternatives - to take it along or not. There will be, however, at least four outcomes - (1) rain and having the umbrella along; (2) rain and no umbrella along; (3) no rain and no umbrella to lug; (4) no rain and an umbrella to lug (and possibly forget on the subway).

Achievements credited to the creators of modern normative decision theory were essentially extensions of the concept of rationality well beyond the "minimal" conceptualization offered above. So it is in the light of these extensions that we must examine to what extent the modern (formalized) conception of rationality has been productive or counter-productive. This we will proceed to do.

Let us consider the first extension, just mentioned, namely including the dependence of consequences on circumstances beyond the actor's control, often called "states of nature". In our umbrella example these states were "rain" and "shine". Sometimes it is possible to estimate probabilities with which such states will occur, in some situations, indeed with great accuracy, as in gambling games or in insurance. Further if preferences can be defined on a sufficiently strong scale, say an interval scale, then the concept of *expected utility* can be defined as the scalar product of the vector of utilities with the vector of probabilities. Now "rational decision" can be defined in a situation of this sort with mathematical precision, namely, as choosing the course of action that maximizes expected utility. Indeed, the origin of decision theory can be traced to the first formulation of this concept in the seventeenth century in the correspondence between Fermat and Pascal stimulated by some questions raised by a philosophically inclined gambler, one Chevalier de Méré.

To illustrate the use of expected utility in a decision problem, let us represent the umbrella problem as a "game against nature". That is, the actor has two choices, T (take umbrella) and L (leave it home). Nature also has two choices, namely, R (rain) and S (shine). Each chooses independently of the other. There are four possible outcomes: (a) I take the umbrella and the weather is fair; (b)

I take it and it rains; (c) I leave it and the weather is fair; (d) I leave it and it rains. The situation is presented by a 2 x 2 matrix:

	R	**S**
T	**0**	**2**
L	**-5**	**10**

Diagram 1

The entries represent my *utilities,* that is, degrees of preferences among the four possible outcomes. I like most to be unburdened by an umbrella when the weather is fine (10), least to have no umbrella when it rains (-5), and so on. The decision is not obvious, since if it should rain, it is better to have the umbrella along, otherwise no. However, if probabilities can be assigned to the two states of nature, a "rational" decision can be arrived at in the sense of maximizing expected utility. Let the probability of Rain be 0.7 hence of Shine 0.3. Then the expected utility having the umbrella is (0)x(0.7) + (2)(0.3) = 0.6, while the expected utility of not having it is (-5)(0.7) + (10)(0.3) = -3.5 + 3 = -0.5. Better take it along. If the probabilities were reversed, we would have (0)x(0.3) + (2)(0.7) = 1.4 as the utility of lugging the umbrella along and (-5)(0.3) + (10)(0.7) = -1.5 + 7 = 5.5, otherwise. Better leave it home.

The development of normative decision theory can be said to have involved a successive generalization of its conceptual repertoire. The concept of expected utility was the first.

A most important generalization was from decision situations involving a single actor to one with two or more actors, whose preferences for the outcomes of their decisions in general did not coincide. This was the beginning of game theory. In the simplest situations there are two actors, whose interests are diametrically opposed. Utilities attached to gains and losses of one of the actors (now called players) correspond to those of the other with the opposite sign. What one gains, the other loses. In this situation, the decision associated with maximization of expected utility is, in general, no longer necessarily "rational".

For it is now assumed that "rational" in this situation entails the supposition that the other player is likewise rational. To see the implication of this assumption assume that Nature, who makes weather, also attaches utilities to the outcomes in the umbrella problem, and that her preferences are diametrically opposed to mine. What is worth 10 to me is worth -10 to her, what is worth -5 to me is worth 5 to her, etc. Then if she has figured out that my decision (based on maximization of expected utility) was, say, to leave the umbrella home, she might make rain just to spite me; if I were to take it along, she might make beautiful weather. This suggests that in playing a "rational" and "hostile" opponent rather than neutral Nature, it behooves one to put oneself into the shoes of the other, as it were, to imagine how the other would decide if she figured out what one was going to do.

Let us again examine the matrix of the umbrella game. Now we suppose that the opponent's payoffs in each entry are equal to one's own with the opposite sign. The matrix now looks like this.

	R	**S**
T	**0, 0**	**2, -2**
L	**-5, 5**	**10, -10**

Diagram 2

The players will now be called Row and Column. Row chooses between R_1 and R_2, Column between C_1 and C_2. As can be immediately seen, Column's choice is obvious, namely, C_1, since her payoffs in that column are larger than in C_2, regardless of Row's choice. This is not true for Row. If Column chooses C_1, Row is better off with R_1, otherwise with R_2. Now if Row assumes that Column is rational, he will expect Column to choose C_1. Therefore Row's better choice is R_1, where he breaks even, whereas choosing R_2 he loses 5.

So in this case we see that rationality includes imagining oneself in the situation of the other, which was not necessary in situations where the actor was not confronted with an equally rational actor.

This extension of the concept of rationality should be of interest to the psychologist. Recall the experiments of Jean Piaget, who demonstrated that the ability to see the world through the eyes of another is attained at a certain stage of maturation. For instance, a small child, whose father is known to have only one brother is asked,

"Does your daddy have a brother?"

The answer is usually, "Yes."

"What is his name?"

"Uncle Pierre."

"Does Uncle Pierre have a brother?"

Up to a certain age the answer is usually, "No." To a small child daddy is not a "brother". Daddy is daddy. Uncle Pierre is daddy's brother. There is no other. Let us now go back to the cost benefit analysis of deterrence. Note that in arriving at a "rational" allocation of available funds with the view of wreaking most destruction on the Russians there was apparently no attempt to carry out a similar analysis from *the Russians'* point of view, so as to anticipate the consequences of not only one's own actions but also those of an equally rational opponent.

In fairness, it must be said that with the advent of the theory of games at the close of World War II, the strategic community did appreciate this further extension of the concept of rationality. In fact, game theory was hailed as a valuable contribution to military science, namely, a powerful mathematical tool of strategic analysis. Generous funds were allocated to the development of the theory.

The type of game that naturally appears as a most appropriate model of a military confrontation is the two-person zero-sum game. It represents a pair of adversaries with diametrically opposed interests: what one wins, the other must lose. Our example illustrated the situation. Recall that the solution of the game was obvious to Column, but Row had to put himself into Column's position to arrive at a rational choice of strategy. In the following game on the next level of complexity both players must go through this exercise.

	C_1	C_2
R_1	**2, -2**	**0, 0**
R_2	**-1, 1**	**4, -4**

Diagram 3

Now we will suppose that the strategist in the role of Row is more mature than Piaget's young subject. In deciding how to choose between R_1 and R_2, he takes into account how Column would choose if he knew how he, Row, would choose. So if Column supposes that Row chose R_1, he would apparently choose C_2, so as to break even instead of losing 2. But if Column supposes that Row would choose R_2, he would choose C_1 so as to win 1 instead of losing 4. Ascribing rationality to his opponent Column, Row assumes that Column, assuming a "worst case scenario", as they say in the military, will choose C_1 in order to guarantee himself the "best of the worst" (the maximin, that is, the maximum of the minima). Hence Row decides on R_1, to maximize his own payoff (winning 2, rather than 1 with R_2). However, still assigning "rationality" to Column, he carries the analysis further, thus assuming that Column has guessed that he will choose R_1. On the basis of that assumption, it is "rational" for Column to choose C_2. But if he does, why should not Row choose R_2 and win 4 instead of just breaking even? Column, being as "rational" as Row, trying to decide between C_1 and C_2 falls into the same trap. We see that this business of assuming the other's point of view has led into a whirlpool of circular reasoning. Is there way out of it?

Game theory showed the way. Again a resolution of a paradox was found by extending further the concept of "rationality". We have seen how in situations involving several different states of nature, on which the outcome of the decision depended, the concept of expected utility was introduced, whereby probabilities were assigned to the possible states. In other words, it was assumed that Nature used some chance device (like tossing a coin or rolling a die) to choose, say, between rain and shine in the umbrella problem. In the zero-sum two-person game, it is each player, who is expected to use a chance device in choosing among his available strategies.

Consider the simple but instructive game of Button-Button. One conceals a button in the right or the left hand, and the opponent tries to guess which. If the game is repeated, it will not do to always hide the button on one or the other hand, since the opponent will catch on and so always guess. Nor will it help to alternate, since the opponent can eventually see through that strategy. It turns out that the best strategy is to let pure 50-50 chance choose the hand, for then the opponent can guess only with probability 1/2. Any other strategy will increase his probability of the opponent's guessing.

The same principle applies to solving games like the one just examined, except that the probabilities with which the one or the other strategy is chosen are calculated in terms of the payoffs of the game. It turns out that Row's optimal strategy is to choose R_1 with probability 5/7 and R_2 with the complementary probability 2/7. Column's best mixture is (4/7, 3/7). If each player uses his optimal mixed strategy, Row's expected gain is 8/7, which is more than 1, his guaranteed payoff in the "worst case scenario". Column's expected payoff is -8/7 (what Row expects to win on the average is what Column expects to lose), which is better than -2, his expected loss in the "worst case scenario". This result is the best each can do.

Once the principle of mixed strategy was accepted, all two-person zero-sum games could be solved in principle. Translating these solutions into real life problems is, of course, a very different matter. Still the game-theoretic approach provided a solid conceptual framework for analyzing the logical structure of situations that could be reasonably modeled by two-person zero-sum games, that is, games of complete opposition. The solutions of such games rested on three cardinal principles: strategic dominance, the maximin, and the equilibrium. All three were illustrated in our examples. In the umbrella game C_1 *dominated* C_2 in the sense that it yielded a larger payoff *regardless of the strategy chosen by Row*. The maximin was illustrated by the "worst case scenario". Each player chose the strategy that contained the "best of the worst", that is the maximum of the two minimal outcomes associated with his two strategies. The equilibrium was illustrated by the mixed strategy solution of the game without dominating strategies. Neither player can improve his expected gain by shifting away from the optimal mixture, provided the other does not shift.

Let us see how these principles fare in a game that is not zero-sum, that is one, in which a player does not necessarily lose what the other wins. In some outcomes, both may win and in others both may lose. A game of this sort that can be a drastically simplified model of deterrence is the now celebrated game of Chicken.

	C_2	D_2
C_1	**1, 1**	**-10, 10**
D_1	**10, -10**	**-100, -100**

Diagram 4
The Game of Chicken

We have changed the notation to adhere to general usage. C now stands for either "careful" or "chicken", D for "daring" or "dangerous". Spirited American youngsters used to play the game as follows (perhaps still do). Two of them get into two cars, perhaps a mile apart and rush at each other, each straddling the middle line. The first driver to swerve so as to avoid the crash loses (is "chicken"). C stands for "swerve", D for "don't swerve". If both swerve the game is a draw awarding a modest payoff of 1 to each. The first to swerve loses 10; the driver with the stronger nerve wins 10. If both are determined to win, both are killed (lose 100).

Herman Kahn, of the American strategic community, who was much concerned with "thinking about the unthinkable", wrote copiously about how to win a nuclear war, preferably without fighting it, but if it became necessary to fight it, how to fight it "rationally". He used the game of Chicken model to illustrate how to intimidate the opponent in a confrontation, that is to make him blink. In playing Chicken, he wrote, yank your steering wheel off and throw it out of the window (making sure the opponent sees this). Now your opponent knows that even if you wanted to swerve, you can't (Kahn, 1965, p.11). Therefore he must swerve, and you have won. Of course (Kahn goes on) your opponent may use the same strategy, in which case you have a problem. Evidently, however, Kahn did not consider this to be *his* problem.

Let us see how the three principles of solving a zero-sum game apply in this case. If both players choose their maximin strategies C (on the basis of a "worst case" scenario), each gets a modest positive payoff. However, this outcome is not an equilibrium. Each can get 10 if he shifts from C to D, provided the other does not shift. There are two equilibria, one at C_1D_2, the other at D_1C_2. The first favours Column, the second Row. But there is no way to show that the one outcome is "more rational" than the other. The game is entirely symmetric (neither player is advantaged by its structure), hence so must the solution be.

There is still another equilibrium in this game, not shown in the payoff matrix, namely, the outcome resulting from mixed strategies. If each player uses the mixture (10/11, 1/11) each gets an expected payoff 0. The solution is symmetric and is an equilibrium in the sense that neither player can improve his payoff by shifting, provided the other does not. But both could have done better if each had chosen his maximin strategy C (each would have received 1 instead of 0). There seems to be no satisfactory answer to the question which is the "most rational" strategy and which the "most rational" outcome.

Attempted applications of mixed strategy solutions in certain situations reveal another, even more conspicuous flaw, especially when applied to games of deterrence or intimidation, of which Chicken is an example.

Consider the following game:

	S_2	T_2
S_1	**1, 0**	**0,1**
T_1	**0, 0**	**-x, -y**

Diagram 5
The Deterrence Game

Here x, y > 0. The payoffs are interpreted as follows. Row wants Column to choose S_2, since if he does, by choosing S_1 Row can get the largest payoff (1). But Row knows that if he chooses S_1, it is in Column's interest to choose T_2 not S_2. What should Row do to induce Column to choose S_2? If Row can manage to send a message to Column before Column makes his choice, he may succeed. The message would be, "If you choose T_2, I am going to choose T_1. Note that the outcome in that case would be T_1T_2, which is worse for both of us than S_1S_2."

Now the question of interest to both players arises, "Is this a credible threat or can it be regarded as a bluff?" Column must decide whether to yield to Row's demand or whether to defy the threat; Row must decide whether to make the threat or not, since it results in a win of 1 if Column yields or in a loss of x, if he does not.

A "solution" is offered by T.C. Schelling (1960) in terms of the payoffs of the game and certain assumed probabilities. Assume that Row estimates that Column will defy the threat with probability p ($0 < p < 1$). Then Schelling recommends that Row threaten to choose T_1 with a certain probability. It turns out that it is worthwhile for Row to make such a threat if the following inequalities hold:

$$(1-p)/px > \pi > 1/(1+y)$$

If the inequalities are incompatible, that is, if $1/(1+y) > (1-p)/px$, then it doesn't pay to make the threat at all. If they are compatible, that is

$$(1-p)/p > x/(1+y)$$

then the best strategy is to threaten T_1 with some probability π.

Now let this game represent the Cuban Missile Crisis of 1962. U.S. is Row, Cuba is Column. S_2 means Cuba removes the Russian missiles; T_2 means she does not. T_1 means U.S. attacks Cuba; T_2 Cuba (or, more likely, U.S.S.R.) attacks the U.S.; -x and -y represent respectively the costs of mutual attack to the respective antagonists. Since the right side of the above inequality increases as x increases, and the left side decreases as p increases, it will not pay to attack when the cost associated with retaliated attack becomes sufficiently large. Therefore it is "rational" to decrease x (say harden one's own missile sites). Also as x decreases, it becomes progressively safer to increase the probability of attacking Cuba. These are common sense conclusions, and from the point of view of a "rationally" thinking strategist it is comforting to arrive at them through sophisticated mathematical analysis, invoking, for example, the concept of mixed strategy, which keeps the opponent uncertain about one's intentions and so presumably anxious and possibly confused.

Let us, however, see how a strategy of this sort can be actually implemented. To attack with probability π means to use a random device, which with probability π implements the command "Attack!" and with probability ($1 - \pi$) "Don't attack!". This device can be calibrated by increasing or decreasing π, and this suggests that the threat can be escalated. If the opponent is not impressed by a low probability threat, one can repeat it with increasing π, "tightening the screw", as it were. Practically this can be done as follows. Construct a missile releasing mechanism activated probabilistically by a bank of buttons. If the first button is pushed, attacking missiles are launched with probability 0.05, if the second, with probability 0.10, and so on. In this way one can escalate or de-escalate the crisis according to the situation. (For further discussion of this sort of "rational thinking about the unthinkable", see Kahn, 1965.)

Undoubtedly the implementation of such threat strategies is technically feasible. However, bothersome questions arise. It occurs, for example, that when one pushes, say, the 0.15 button, one doesn't release a 15/100ths of a nuclear war. One either releases a full blown nuclear attack or none at all. In the latter case, if the opponent has called your bluff, you lose credibility. In the former case you and your opponent both lose the game. In fact your chances of "winning", that is, "forcing your opponent to do your will", as Clausewitz defined the objective of any war, is slightly better than "winning" at Russian roulette (which is 5/6). The question reduces to the question of how rational it is to play Russian roulette.

The most dramatic paradox of formal decision theory has been repeatedly demonstrated by the most famous of so called "dilemma games", namely Prisoner's Dilemma represented by the following payoff matrix.

	C_2	D_2
C_1	**0, 0**	**-10, 10**
D_1	**10, -10**	**-1, -1**

Diagram 6
Prisoner's Dilemma

Note that in contrast with Chicken the different principles of rational choice are not at odds with each other. In Chicken there was no dominating strategy. The intersection of the maximin strategies, was not an equilibrium. There were three equilibria, which raised the problem of choosing among them. In Prisoner's Dilemma each player has a dominating strategy, namely D, the defecting strategy. The maximin strategies likewise dictate D. Finally, there is a single equilibrium, the outcome D_1D_2. Thus all three principles dictate the same outcome as the "rational" outcome of the game. The paradox arises from a different circumstance, namely that both players are better off in the cooperative outcome C_1C_2 than in the supposedly "rational" defecting outcome D_1D_2.

Again a generalization is required to resolve the paradox. Resolution can be effected by refining the concept of "rationality" namely, by distinguishing between individual and collective rationality. The former dictates D, the latter C. Sticking to individual rationality results in an outcome bad for both; guiding one's decision by collective rationality leads to an outcome good for both. It isn't a matter of sacrificing individual interest for collective interest. It is a matter of realizing one's individual interest by acting in the collective interest.

To act in collective interest means to follow a variant of the Golden Rule or of Kant's dictum: act as you would want every one to act. Innumerable real life situations illustrate the principle. If a fire breaks out in a crowded theatre, individual rationality dictates getting most quickly to the nearest exit. If every one acts "rationally" in this sense, every one may perish. If military superiority is thought to be a key to national security, it pays to keep increasing one's military potential vis-a-vis real or imagined enemies. If all nations do this, none is secure. It seems rational to develop commercial fishing technology with the view of maximizing catch. But if every fishing fleet maximizes its catch, the fish ultimately disappears, and every one loses. Need we go on?

The examples illustrate how application of the rational choice paradigm can be counterproductive. It remains to explain how these examples fit into the theme that I was asked to discuss at this meeting, namely, "Counterproductive consequences in the application of rational models to the quest for peace and social justice."

It may occur to some that I never mentioned the quest for peace and social justice. All I talked about was the application of rational models to what military establishments and their civilian advisers regard as national security. However, I was merely giving them the benefit of the doubt. I take seriously the claims of these people that security, as they understand it, is the key to the maintenance of peace. That is, I believe that many of them actually believe it. They take seriously the motto of the American Air Force, "Peace is our profession". They see nothing obscene in christening an intercontinental ballistic missile "The Peace Keeper". In fact, practically every one is for peace. All the ministries of war in the world have become ministries of defence, implying that what is being defended by threat of total destruction is peace. Even Hitler was for peace before 1939. Provided, that is, all territories where Germans lived were incorporated into the Third Reich and provided there was enough *Lebensraum* for the Master Race to increase and multiply. Even the present Israeli prime minister is for peace, provided the land given by Jehovah to his chosen people remains under their control in perpetuity and provided Arab states become democracies. Not long after the Israeli election I read an interview with Mrs. Netanyahu, the Israeli First Lady. She, it appears, is also solidly for peace. But what kind of peace is it, she pointed out, when one is afraid to take a bus? She isn't for *that* kind of peace, one that her husband's predecessor was ready to settle for!

So the main thrust of my argument was aimed at the rationales of those who identify peace with security, as security is understood in the context of the so called "realist" school of international relations, that is, when security is identified with power and power with the potential to wreak violence. The main hallmarks of this approach are, first, egocentrism. "Rationality" is understood as the "rationality" of an single actor, who may be an individual, a corporation, a state, or a gang – any entity with well defined individual interests. The second hall mark is the use of sophisticated formalism, usually mathematical, which provides an aura of "scientific rigour" to the arguments and also justifies a dismissal of misguided idealism, sentimentality, and the like. These are the hallmarks of those, who regard themselves as belonging to the man's world. Note that the most prestigious occupations in this world are competitive business, competitive politics, and war. From this point of view the rationality of the D_1D_2 outcome in Prisoner's Dilemma is unassailable.

When the quest for peace and social justice is considered from the point of view of organizations and movements motivated by rejection of the "peace equals national security" assumption, rational analysis is no longer counterproductive, since not individual but collective rationality is the point of departure. From this point of view it is the C_1C_2 outcome that is the rational one in both Prisoner's Dilemma and in Chicken. Moreover peace activists are not usually addicted to prestige-enhancing mathematical formalism that leads arguments into sterile channels, as we have seen in the threat game. (Peace activists cannot hope to impress the power elite with intellectual high tech). Terms excluded by a cost-benefit analyst but important to a psychologist or a social activist, like "compassion", "justice", "self realization" often enter analysis constructed on the basis of collective rationality. I hope to have shown that a rigorous analysis of the most important problematics of our day, the problematics of peace and social justice of our day can be constructed also on that basis.

REFERENCES

C.J. Hitch/R.N. McKean, "The criterion problem", in: M. Berkowitz and P.G. Bock (Eds.), *American National Security*. New York: The Free Press 1965.

H. Kahn, H.. *On Escalation. Metaphors and Scenarios*. New York: Frederick A. Praeger 1965.

T.C. Schelling, *The Strategy of Conflict*. Cambridge, Mass.: Cambridge University Press 1960.

38 Winchwood Park
Toronto ON M6G 2V5
Canada

Thomas C. Schelling

Rationally Coping With Lapses From Rationality

A man gave up smoking three months ago. For the first six or eight weeks he was regularly tormented by a desire to smoke, but the last three or four weeks have been less uncomfortable and he is becoming optimistic that he has left cigarettes behind for good. One afternoon a friend drops in for a business chat. The business done, our reformed smoker sees his friend to the door; returning to the living room he finds, on the coffee table, an opened pack of cigarettes. He snatches up the pack and hurries to the door, only to see his friend's car disappear around the corner. As he will see his friend in the morning and can return the cigarettes, he puts the pack in his jacket pocket and hangs the jacket in the closet. He settles in front of the television with a before-dinner drink to watch network news. Twenty minutes into the news he walks to the closet where his jacket hangs and takes the cigarettes out of the pocket, studies the pack for a minute, and walks into the bathroom, where he empties the cigarettes into the toilet and flushes it. He returns to his drink and his news.

What have we witnessed? I think we can confidently guess that our subject came to anticipate that in the presence of the cigarettes something might occur that he did not want to happen; by disposing of the cigarettes he has made it not happen. Wasting a dollar's worth of his friend's cigarettes was an inexpensive safeguard. He has coped rationally with the risk that he would do something he did not – at the moment of flushing the cigarettes – want himself later to do.

My usual interest is in how people actually exercise strategy and tactics, successfully or unsuccessfully, in constraining their own future behavior. Often the ways people try to constrain their own future behavior are like the ways they would try to constrain someone else's behavior; they appear to be treating their "future self" as if it were another individual. If our man had never smoked but his wife had, and she had recently with great discomfort foresworn cigarettes and was not yet confidently weaned, and his business friend had left cigarettes behind, he would surely dispose of the cigarettes before his wife came home. So whether we want to say that he treats his ten o'clock self as if it were "another self", or only that he treats it as he would "another's" self, makes little difference.

Most literature on this subject in economics and philosophy concerns what is usually described as an apparent change in preferences. At five o'clock the man does not want to smoke; at five o'clock he does not want to smoke at ten o'clock; at ten o'clock he may want to smoke, remembering perfectly well that five hours ago he did not want himself to smoke at ten o'clock, remembering that three months ago he did not want himself to smoke at any time. Just describing what the man is doing if he lights the cigarette that even a few moments ago he may have hoped he wouldn't smoke is not easy. Whether it is "rational" that

W. Leinfellner and E. Köhler (eds.), Game Theory, Experience, Rationality, 49–53.

he satisfies an urge to smoke, exercising his unalienable sovereignty at ten o'clock, may not be answerable within the classical paradigm of rational choice; neurologically there may be a resolution of the question, but I'll settle for calling it *nonoptimal as of five o'clock*; I'm willing to let the man refer to his anticipated lapse as "irrational" if he wants to call it that; at least, it so appears to him as of five o'clock.

Just to be clear: I do not consider the injection of heroin or the smoking of nicotine to raise any issue of rationality. It is only when the user of heroin or nicotine makes a serious attempt to stop and has difficulty doing so, suffering occasional relapse or suffering torment on the verge of relapse, perhaps attempting to restructure his or her environment or his or her incentives, that the issue arises whether some preferences are "true" and some are interlopers, whether fulfilling one preference is rational and fulfilling an opposing or alternating preference not.

There are many behaviors that display the paradox that a person, quite uncrazed, fully conscious, apparently "voluntarily" does what the person is simultaneously fully aware one shouldn't do. Scratching hives, poison ivy, or chicken pox almost always aggravates the burning and itching, the person scratching knows that it does, and as he or she who scratches can be amazed at what he or she is doing. The boy who believes masturbation is evil, harmful, or contrary to God's command, continues to believe it as he watches himself masturbate. The man or woman who is persuaded that sexual infidelity is fraught not just with risk but with the near certainty of awful consequences continues with the persuasion while engaging in the act. This phenomenon is utterly tantalizing: one sees oneself doing what one knows one should not be doing; there is no loss of awareness of consequences, only – this is one way to say it – loss of command from one part of the brain to another.

Return now to our friend who flushed his friend's cigarettes away at the beginning of this paper. I averred, and expected no dispute, that the man's action was interpretable as a rational act, probably taken in anticipation of some lapse in his later rationality. This anticipatory pre-emption is easy to describe. Harder to describe is what that later action, the one he wanted to avert, would represent, what the state of his "rationality" would have been, at the later time, had he later engaged in smoking, or perhaps what the state of his rationality is still going to be at that later time when he might have engaged in smoking but does not. (Actually, maybe he will; his preclusive act may not have been decisive.)

An interesting question is whether, at the time he disposed of the remaining cigarettes, he had a good understanding of what his later state of mind was going to be. Might he, if we could ask him, be able to tell us just what the mental or emotional circumstances were going to be that he wanted to guard against?

There are a number of possibilities.

ALTERNATIVE SCENARIOS

One possibility is that he anticipates drinking alcohol and knows by experience that drinking interferes with his reasoning so that he succumbs to the temptation of cigarettes if cigarettes are readily available. If the individual believes that what alcohol does is to distort or to anesthetize some of the functions or characteristics that go under the name of rationality, we can call this situation "anticipated irrationality". We discussed it earlier.

This drinking contingency can be divided in two. The simpler case is that he enjoys relaxing in the evening with a few drinks even though it impairs somewhat his "rationality", because nothing he plans for the evening places demands on that part of his rationality that is subverted by alcohol. He could keep his smoking under control by going without alcohol, but for the price of a pack of cigarettes he can eliminate the danger and have his drink, and that is what he has chosen.

The second is that he wishes not only not to smoke but also not to drink, but he has a drinking problem that he knows he may not be able to control this evening: to be on the safe side – with respect to smoking, not to drinking – he destroys the cigarettes. We need here some reason why he doesn't pour the liquor down the toilet too. Maybe he is going out for the evening where he will be confronted by an opportunity to drink, expects to succumb against his (currently) better judgement or expects to feel obliged to drink among his friends, and fears coming home with impaired rationality to find cigarettes on the coffee table. (Our story might have been more tantalizing if, upon discovering the cigarettes his friend left behind, he had studied the pack momentarily, had then put it back on the coffee table and gone to the pantry and poured his whiskey down the sink!)

Counting the alcohol contingencies as the first two, a third possibility is that cigarettes by their physical presence – the sight of them, perhaps the smell of them, or the mere unforgettable knowledge of their presence – stimulate a craving, a craving that is a serious discomfort and distraction unless satisfied by the smoking of a cigarette. This contingency in turn breaks down into three.

One is that the craving is such a discomfort and distraction that it will produce an unproductive and disagreeable evening; at the risk of complete relapse it would be rational to avoid such a dismal and unproductive evening by going ahead and smoking. The presence of the cigarettes is an irritant that if not removed will make it rational to smoke. (An alcoholic might be considered rational to drink before undergoing some exceedingly painful procedure, the risk of relapse being a lesser evil than the pain.)

A fourth possibility, the second of these craving contingencies, is that he knows he will *not* succumb – will not suffer a lapse from rationality – but will be sufficiently distracted by the presence of cigarettes to make it worthwhile to dispose of them. In the same way he may, when it is time to get to work, turn off the TV, not because he expects to succumb to it but just because the flickering screen continually disturbs his concentration.

The third craving contingency – the one I find hardest to understand, though not to recognize – is that he expects to succumb "irrationally", not merely to be distracted, and not to be so distracted that it would be better to go ahead and smoke, but simply to be unable to maintain his resolve not to smoke. He wants now not to smoke later no matter how strong the craving, and he cannot trust his brain to function "rationally" under the stimulus of cigarettes present. (The case of extreme thirst may be an exemplar.) He would rather handcuff himself to the radiator across the room than be free to succumb. He doesn't need to; it is the presence of the cigarettes, or the knowledge of their presence, that will influence his brain chemistry, and *that* he can take care of in the toilet. Less poignantly, people often remove the peanuts, or the Danish, from the conference table or the nearby buffet, either in fear of succumbing or just to remove the distraction. (Clinicians have told me that recovering heroin addicts suffer much less craving when in a "clean" place where they know there is no chance to obtain heroin than when they believe, rightly or wrongly, that there is some possibility of obtaining the stuff.)

A sixth possibility is that the craving is independent of the presence of cigarettes. He knows that late-night fatigue (or late-night awakening) or some circadian change in brain chemistry – he needn't know what – will produce an irresistible craving for tobacco. It is the same whether the cigarettes are in open view, stowed in a cupboard, or five miles away in a vending machine. Destroying the cigarettes has no effect on his choice to smoke, only on his ability.

Here, too, we can distinguish two cases. One is that there will be no other cigarettes available at reasonable cost. No stores will be open, no one will be on the street from whom to bum a cigarette, or perhaps the man lives out of town without a car and no busses will be running. Tossing the cigarettes effectively denies the wayward choice.

The other – the seventh case – is that the craving is the same whether or not cigarettes are present, but their absence affects his ability to avoid surrender. I can think of two reasons. One is that an impetuous urge might control his behavior for a minute or two, but not for an hour, and if he grabbed his car keys and drove away his better judgement would overtake him within the time it would take to get to the store and park the car, and he'd be safe. (Maybe knowing that he'd return without cigarettes suffices to keep him from getting the car.) Alternatively, in self discipline some lines are harder to cross than others: he might allow himself "just one cigarette" if this act of God offers the opportunity, whereas getting the car and driving in search of cigarettes would be a flagrant violation of the regime he has imposed on himself, and the prospective loss of self respect would suffice to deter.

An eighth case is that the individual simply believes that if the cigarettes are there he will smoke. It may not matter where his belief came from. If instead his problem were alcohol, it is easy to believe he may have been authoritatively told that if there is liquor in the house he will drink it, or is likely to. Therapies to cope with cigarette addiction have not institutionalized any such lore about the inexorable consequences of spending an evening in the company of cigarettes, but our man could have received strong advice to that effect and takes it seriously even without a theory of his own of how the breakdown occurs in the presence of cigarettes. If his belief is based on his own experience with earlier attempts to quit, it may be a belief in behavioral phenomena that he can analyze in retrospect.

His belief could be true, and it serves his purpose that he destroyed the cigarettes. His belief could be false, and he has wasted a dollar for his friend. And there is a third possibility.

The proposition that if cigarettes are available he will ineluctably smoke before the evening is over may be the kind that is true of anyone who believes it and false of people who do not. Anyone who "knows" that with cigarettes around he will smoke before midnight knows that as midnight approaches he will have an irrefutable argument for going ahead and not waiting for midnight.

Somewhat akin to belief is suspense. Suspense produces discomfort, anxiety. If one spends an evening watching oneself, wondering whether one is going to succumb to temptation, two things can happen. One is that the evening can be exceedingly disagreeable; the second is that the suspense goes away once a cigarette is smoked. Until the cigarette is lit one is uncomfortably apprehensive that he may light a cigarette; lighting the cigarette is an escape. Is there something irrational here? There is a painful uncertainty whether I shall survive the night without smoking, an uncertainty that is dispelled upon smoking. Can the certainty that failure has occurred bring relief from uncertainty whether it will occur? (Again if it were alcohol, the anaesthesia might enhance the relief.)

We have so far, I believe, somewhere from nine to eleven distinct scenarios, depending on how we count subcases. But they are all scenarios of what the person has in mind when he "rationally" disposes of the cigarettes. That is an important part of our subject, and in several of the cases may be an adequate diagnosis of just what later would happen if the cigarettes were not made to disappear.

But the most tantalizing case, in terms of sheer description, is the third "craving" contingency: no alcohol or other exogenous chemical influence, no brainwashing about inevitability, just the anticipation that, either gradually or impetuously, the resolve not to smoke will be, or may be, replaced or overwhelmed by the desire to smoke. This is often a correct anticipation; it is the anticipation of something that actually occurs. But what "it" is that occurs continues to defy description.

JOHN D. HEY*

DO RATIONAL PEOPLE MAKE MISTAKES?

I. INTRODUCTION AND OVERVIEW

Unlike many of the papers at this Conference, this paper concerns games against nature rather than games against strategically-motivated human (or computer) opponents. Nevertheless, games against nature are relevant to games against opponents - as the literature makes clear. Indeed, one of the seminal pieces of that literature, John von Neuman/Oskar Morgenstern, *The Theory of Games and Economic Behaviour* (1947)[1], whilst ostensibly addressed to the theory of strategic games, derived one of the key elements of the theory of rational decision making against nature - namely Expected Utility theory - as a key component of the theory of strategic games. Moreover, many of the interesting new developments in Game Theory are pursuing the implications of non-Expected-Utility behaviour in games. Contrariwise, some recent developments in individual decision-making, such as those related to intertemporal decision-making, are using results from the theory of games. Clearly, developments in each branch of the literature are relevant to the other branch.

This paper is concerned with the tidal wave of new theories of decision-making under risk (and, of necessity, under uncertainty) that have been proposed as 'improvements' over Expected Utility (EU) theory. The paper is concerned with the fact that most (if not all) of these new theories (and, of course, EU itself) imply deterministic behaviour on the part of the decision maker, or (but only with a small subset of those theories) imply a stochastic component (of decision-making) which is deliberately (or deterministically) chosen by the decision maker. The idea of 'error' in decision-making is relatively unusual (though it does creep into certain theories of strategic decision-making - such as those involved with 'trembling hands' - a concept associated with one of those we honour at this Conference). Indeed, it might well be argued that 'error' must definitionally be 'irrational'.

However, the experimental evidence suggests that there is a stochastic component in decision-making, and, moreover, one that does not appear to be the product of deliberate 'rational' choice. I shall try and argue that this component is indeed what one might call 'error' or 'mistakes' - and that it is rational for decision-makers to display such behaviour. Furthermore, I will try and argue that behaviour in relatively simple choice contexts might better be described by some simple decision rule (such as EU) *plus error*, rather than by one of the more complicated decision rules recently proposed by decision theorists.

W. Leinfellner and E. Köhler (eds.), Game Theory, Experience, Rationality, 55–66.

I present the theoretical background in section II and give empirical evidence concerning the existence of 'error' in section III. In section IV, I interpret this evidence and suggest that it may indeed be properly termed error. I then ask, in section V, whether such 'error' and the behaviour which gave rise to it, can be considered *rational*. Section VI concludes.

II. THEORETICAL BACKGROUND

We should distinguish at the outset between *preferences* and *choices*. We must also distinguish between *deterministic* and *stochastic* preferences and choices. What is popular in decision theory concerning whether preferences and choice are deterministic or stochastic seems to vary over the years and over the disciplines. In economics, the current fashion is to view preferences as deterministic; 40 years ago the fashion in psychology was to regard preferences as stochastic. Let me briefly review the main components.

In economics at the present time, the mainstream regards EU as the (correct) theory of decision-making under risk (and under uncertainty). Those economists who are interested in decision theory and who have studied the (economic and psychological) literature on experiments in decision theory regard EU theory with some suspicion - and now put their faith in one or more of the many generalisations of EU theory. A partial listing of the members of this set includes Anticipated Utility theory, (Cumulative) Prospect theory, Disappointment (Aversion) theory, Implicit EU theory, Lottery Dependent EU theory, Machina's Generalised EU theory, Prospective Reference theory, Quadratic Utility theory, Rank Dependent EU theory, Regret theory, SSB theory, Weighted EU theory and Yaari's Dual theory. All of these[2] include EU theory as a special case and all have been developed in the light of experimental tests of the (axioms) of EU theory. All are *deterministic* theories of preference (with the exception of *Regret theory* which is a theory of choice not of preference). And they all imply (with an exception noted below) that choice itself is deterministic.

To be specific, let P and Q denote risky choices defined over a vector $\mathbf{x} = (x_1, x_2, \ldots, x_n)$ of final outcomes, where under P and Q outcome x_i occurs with probability p_i and q_i respectively. Denote by $V(.)$ some *preference function* as given by one of the above theories. So, for example, under EU, we have:

$$V(P) = \sum_{i=1}^{i=n} p_i \, u(x_i) \tag{1}$$

Now suppose our individual is asked to choose between P and Q; then, according to any of these theories, we have the following description of preference *and* choice:

> P is preferred to Q, *and hence is chosen from the pair* $\{P,Q\}$, if and only if $V(P) \geq V(Q)$

Note that this is a deterministic theory of choice. It is so partly due to the fact that the underlying preferences that gave rise to the choice are also deterministic.

However, it could be the case that the underlying preference function is deterministic yet (optimal) choice is stochastic. If an individual is offered a choice between P and Q, it may be the case that the individual prefers to randomise – that is, instead of choosing one of P or Q, to choose instead the *mixture* $M_\lambda = [P,\lambda;Q,(1-\lambda)]$, by which notation I mean a mixture which gives a probability λ of getting P and a probability $(1-\lambda)$ of getting Q. So, if λ is 0.5, the individual tosses a fair coin to decide whether to choose P or to choose Q. The mixture M_λ itself is, of course, a gamble over the vector **x** in which outcome x_i is received with probability $\lambda p_i + (1 - \lambda)q_i$. Hence the preference function $V(.)$ can be applied to it. The best such mixture is given by λ^* where λ^* maximises $V(M_\lambda)$. The necessary condition that such a best mixture exists (with λ^* properly between zero and one) is that $\partial^2 V(M_\lambda)/\partial\lambda^2 < 0$ where $\lambda = \lambda^*$. In other words, the preference function must be concave in the probabilities (which means that the indifference curves in the Marschak-Machina Triangle are convex[3]). In this case, the individual optimises by choosing P and Q randomly; namely, a λ^* probability of choosing P and the residual probability of choosing Q.[4]

There is also an older theory of random choice with deterministic 'preferences' that was popular amongst psychologists in the 1950's[5]; here 'preferences' are represented by a 'utility function' $U(.)$ over choices. Consider now a pairwise choice problem between choices A and B. This theory asserts that A is chosen 'in preference to' B with probability $U(A)/[U(A)+U(B)]$ and B 'in preference to' A with the residual probability. However, in contrast to the theories described above (both for deterministic choice and stochastic choice), there is no axiomatic basis to this theory. A similar property 'holds' with respect to the theories described below.

The theories above have *deterministic* preferences. There is an older set of theories – mainly appearing in the psychological literature of the 1950's – in which the preferences themselves are *stochastic*. The crucial feature of these theories is that the decision-maker is perceived *not* to have a well-defined preference functional – but rather one with a stochastic component. Thus, the actual functional used in any decision-problem varies from decision-problem to decision-problem. So, for example, in *Stochastic* Expected Utility theory the decision-maker is perceived as having the functional (1), but the utility function $u(.)$ is thought of as having a stochastic component. So the decision-maker knows that he or she is an expected-utility maximiser, but does not exactly know his or her utility function. It varies from problem to problem in a stochastic fashion – though possibly subject to the restriction that whichever function is used it is always monotonically increasing. The great advantage of this is that the decision-maker never violates dominance – though he or she may violate other perceived canons of rationality – such as independence.

Of course, each of the theories (of deterministic choice with deterministic preferences) described at the beginning of this section have their corresponding stochastic version. The main difficulty, however, from the point of view of the economist, long-steeped in the tradition of intuitively-normatively appealing axiomatically based theories, is that these psychologically-based theories have no theoretical appeal of the type normally sought-for amongst economists. Rather obviously, the appeal of the (Neumann-Morgenstern) axioms underlying (deterministic) Expected Utility theory tend to lose some of their charm when transferred to the world of the *Stochastic* Expected Utility 'theory'; indeed the axioms hardly make sense in such a world – for the simple reason that the axioms imply deterministic preferences and deterministic choice.

So we have a choice (except for Machina's theory of deterministic preferences and stochastic choice) between axiomatically-based theories of deterministic preferences and deterministic choice and non-axiomatically based theories of stochastic preferences and stochastic choice. The former seem to be preferred by the economic theorist – but they run into problems when it comes to the interpretation of data, as we shall see in the next section.

III. EMPIRICAL EVIDENCE

All the direct (experimental) empirical evidence on the actual behaviour of people when taking decisions under risk suggest that there is a stochastic component to that behaviour. Such evidence is either *descriptive* or *analytical*. The descriptive evidence comes from experiments in which people were asked the same question twice (or, more unusually, more often) either on the same experimental occasion or on separate occasions. An example of the latter is reported in Hey/Orme (1994)[6], in which 80 subjects were asked on two separate occasions (separated by an interval varying between 3 and 5 days) the same 100 pairwise choice questions[7]. The consistency rate varied from a low of less than 60% to a high of over 90%; clearly subjects were not answering randomly (for then a consistency rate of just 50% would be observed) but neither were they answering in a completely consistent manner (when a consistency rate of 100% would be observed). Whether this was error or something else remains to be discussed. In Carbone/Hey (1995)[8], subjects were asked *on the same occasion* the same 42 pairwise choice questions (again with the order randomly varied and left and right interchanged) as well as 8 questions testing for violations of dominance. The consistency rate on the repeated questions again varied across questions and across subjects; see Table 1. However, dominance was violated just once (out of 320 possible occasions); the details can be found in Carbone/Hey (*op.cit.*), ibid. A final example (there are others) is given by Ballinger/Wilcox (1996)[9], where the 'within sets' switching rates were well above zero (even though well below 0.5).

Analytical evidence on 'error' comes from attempts to fit models (of deterministic preferences and deterministic choice) – that is, to estimate preference functionals – to experimental data. How such fitting is done depends on the numbers of questions asked to experimental subjects. Two main approaches can be identified: first, when large numbers of subjects are asked a relatively small number (3, 4 or 5 are typical values) of pairwise choice questions and where the estimation or fitting is done *across* all subjects; second, when an immaterial number of subjects are asked a relatively large number of questions (94 or 100 pairwise choice questions are typical values) and when the estimation or fitting is done subject by subject. In both approaches it is almost invariably true that no one model, no one preference functional, fits the data exactly: with the first approach, there are some subjects whose behaviour *is* consistent with the model being fitted and some subjects whose behaviour is *not* consistent; with the second approach, there are, for each subject[10] and for each estimated preference functional, a subset of the subject's answers that are inconsistent with that estimated functional. In other words, for neither approach is there a (relevant) preference functional that fits the data exactly.

Thus fitting, or estimation, requires some stochastic specification – specifying where the randomness in the data is coming from. Some of the early studies had some hidden assumptions – they were not explicitly stated – and sometimes it was impossible to reconcile the statistical test being used with the preference functional being estimated/fitted/tested. For example, in tests of EU using Common Ratio or Common Consequence (or 'Allais Paradox') type questions, subjects were presented with 2 pairwise choice questions, first between C_1 and C_2 and then between D_1 and D_2, where the risky prospects were chosen so that, under EU, C_1 would be preferred to C_2 if and only if D_1 was preferred to D_2. To test whether EU was being violated or not, a standard binomial test of the difference between 2 proportions (the proportion choosing C_1 in preference to C_2 and the proportion choosing D_1 in preference to D_2,) was usually carried out. But this test relies on the assumption that, for each observation (each subject) there is a constant probability of C_1 being chosen out of C_1 and C_2 and the same constant probability of D_1 being chosen out of D_1 and D_2 – and that the realisations on the 2 pairs are independent. But this is not what EU says – on the contrary it says that for each subject (each observation) the 'probability' of choosing C_1 *and* choosing D_1 is either zero or one – depending upon the risk aversion of the subject. The only randomness that EU allows is in the distribution of the degree of risk aversion *across* subjects. Unfortunately, if the *same* subjects are asked the two pairwise choice questions (as is usually the case[11]) then the observations are *not* independent (across the 2 pairs) and exactly the same proportion of subjects should report a preference for C_1 as report a preference for D_1. No statistical test is necessary – any difference in the 2 proportions is a violation of EU.

Two recent studies which tell a consistent story are Harless/Camerer (1994)[12], and Hey/Orme (1994), *op.cit*. They differ fundamentally in the way they tell that story. Harles/Camerer (1994) *ibid.*, when fitting a particular preference functional, assume that there is a distribution of functions (within that functional) across the population of subjects – as in the story above. So, for example, with EU there is a distribution across the subjects of the utility values $u(x_i)$ (i = 1, ..., n). However, this is invariably insufficient to explain the data – as there are responses that are inconsistent with *any* specific preference function. How do Harless and Camerer cope with this? By assuming that these responses are the consequences of *error*: the subjects were trying to respond with their true preferences but for some reason reported incorrect preferences. Harless and Camerer need some stochastic specification for this error: they make the particularly simple assumption that on any pairwise choice question there is a given probability that the subject makes a mistake – and reports, for example, that they prefer C_1 when they in fact prefer C_2 (or *vice versa*). This enables them to 'explain' the data.

Hey and Orme also assume that subjects make mistakes, though their story is rather different; in essence, they assume *measurement error*. Subjects have true preferences – as given by their true preference functional, but they use it with error. Instead of using the true value of the functional, $V(P)$ when evaluating a risky prospect P, they use instead $V(P) + \epsilon$ where ϵ is a white noise error term. This inevitably means, in contrast to Harless and Camerer, that the probability of making a mistake on a pairwise choice question is not constant, but instead depends on the difference between the two valuations, $V(P)$ and $V(Q)$, where P and Q are the two prospects. It is clear[13] that this is a better approximation to the truth[14].

Other stories are clearly possible, but what is important is the fact that none of the 'deterministic preferences and deterministic choices' theories explains the data exactly. So my conclusion at this stage of the argument is that, if theories with deterministic preferences and deterministic choices are what people are trying to follow, *then people must be making mistakes.*

IV. ALTERNATIVE INTERPRETATIONS OF THE EVIDENCE

Of course, there is an alternative interpretation: namely that people are 'deliberately' randomising. This could arise either with 'deterministic preferences and stochastic choice' or with 'stochastic preferences and stochastic choice'. There are several variants of each of these. Let me begin with the first.

The most satisfying story (to economists) within the 'deterministic preferences and stochastic choice' framework is that of Mark Machina[15]. Here we have the story discussed above – where indifference curves in the Marschak-Machina Triangle are convex – and hence mixtures (that is, randomisations) *may* be preferred to either of the two choices in a pairwise choice problem. The difficulty is the word 'may', since it is not *guaranteed* that mixtures are preferred. In such a case the probability of choosing one or other of the two choices must be zero – from which it follows that the likelihood takes the value zero.

To operationalise this approach one needs to take a particular functional form for the (convex) preferences; a natural contender is the *Quadratic Utility* theory of Chew, Epstein and Segal (1991).[16] This gives quadratic indifference curves in the Marschak-Machina Triangle. This approach was followed in earlier work of mine with Enrica Carbone[17], but the results (as far as Machina's story was concerned) were disappointing. The model was fitted on part of the data reported in Hey/Orme (1994) – where 80 subjects were asked 100 pairwise choice questions. In this experiment, subjects were allowed to express indifference – indeed on one repetition of the experiment, 36 subjects did so on at least one of the 100 questions. This creates problems in the analysis of the data[18] – particularly so for Machina's model, since in his model they will be genuinely indifferent for just a small discrete set of parameter values. We discarded these 36 subjects from our subsequent analysis.

As is reported in Hey/Carbone (1995), of the remaining 44 subjects, we were unable to find any parameter vector for which the log-likelihood was finite for 40 of them. This leaves 4 subjects (out of the original 80) for which Machina's model fits. Of these 4, there were 2 subjects for whom one of the 'deterministic preference plus mistakes' stories fitted significantly better. Of the remaining 2, neither subject had estimated parameter values which respected the monotonicity requirement of the theory – though for one of these two the monotonicity requirement was approximately satisfied. So, out the original 80 subjects, we end up with just one for whom Machina's model fits, fits better than other models, and fits in a manner consistent with the theory itself. Hardly an overwhelming endorsement of the theory!

The alternative story – that both preferences and choices are stochastic – has been suggested by Graham Loomes and Bob Sugden[19], and has been empirically explored by Enrica Carbone[20]. She fits the Stochastic EU story to the data in Carbone/Hey (1995). She finds that it does indeed do better than 'EU plus mistakes' for a significant number of the 40 subjects – though there are also a large number of subjects for whom 'EU plus mistakes' fits better.

Recall that Stochastic EU has the great attraction that it respects dominance – as do the vast majority of subjects in experiments. Indeed, when dominance is 'obvious', it is very rarely the case that dominance is violated by subjects (though it is more often violated when dominance is not so obvious). However the other attractions of Stochastic EU are not so clear. First, there is no axiomatic base to the theory – in contrast to most of the theories in the 'deterministic preferences and deterministic choice' set of theories. Moreover, while the basic idea of the theory – that subjects do not quite know their utility function, so that it appears to vary over time – is appealing, the idea that it varies from choice to choice in a sequence of (almost contemporaneous) pairwise choices is somewhat less appealing. One can understand it changing from day to day, but hardly from minute to minute.

Which brings me back to 'deterministic preferences plus mistakes'.

V. RATIONAL MISTAKES?

The argument above suggests that contenders for a satisfactory explanation of the data are two: 'deterministic preferences plus mistakes' and 'stochastic preferences and stochastic choices'. Whilst there may be difficulties in empirically distinguishing between these two contenders (though I have offered some remarks on how this might be done in the section above), the former has greater theoretical appeal. There is also an additional theoretical argument that I would now like to explore. In essence, it asks: "is it rational *not* to make mistakes?".

Consider the costs and benefits of not making mistakes. The benefits are clear: it is getting the truly preferred choice rather than one less preferred. The magnitude of this benefit is related to the 'distance apart' of the two choices – the further apart (in terms of the preferences of the subject), the greater the benefit of not making a mistake. If the individual is indifferent then the benefit is zero. A measure of the benefit, when choosing between P and Q can be found by finding the value of a such that $V(P+a) = V(Q)$, where by '$P + a$' I mean the risky choice with the same probabilities as P but wherein all the outcomes have been augmented by the constant amount a.[21] Clearly in general, the larger is $V(P) - V(Q)$ the larger will be a; moreover, if $V(P) - V(Q) = 0$ then a will equal zero.

The costs of not making a mistake depend on the expenditure of time and effort sufficient to ensure that a mistake is not made.[22] This depends on how difficult the particular choice problem is. There is very little good theory to guide us here, though one could argue that the further apart are the two choices, then the easier the decision problem is, and hence the lower the costs of not making a mistake. If the two choices are very close (but not actually indifferent) then the costs of not making a mistake could be rather high.

The details can be left to another occasion, but what is clear from the above decision is that there *are* costs and benefits of not making mistakes, and it is by no means obvious that the 'best' decision involves making no mistakes at all. Indeed, on the face of it, the contrary conclusion seems to be the correct one: that it is unlikely to be 'optimal' to make no mistakes. Furthermore, the above discussion suggests that, the closer together are the two choices, then the more 'sensible' it is to make a mistake. But note also: the whole idea of choosing mistakes *optimally* sounds a thoroughly misguided one.[23]

VI. CONCLUSIONS

There is one final point to be tied up before concluding: some may argue that *all* the theories are absolutely wrong (which is manifestly the case) and *hence* that we should seek further theoretical explanations – ones which encompass the present explanations as well as adding more. The problem with this is that is will further weaken the predictive power of our theories. Worse, such theories will simultaneously be more difficult to apply as well as being more implausible as theories of real decision making in simple risky-choice problems. For after all, the choice problems being considered are extremely simple choice problems; is it plausible to suggest that human beings tackle such simple choice problems with horrendously complicated decision rules? Is it not more plausible to suggest that people use a relatively simple decision rule (such as EU) but make mistakes? This is what the data seems to be suggesting. To that I finally turn.

My conclusion concerning the interpretation of the data is *either* that people have deterministic preferences but make mistakes in the application of their preferences, *or* that people have stochastic preferences and make stochastic choices. Whilst empirically these two approaches may be difficult to disentangle, theoretically the first of the two has greater appeal, particularly when it is appreciated that there are costs involved with implementing ones preferences (whether deterministic or stochastic) *exactly* and hence it may well be rational to make mistakes: my answer to the question in the title of this paper must be 'yes'.

TABLE 1: CONSISTENCY IN THE ANSWERS OF THE REPEATED QUESTIONS

```
 1                                      40
 1 ******************************1*********
 2 0****0**1*****0**********0****0*****0***
 3 **********1*0******1*****1***0*******0**
 4 ***********1*0***1**1*****1***11*1***0**
 5 *************1************1************1
 6 ***0**1***0****0*0*0*******0************
 7 *************************************1**
 8 ****************************************
 9 ***0**0****1******10*******0***1********
10 *************1*0***********0*****1******
11 ******0*************************11******
12 1*0********1*****************0***00****1
13 *******************1*******11***10***1**
14 ****************************************
15 ********1****1************1*************
16 *****************01*1*********0*******00
17 ****1*************0*0******0****1****0**
18 *******1***0**1**********************0**
19 1*******0*00*******0*********1**0***1***
20 *****1**0*1**********1**1***********0***
21 ***********1*0*****************1*******0
22 ******0***01*0******0*********0****1*0**
23 *******************************0********
24 ****************************************
25 ****0****0*0*****0000******0****0***01**
26 ******1*1**1*****0*10*1**************1**
27 ****0*1*0**********0*********0**1*******
28 **********1******************01**0******
29 0**0*********1***********0*******1******
30 **0******0*******000**0**************00*
31 ***********0****************************
32 ***0******11*0****************1110**1**0
33 ***********0*************************0**
34 ***********************************1****
35 *************************************0**
36 **0*************************************
37 ****************************************
38 ***0*0**00**************1**1*****1*00***
39 ******10110**1**0*1*1******11*****11*1**
40 0*0*0000*0000000000000*0000*0000000*0*00
41 ***************************1************
42 *******0********************************
43 ***************************0*1*****0****
```

Key:
* if the answer is consistent
0 if the answer is left-left
1 if the answer is right-right

Note: (1) the columns go from subject number 1 to subject number 40 the rows go from question pair 1 to question pair 43
(2) there was a data entry mistake in question pair 40.

NOTES

* I am grateful to the organisers of the Conference for inviting me to Vienna, and for organising an extremely interesting conference in which the most fruitful feature was the bringing together of experts from different disciplines.

1. John von Neuman/Oskar Morgenstern, *The Theory of Games and Economic Behaviour*, Princeton: Princeton University Press, 1947
2. With the exception of Yaari's Dual theory and the possible exception of the (original) version of Prospect theory.
3. This can be so under Quadratic Utility theory (see Soo Hong Chew/Larry Epstein/Uzi Segal, "Mixture Symmetry and Quadratic Utility" in *Econometrica* 59, 1, 1991, pp.139-163).
4. There is an interesting philosophical point here: suppose the individual recognises all of this, carries out the randomisation - and that P is the consequence. Could one argue - as Machina invites us to do - that the individual ought to carry out the randomisation again, since the individual prefers the mixture to P? This line of argument seems to lead to inaction rather than action!
5. See, for example, Duncan Luce, "A Probabilistic Theory of Utility", in: *Econometrica* 26, 1, 1958, pp 193-224.
6. John Hey and Christopher Orme, "Investigating Generalisations of Expected Utility Theory Using Experimental Data", in *Econometrica*, 62, 6, 1994, pp.1291-1326.
7. The order of the questions was randomly chosen on each occasion, and the position of the left and right choices was reversed on the second occasion.
8. Enrica Carbone/John Hey, "A Comparison of the Estimates of EU and non-EU Preference Functionals Using Data from Pairwise Choice and Complete Ranking Experiments", in: *Geneva Papers on Risk and Insurance Theory* 20, 2, 1995, pp.111-133
9. Parker Ballinger/Nat Wilcox "Decisions, Error and Heterogeneity", University of Houston discussion paper, 1996.
10. Except in very rare cases.
11. Though was not the case in an early experiment by Kenneth MaCrimmon and Stig Lars, "Utility Theory: Axioms versus 'Paradoxes'", in Marice Allais/Ole Hagen, *Expected Utility and the Allais Paradox*. Dordrecht: Reidel, 1979, pp.333- 409.
12. David Harless/Colin Camerer "The Predictive Utility of Generalized Expected Utility Theories", in: *Econometrica* 62, 6, 1994, pp.1251-1290.
13. See John Hey, "Reconciling Harless and Camerer and Hey and Orme", discussion paper, University of York, 1996.
14. An even better approximation is when the variance of the error term ϵ is not assumed to be constant, but rather a function of what appears to be a measure of the complexity of the decision problem facing the subject. See John Hey, "Experimental Investigations of Errors in Decision Making under Risk", in *European Economic Review* 39, 4, 1995, pp.633-640.
15. Mark Machina, "Stochastic Choice Functions Generated from Deterministic Preferences over Lotteries", in: *Economic Journal* 95, 379, 1985, pp.575-594.
16. See endnote 3.
17. John Hey/Enrica Carbone, "Stochastic Choice with Deterministic Preferences: An Experimental Investigation", in: *Economics Letters*, 47, 2, 1994, pp.161-167.
18. Indeed so much so that we now wonder why we gave the subjects the option: it clearly does not increase their utility, since, if they are genuinely indifferent, there is no advantage to them in saying so; and it certainly does not help us because we cannot be sure - for the reason noted above - that the cases when they reported indifference were indeed the only cases when indifference was experienced. Furthermore, there is a serious problem in terms of most of the stories we are trying to fit - in that the probability of them being *exactly* indifferent is clearly zero.
19. Graham Loomes/Bob Sugden, "Incorporating a Stochastic Element into Decision Theories", in: *European Economic Review* 39, 4, 1995, pp.641-648.

20. Enrica Carbone, "Estimation of Stochastic Utility Theory using Experimental Data", paper presented at the Naples Experimental Economics conference, 1996.
21. An alternative measure is the value b such that $V(P) = V(Q+b)$. Clearly $a > 0$ if and only if $b < 0$.
22. Actually, a proper 'theory of errors' would recognise that error-making is (necessarily) stochastic – which means that the above discussion should be formulated in stochastic terms. However, this rather informal treatment is sufficient for my present purposes.
23. See Vernon Smith/James Walker, "Monetary Rewards and Decision Cost in Experimental Economics", *Economic Inquiry* 31, 2, 1993, pp.245-261, for an apparently conflicting view.

Department of Economics
University of York
York YO1 5DD
UK

DANIEL M. HAUSMAN*

RATIONALITY AND KNAVERY

This paper makes a modest point. Suppose one wants to evaluate alternative policies, institutions or even constitutions on the basis of their consequences. To do so, one needs to evaluate their consequences and one needs to know what their consequences are. Let us suppose that the role of economic theories and game theory in particular is mainly to help us to use information we already possess or that we can acquire at a reasonable cost to judge what the consequences will be. We do not necessarily need true theories or theories that provide perfectly precise predictions. What sort of accuracy and precision we need depends on how the consequences of alternative policies are evaluated. The worth of an economic theory in this context of policy assessment depends on the accuracy of the predictions the theory permits, the costs of gathering the needed information, and the costs of learning and using the theory. My only excuse for uttering these truisms is that well known economists and philosophers have denied them.

My title derives from the following remarks of David Hume:

> Political writers have established it as a maxim, that, in contriving any system of government, and fixing the several checks and controls of the constitution, every man ought to be supposed a *knave*, and to have no other end, in all his actions, than private interest. By this interest we must govern him, and, by means of it, make him, notwithstanding his insatiable avarice and action, cooperate to public good....
>
> It is, therefore, a just *political* maxim, *that every man must be supposed a knave*; though, at the same time, it appears somewhat strange, that a maxim should be true in *politics* which is false in *fact*. (Hume 1741, pp.40-42)

When Hume says one should suppose that people are knaves, he means that one should suppose both that they are motivated exclusively by private interest and that they will pursue their private interests competently. He has in mind in particular office holders, who, faced with a conflict between private and public interest cannot be trusted to act in the public interest. But the axiom is stated in a perfectly general way, and one can take it to apply equally to those who govern and those who are governed.[1] If one seeks to design institutions to secure the public interest, one must incorporate incentives and sanctions that will lead individuals who pursue their private interests to serve the common good. The most obvious way legislators accomplish this is, of course, by the use of rewards and punishments. For example, suppose (as seems likely) it is in the common interest if individuals are secure in the possession of their property. Individuals seeking their own advantage will not automatically cooperate to secure this good. One can lessen the conflict between the common good and the pursuit of individ-

W. Leinfellner and E. Köhler (eds.), Game Theory, Experience, Rationality, 67–79.

ual interest by making theft more difficult (with the help of security procedures) and less advantageous (with a system of sanctions).

This idea has a long lineage. St. Augustine held that in politics everyone ought to be considered not merely a knave, but a sinner doomed to eternal damnation. Although a few people are saved by God's merciful grace, there is no way to tell who is damned and who is saved, and political institutions need to control the damned. By a fortunate dialectical twist worthy of a deity, the central characteristic of the damned – their egoistic attachment to earthly goods – renders them governable and makes possible peace and imperfect earthly justice (Deane 1963).

This reading of Hume's axiom should be somewhat alarming to liberals, who do not like state and social coercion. Although liberals recognize the need for sanctions to protect individual rights, they would like to circumscribe the functions of government strictly and to permit individuals to enjoy as much freedom as is consistent with the protection of rights. When individual rights, such as the right to property, are not at stake, how can the pursuit of private interest be made to serve the common good without intolerable interferences with individual freedom?

Fortunately, institutions can sometimes lead the pursuit of self-interest to serve public interests in more subtle ways. Competitive markets are the most famous example. As Hume's young friend, Adam Smith, maintained, sometimes the invisible hand will do the trick (1776, Book IV, ch. 2). A liberal might put Hume's and Smith's thoughts together as follows: Since we must suppose individuals to be knaves in our design of constitutions, let us rely on markets for the organization of production and distribution. For markets harmonize private and public interests, and at the same time they minimize coercive state actions. Arguments like this one send a warm glow to the heart of every liberal.

But there are well-known difficulties. The most obvious is that Smith's argument that the pursuit of private interest serves the public is unsound. Nobody with any training in economics would have trouble punching holes in it. The problem is to devise rules for a game that will have the outcomes one wants. When economists have repaired Smith's argument, they have evaded rather than solved this problem. Rather than proving that the outcomes people will achieve meet relevant normative criteria, they have proven that unusually rational and self-interested agents might possibly achieve an efficient outcome in unusual circumstances.

Economists have a great deal to say about the virtues of markets that do not depend on fancy general equilibrium theorems. Markets economize on information. They are highly flexible and encourage individual innovation. Although these considerations do not constitute a demonstration that markets harmonize private and public interest, they give one good reason to believe that markets are a good alternative to coercion. The historical record also provides evidence that markets typically coordinate individual pursuits better than the alternatives.

I. CAN ECONOMICS AND RATIONAL CHOICE THEORY BE OF VALUE EVEN IF THEY ARE PREDICTIVELY EMPTY?

Some thinkers have held that rational choice models like those in economics are of value regardless of their predictive worth. In his often-cited essay "Rational Fools", Amartya Sen asks why Edgeworth portrays individuals in his economics as self-interested, despite believing that individuals are not in fact always self-interested. One answer is that Edgeworth believes that in particular domains (war and "contract") individuals are, to a good approximation, self-interested (1977, p.318). Another answer, Sen maintains, is that Edgeworth - like many other economists - is interested in the abstract question of whether egoistic behavior serves the general good (1977, pp.319-22). General equilibrium theory answers this theoretical question (with respect to idealized circumstances). It does not matter if people are in fact egoists.

From a very different perspective, Alexander Rosenberg also attempts to separate the worth of rational choice theories from the truth of their axioms concerning preferences and choice and from their predictive adequacy. In a series of books and articles over the last fifteen years, culminating in his *Economics: Mathematical Politics or Science of Diminishing Returns* (1992), Rosenberg has consistently criticized the predictive value of economics. Many of his comments are focused on general equilibrium theory, but his critique applies to all rational choice models and to game theory. I believe that this critique cannot be sustained (see Ross 1995, Rappaport 1995, and Cottrell 1995), but I shall not challenge it here.

I do not accept Rosenberg's indictment of standard economic theory, but if one does, then one needs to explain why so many people respect standard economic theory, work on it, and apply it to new domains, indeed even to new disciplines. One answer Rosenberg gives is that economists (and their fellow-travellers among political scientists, anthropologists, and sociologists) are interested in economic theory for its mathematical structure (1992, ch. 8). Like mathematicians concerned to prove theorems in some system of geometry and uninterested in the true structure of space, rational choice theorists, including economists, are concerned to prove theorems employing axioms of rationality, and they are uninterested in whether individuals conform to those axioms. Undoubtedly Rosenberg is right that some economists are really mathematicians.

Rosenberg's second reason why rational choice and economic models are of interest, even if they are empirically empty, is more relevant here. Rosenberg claims that economics forms part of a *normative* argument for the market.

> But at least now we can understand why economists continue to lavish attention on general equilibrium theory. It is not because they think it can be improved in the direction of a descriptively and predictively accurate explanation of economic activity, but because they believe it is already part of the best contractarian[2] argument for the adoption of the

market as a social institution and, more important, stands a chance of becoming an even better one, as its assumptions are weakened, changed, and varied. (1992, pp.219-20)

Rosenberg's assertions here are incredible, if they are read as claims about the conscious beliefs of economists who value orthodox economic theory, including game theory. Presumably the real story has to involve a considerable element of false consciousness, and the account looks much like an ideological critique, though Rosenberg never says whether this reason for lavishing attention on general equilibrium theory is a mistake.

Although the theorems of general equilibrium theory and welfare economics establish conclusions concerning the existence and Pareto optimality of equilibrium (given idealized agents and circumstances), the political argument for the market surely rests as much on informal arguments for the virtues of markets and on claims concerning the *freedoms* markets permit and protect. Rosenberg does not mention this, in part because he tends to overemphasize the centrality of general equilibrium theory and in part because economics and rational choice theory in general contribute little to our understanding or appreciation of the links between markets and freedom. The normative contribution of economics and rational choice theory apparently lies in their account of the consequences of markets for welfare and growth. Since Rosenberg is interested in the normative role of markets, he accordingly focuses exclusively on welfare arguments for markets.

Rosenberg maintains that rational choice theories can contribute to welfare arguments concerning markets, even if they are largely without empirical content. He cites Brennan's and Buchanan's contrast between the choice *of* constraints (in a constitutional context) and choices *within* constraints (in the ordinary course of economic life) (Buchanan 1989b, pp.44-5, Brennan and Buchanan 1983, p.4; 1985, p.xii). Even though rational choice theory and game theory cannot predict very well how individuals choose within constraints, they can still guide our choice of constraints.

Notice that this argument for the worth of economic theories can be made just as well by someone who questions whether they underwrite *laissez-faire* policies. General equilibrium theory does not in fact tell us that knaves interacting on unfettered markets in circumstances involving externalities and public goods achieve optimal outcomes. Frank Hahn goes so far as to argue that one of the main virtues of general equilibrium theory is that it enables one to see through shoddy arguments for *laissez-faire* policies (1973, p.324). Hahn's point (with which I think Rosenberg might agree) does not in any way undercut the normative importance of economics, which Hahn emphasizes as strongly as Rosenberg. Hahn is making a claim about what the normative significance of general equilibrium theory is. He is not denying that it has normative significance. Hahn thinks that economics underwrites the normative conclusion that *laissez faire* is not always good policy.

II. AN OBVIOUS DIFFICULTY

On Rosenberg's view, economic theory is supposed to guide the choice of constraints that structure the production and distribution of goods and services. It is supposed to guide this choice by providing an account of the consequences of alternative sets of constraints for welfare and growth. But if rational choice and game theoretic models have little predictive value, how can they successfully tell us what the consequences of markets will be for welfare and growth? If they cannot tell us what the consequences of markets will be for welfare and growth, what role can they play in normative arguments concerning institutional design?

Rosenberg notes the difficulty and asks, "Can a theory that is, as I have argued, predictively weak, carry the normative burden that general equilibrium is being saddled with by this interpretation?" (1992, p.224) His answer is, in effect, that a theory may be "true in politics which is false in fact". The fact that individuals are not perfectly self-interested, intelligent, or rational and that their information is often biased and mistaken undercuts the predictive worth of game theory and rational choice models.[3] But it does not undercut their normative worth. For what matters when we are considering alternative "constraints" is not what their consequences would in fact be, but what their consequences would be *if all men were knaves* – that is, if everybody were rational, intelligent, self-interested, and well informed.

(A word about terminology: "Rationality" here involves having complete and transitive preferences and choosing among feasible alternatives what one most prefers. "Intelligence" means being able to figure out anything that the theorist can. "Self-interested" means that the *objects* of one's preferences ultimately contribute to one's own well-being. To be "well-informed" is to have information that is at least as good as the theorist's or policy maker's. Let us dub someone who is rational, intelligent, self-interested, and well-informed "economically rational". I shall take economic rationality as an explication of Hume's notion of a knave.)

Rosenberg argues that if game theory can tell us what the consequences of institutions and policies would be if everyone were economically rational, then it is of great normative importance, even if it cannot tell us what the actual consequences of institutions would be. Policy-making and the choice of constraints should be guided by knowledge of how ideally rational and self-interested agents *would behave*. Actual individuals may not choose that way, and a set of institutions designed to work optimally with knaves may not work optimally with actual individuals. With John Stuart Mill as absolute monarch, absolute monarchy might be best. But we cannot be assured that our absolute monarchs will always have sterling characters. As public spirit waxes and wanes, our institutions must be robust. Institutions must be designed to survive the "worst case" in which individuals are driven by nothing but their private interests and are ruthlessly rational in pursuing them. Thus Hume's axiom and thus the oddity "that a maxim should be true in *politics* which is false in *fact*".

III. FURTHER ARGUMENTS FOR THE NORMATIVE RELEVANCE OF ECONOMICS

In developing this argument for the independent normative significance of models of interactions among knaves, Rosenberg cites the work of Geoffrey Brennan and James Buchanan. In their 1985 book and in a joint essay, they offer a lengthy defense of taking economic rationality to be the appropriate model of human behavior for the purposes of "constitutional economics". "Constitutional economics" is the attempt to assess general rules within which economic and political actors pursue their objectives.[4] Like Rosenberg they argue that the appropriateness in the context of constitutional economics of taking individuals to be economically rational – that is, rational, self-interested, intelligent, and reasonably well-informed – is independent of the extent to which these claims about individuals are true.[5]

Brennan and Buchanan's first argument (1985, pp.48-50) is that if agents are modeled as knaves in their economic behavior, then political behavior should be modeled the same way. This argument assumes what Rosenberg wants to show, that economic rationality is the right model in this context for economic behavior, and is of no use in establishing Rosenberg's conclusion. Brennan and Buchanan are right to be dissatisfied with theories that posit different models of behavior in different domains, but wrong to suggest that this methodological commitment takes precedence over evidence. If political theory T1 models people as altruists, while T2 models them as knaves, then those who think that economic rationality is the right model of economic behavior have reason for dissatisfaction with T1. But this reason for dissatisfaction is no argument for accepting T2, if T1 fits the data much better.

Second, Brennan and Buchanan argue that their inquiry concerns how institutions can reconcile or harmonize conflicting interests, and that it thus changes the subject to assume away conflicts of interest. "If we want to discover how institutional rules can turn conflict into cooperation, we cannot simply assume that persons who operate within those rules are naturally cooperative" (1985, p.53). But to deny that people are economically rational is not, of course, to assume that people's aims and interests are in harmony.

Third, Brennan and Buchanan argue that the truth about people's actual interests and capabilities influences only the *significance* of the enterprise of constitutional economics, not what the appropriate "analytical method" is (1985, p.53). It appears that they *define* constitutional economics as an inquiry into what institutional frameworks are best for a society full of economically rational agents. If constitutional economics is defined this way, then this inquiry must of course model individuals as knaves. But this trivial conclusion is irrelevant to the question of whether one ought to assume that individuals are economically rational when one is addressing problems of constitutional design.

Fourth, Brennan and Buchanan make the assertion (which they bolster by means of an extended example) that models that best fit the data "will be systematically biased in the direction of inadequate constraints". (1985, p.55). Their 1983 essay is devoted mainly to this argument. Suppose that actual behavior involves a mixture of economic rationality on the part of some agents and public spiritedness on the part of others. If the welfare loss connected with a failure of public spiritedness is more than a linear function of the divergence from public spiritedness, then one will underestimate the loss if one supposes that every agent pursues the same partly public-spirited policies. Using the best estimate of average behavior leads one systematically to underestimate the welfare loss. One might do better by modeling everyone as a knave.[6]

Even if one concedes that welfare losses are generally convex functions of divergences from public-spiritedness, this argument fails. First, it shows only that one *might* do better using a model of individuals as economically rational. It does not deny that one might do worse. How is demonstrating that procedure 1 might be better than procedure 2 (though it might be worse) an argument for procedure 1? Second, as Brennan and Buchanan note, it seems that they have offered a critique of one particular misuse of data rather than an argument that an *a priori* model is better than one informed by data (1983, p.13). In response Brennan and Buchanan point out that information about the distribution of different modes of behavior may not be available, and they seem to suggest that one is forced to choose between fitting average behavior and supposing everyone is a knave. But if one really has no idea about the distribution of different behavior, why should we assume that everyone is economically rational rather than, for example, assuming some uniform distribution over different motives. (Clearly the reason is empirical!)[7]

Finally Brennan and Buchanan argue that without institutions that constrain the actions of knaves, altruistic behavior may be driven out.[8] Why this argues for modeling behavior as entirely self-interested is never explained.

IV. PROBLEMS WITH THIS ARGUMENT AND HUME'S AXIOM

Rosenberg takes Hume's axiom as a maxim of caution: our institutions ought to hold up even if people were not at all altruistic or public spirited. This argument does not, I think, fare any better than Brennan and Buchanan's. I suggest that game theoretic models of the interactions of knaves have normative significance concerning some domain only if they predict the outcomes people will reach. If a model goes predictively awry, then one should not rely on its advice concerning institutional design.

There are three decisive problems with using Hume's axiom to pull a normative rabbit out of an empty predictive hat. First, institutional choice should not be made on the assumption that every man is a knave. Second, actual economic theories typically fail to model agents as economic rational, and the attempt to repair this deficiency leads to deep problems. Third, economics is not much better at predicting what outcomes knaves will reach than it is at predicting what outcomes actual individuals will reach.

Should We Suppose That Everyone Is a Knave?

Suppose it is the case that the optimal institutions (the institutions that best promote the public good, whatever that may be) for real individuals are not the same as the optimal institutions for knaves. Then there are costs to employing institutions and policies designed for knaves.[9] There were costs attached to governing Victorian England by means of imperfectly representative institutions rather than handing over the power to John Stuart Mill. Once one recognizes such costs, one needs to ask whether the benefits in security, resilience or whatever are worth the costs. If one is an expected utility maximizer, one also needs to consider the probabilities of the possible outcomes of different institutions. Even if some institutions and policies would be disastrous if everyone were a knave, they may be superior in all actual circumstances. If the outcomes of institutions designed for knaves are much worse than the outcomes designed for actual individuals and the odds of everybody being a knave were low, then it would be foolish to choose the institutions designed for knaves.

One might argue that if institutions rely upon altruism or public spirit, they will be exploited by those individuals who are in fact knaves, and the exploitation will lead to the same outcomes that a society full of knaves would achieve. For example, individual entrepreneurs might act in a public spirited way by refusing to dump their wastes in adjoining marsh, paying higher wages, refusing to lay off workers in hard times, and so forth. Similarly individual entrepreneurs might be foolish and ill-informed. But public spirited, irrational, and ignorant actions reduce profits, and firms cursed with moral or stupid managers will be unable to compete with firms whose leaders are less squeamish and more intelligent in pursuit of profits. To preserve altruism and public spirit, institutions must place no reliance on them.

This argument relies on questionable empirical assumptions (Hausman 1989). Moral or public-spirited behavior on the part of firms is not always costly, and indeed it can be advantageous. Firms that are known to be "moral" face lower labor costs and enjoy the good will of customers and suppliers (Hausman and McPherson 1996, ch. 3). Furthermore, even if deviations from economic rationality reduce profits, non-profit maximizing behavior will not necessarily be driven out. Markets must be sufficiently competitive and there must be a sufficient amount of profit-maximizing behavior already there. Altruistic behavior can spread, too (Becker 1976).

Suppose one rejects expected utility maximization as the best way to choose institutions and employs a maximin principle instead. If in addition one holds that the worst outcomes would obtain if everyone were a knave, then it would follow that in institutional choice one should suppose that everyone is a knave. This may be Rosenberg's view. Maximin is, of course, a dubious principle of choice. It could condemn all mankind to stunted lives to protect against a one in a billion chance of a society of knaves.

Even if one accepts maximin, the argument does not go through unless the worst outcomes occur if everyone is a knave. But that claim only needs to be made explicit to be doubted.[10] The worst social disasters -- Nazi genocide, Serbian "ethnic cleansing", slaughters of Hutu's and Tutsi's in Rwanda - would never occur in a game played by economically rational individuals. If one ought to design institutions for the worst case, then one ought to prepare for something much worse than universal knavery. In the waning years of the twentieth century, it seems overly optimistic to suppose that everyone is a knave.

Robertson's suggestion (1956) that economists economize on "love" is more moderate and sensible. Instead of supposing falsely that there is no altruism or civic spirit, one should recognize that its quantity is limited, and that institutions that limit conflicts between private and public interest work better.[11] Yet even this more moderate proposal is dubious, because it falsely supposes that the extent to which individuals are economically rational is independent of institutions and policies. But, as Albert Hirschman (1985) points out, altruism and public spirit are not given in fixed quantities like iron ore or petroleum. Their extent depends upon institutions and policies. If there are too many possibilities for free-riding - if one overestimates the extent to which people will be public-spirited and moral -, altruism and morality may be undermined. If, for example, one has a public transit system with too little enforcement of fares and prevalent cheating, those who pay their fares will feel like "suckers", and the system can undermine the public-spiritedness upon which it depends. But a system with too *few* possibilities for free-riding can undermine public-spiritedness and moral commitments as well. Workers who have to punch a time clock may be more likely to leave when they have put in their eight hours than workers who are trusted to fulfill their responsibilities. People can become what they are assumed to be, and with too much regulation people may not be able to make trust-inducing overtures to one another (Pettit 1995, p.225).

Economic Models Are not Populated by Knaves

If one takes a knave to be someone who rationally and effectively pursues his or her material self-interest, individuals as depicted in most economic models are not knaves. First of all, economic theories typically suppose that firms maximize net returns. Just how the internal structure of a firm is supposed to insure that a set of knaves acts so as to maximize the net returns for the firm is deeply mysterious. Second, standard theory says nothing about how knaves are supposed to reach equilibrium and to restore equilibrium after various "shocks". Instead the theory offers an entirely fictitious story. If I've got some extra beaver and you've got some extra deer, I tell an auctioneer how much beaver I'm willing to part with at the price the auctioneer announces, and you tell the auctioneer how much beaver you're willing to buy and the auctioneer adjusts this price (and all other prices) up or down until quantities demanded and offered equalize.

Third, economic theories generally suppose that agents obey the law. You and I patiently keep the auctioneer filled in on how much we offer and demand. You don't just shoot me and take my extra beaver. Why not? Of course there are the sanctions of the state. You might get caught. But people can get away with murder, sometimes literally and often figuratively. Opportunities abound to take advantage of the inevitable incompleteness of contracts. Over the past generation economists have begun to take all these questions seriously, but as game theorists have made clear, avoiding these short cuts and insisting on modeling how outcomes are reached pose enormous theoretical challenges. The very characterization of rationality is controversial. I would argue that game theory has taught economists modesty. It is not easy to say what knaves are or what games they are playing.

The Predictive Problems of Economics Do not Derive Mainly From Assuming People to Be Knaves

Suppose, as Rosenberg alleges, that economic theories in general have little predictive value. Then I maintain that they will be of little help in designing institutions, even if everyone were a knave and these models correctly characterized what a knave is. Suppose there were no difficulties with the axioms concerning individual rationality (that people's preferences are complete, transitive and continuous and that choice follows preference) and self-interest (that individuals care only for their own consumption bundles). Would the predictive weaknesses that concern Rosenberg then disappear? Surely not. The predictive problems (which, to repeat, I do not think as serious as Rosenberg urges) do not result entirely or even mainly from the assumptions that people are economically rational. There are still incomplete futures markets, increasing returns to scale, indivisibilities, non-equilibrium trading, and so forth. Furthermore, even the best informed individuals will not have perfect knowledge. One might, of course, maintain that, unlike the assumptions about rationality and self-interest, these other assumptions are reasonable approximations that permit one to make accurate predictions concerning the outcomes of the interactions of knaves. But Rosenberg offers no argument for this view. Why should one believe that it is a reasonable approximation to assume that there are complete futures markets, but unreasonable to assume that people's preferences are transitive? The assumptions about rationality and self-interest are apparently *better* rather than worse approximations than are many of these additional assumptions. If Rosenberg were right about the predictive weakness of economic theories, there would be little reason to rely on them when designing institutions even if everyone were a knave and the theories correctly characterized knaves.

V. CONCLUSIONS

The moral I draw is that economics has normative value in the context of institutional design only if it has predictive value concerning individual behavior within particular institutions. If one believes, as Rosenberg does, that the theories have little predictive value, then one should conclude that they have little normative value as well, and the attention economists and other social theorists lavish on them reflects either mistake or their purely mathematical interest. If one has a less negative appraisal of the empirical content of the theories, one need not have such a negative view of their normative importance or of their overall worth. The bottom line however is empirical content. To what extent do the theories have true implications concerning individual behavior and its social consequences?

Let me emphasize that this paper is no brief for Polyannaism. I maintain merely that when thinking about institutional design, we should make use of the best knowledge we have of human motivations and capacities and of how they are affected by institutions and policies. What little knowledge we have of human beings does not justify a particularly rosy view of their character. Indeed it may be economists who are the Polyannas.

NOTES

* I am grateful for criticisms from Jeffrey Johnson, Julian LeGrand, and audiences at the University of Wisconsin, Duke University, the University of Exeter, and the Conference on Game Theory, Experience, and Rationality sponsored by the Institute Vienna Circle.

1. If Albert Hirschman's reading of the history in *The Passions and the Interests* (1977) is correct, then it is a misreading of Hume to suppose that he is saying that institutions ought to be designed to withstand a worst-case scenario, in which all citizens pursue their own interests without altruism or public spirit. Hirschman argues that during the 17th and early 18th century political philosophers distinguished between, on the one hand, the passions, such as lust, political ambition, the pursuit of honor or revenge and, on the other hand, the *interests*, which were identified with material interests. Insofar as institutions enabled individuals to become wrapped up in the pursuit of their interests, they were thought to promote stability. What had been regarded as the vice of avarice came to be seen as a feature of human beings that made their behavior predictable and that reined in their passions. This essay is concerned mainly with this modern misreading of Hume, not with exactly what Hume meant.
2. Rosenberg maintains that general equilibrium plays its normative role as part of a specifically contractarian argument in defense of the market. But the only connection I can see to contractarianism is the irrelevant fact that both microeconomics and contractarian theories portray individuals as rational. The argument that markets are more efficient than the alternatives would be just as relevant to a utilitarian policy-maker as it would be behind a veil of ignorance.
3. This is an empirical claim, not the fallacious assertion that false premises have only false conclusions.

4. Brennan and Buchanan argue explicitly that one cannot evaluate sets of rules by evaluating their outcomes, on the ground that "there is no external standard or scale through which end states can be valued" (1985, p. 45). I shall, however, ignore this claim both on account of the difficulties in interpreting it and because Brennan and Buchanan's own arguments conflict with it.
5. "In any evaluation of alternative institutions, therefore, *Homo economicus* is a uniquely appropriate caricature of human behavior, not because it is empirically valid but because it is analytically germane" (1985, p. 53).
6. Sometimes it seems that Buchanan is only arguing for models that treat individuals as *less* altruistic, trustworthy, or public spirited rather than for models that treat them as economically rational, "In constitutional choice, therefore, there is an argument for incorporating models of individual behavior that presume more narrowly defined self-interest than any empirical record may warrant" (1987, p. 67).
7. And in his 1989b, after quoting Hume's axiom that every man ought to be supposed a knave, Buchanan writes, "Let me not be misunderstood here. The argument does not defend the use of the narrowly restricted economic model of behavior independently of its descriptive qualities" (p. 48). This is puzzling, because the arguments Buchanan and Brennan make purport to offer a defense of the use of "the narrowly restricted economic model" regardless of how well it fits the data.
8. As their examples show, Brennan and Buchanan are not very precise about what they are assuming about individual behavior. Thus they argue for the claim that individuals ought to be modeled as knaves by pointing out that without social sanctions malevolent behavior is as likely to spread as is altruistic behavior (1985, p. 61). But one who is economically rational is of course neither malevolent nor altruistic. If one is concerned about malevolence, one should not model people as self-interested.
9. Brennan and Buchanan note this point, "In particular, we have not discussed the implications of the elementary fact that more restrictive rules will not only help to prevent the occurrence of disaster but also often preclude actions that may be intended to promote desirable outcomes" (Brennan and Buchanan 1985, p. 54).
10. In their 1985 work, Brennan and Buchanan sometimes equate a society consisting exclusively of economically rational agents with the "worst case" (1985, p. 55), while in their 1983 essay, they write, "Homo economicus by no means represents the worse imaginable character for the social drama" (1983, p. 19). They then argue that the economically rational agent may wind up being only "somewhat worse" than the representative agent, which would be a virtue of their method.
11. Oddly enough, this proposal, which is in many circumstances inconsistent with assuming that everyone is purely self-interested, is cited in support of that assumption by Brennan and Buchanan (1985, p. 63).

REFERENCES

Gary Becker, "Altruism, Egoism and Genetic Fitness", in: *Journal of Economic Literature* 14, 1976, pp.817-26.

Geoffrey Brennan and James Buchanan, *The Reason of Rules*. Cambridge: Cambridge University Press, 1985.

Geoffrey Brennan and James Buchanan, "Predictive Power and Choice Among Regimes", in: *Economic Journal* 93, 1983, pp.89-105; rpt. and cited in Buchanan (1989a), pp.3-23.

James Buchanan, "Constitutional Economics", (1987) from the *New Palgrave*, rpt. and cited from

Buchanan (1989a), pp.57-67.
James Buchanan, *Explorations into Constitutional Economics*. College Station: Texas A&M Press, 1989a.
James Buchanan, "Rational Choice Theory", (1989b), in: Buchanan (1989a), pp.37-50.
Allin Cottrell, "Intentionality and Economics", in: *Economics and Philosophy* 11, 1995, pp.159-76.
Herbert Deane, *The Political and Social Ideas of St. Augustine*. New York: Columbia University Press, 1963.
Frank Hahn, "The Winter of Our Discontent", in: *Economica* 40, 1973, pp.322-30.
Daniel Hausman, "Arbitrage Arguments", in: *Erkenntnis* 30, 1989, pp.3-22.
Daniel Hausman and Michael S. McPherson, *Economic Analysis and Moral Philosophy*. Cambridge: Cambridge University Press, 1996.
Albert Hirschman, *The Passions and the Interests*. Princeton: Princeton University Press, 1977.
Albert Hirschman, "Against Parsimony: Three Easy Ways of Complicating Some Categories of Economic Discourse", in: *Economics and Philosophy* 1, 1985, pp.7-21.
David Hume, "Of the Independency of Parliament", (1741) in *Essays Moral, Political, and Literary*. Rpt. Oxford: Oxford University Press, 1963, pp.40-47.
Philip Pettit, "The Cunning of Trust", in: *Philosophy and Public Affairs* 24, 1995, pp.202-25.
Steven Rappaport, "Is Economics Empirical Knowledge?" in: *Economics and Philosophy* 11, 1995, pp.137-58.
Dennis Robertson. "What Does the Economist Economize?" in *Economic Commentaries*. London: Staples Press, 1956, pp.147-55.
Alexander Rosenberg, *Economics - Mathematical Politics or Science of Diminishing Returns*. Chicago: University of Chicago Press, 1992.
Alexander Rosenberg, "What Is the Cognitive Status of Economic Theory?" in Roger Backhouse, ed. *New Directions in Economic Methodology*. London: Routledge, 1994, pp.216-35.
Don Ross, "Real Patterns and the Ontological Foundations of Microeconomic" in: *Economics and Philosophy* 11, 1995, pp.113-36.
Amartya Sen, "Rational Fools", *in: Philosophy and Public Affairs* 6, 1977, pp.317-44.

35 Bellaire Court
Appleton WI 54911
USA

JULIAN NIDA-RÜMELIN

STRUCTURAL RATIONALITY IN GAME THEORY

I. PRELIMINARY REMARKS

When I began to think about normative questions of rationality and morality, I soon became a consequentialist. This is the natural result of questioning socially established institutional and moral rules and their metaphysical underpinnings. A critical mind cannot accept a duty to do this or that just because it is told that this duty is implied by some general principle. The immediate question is: Where does this principle come from and how is it justified? And if this question cannot be answered satisfactorily, some critical minds become normative sceptics and others search for rational grounds. Those who search for rational grounds easily end up with one form or other of consequentialism. A rational person does not act by established convention or acquired habit but chooses her action as a means to cause the preferred effects. Thus it seems that the rational person tries to optimize, and this seems to imply that moral action as a variant of rational action should optimize results from a universal point of view, e. g. by maximizing the sum of individual utilities.

In my case the story did not end there. When I became aware of the conflict between aspects of distributive justice and this form of consequentialism, I had second thoughts about moral reasoning in general. Questions of justice have at least equal weight as questions of utility-efficiency. There is a conflict between the simple extension of individual rationality to social or moral rationality on the one hand and the central moral intuition which does not allow us to disregard aspects of distributive justice. But if one takes this argument seriously, the door is open to include further considerations of different moral intuitions. The initial effort to rationalize normative discourse eventually resulted in dismissing almost all of our normative convictions, which is the dead end of normative theory. A good normative theory has to be compatible at least with the most central and immanently coherent normative convictions we share. This is the basis of the critique of consequentialism not only as an ethical but also as a theory of practical rationality (cf. Nida-Rümelin 1993, 1995).

The critique of consequentialism has shown that normative evaluation cannot be external: what we value as good or bad, right or wrong, depends on a complex structure of good reasons for actions. These reasons can only partly be reconstructed by the consequentialist device. To act irrationally is to act so that the balance of good reasons is not in favor of that act. The accepted reasons represent a form of social community, i.e. they constitute social interactions as legitimate or illegitimate. Likewise the person in choosing her favorable good

W. Leinfellner and E. Köhler (eds.), Game Theory, Experience, Rationality, 81–93.

reasons for action chooses an individual form of life. To act irrationally and to act immorally means not to comply with these (normatively acceptable) interpersonal and intrapersonal structures.

If the divergence between normative and empirical analysis is endemic to all social analysis, one should have second thoughts about the adequacy of the normative theory. There can be no doubt that there are systematic misrepresentations of decision situations caused by our limited capacity to deal with complex strategic situations and probability distributions, but there are simple cases where this kind of reconciliation is not convincing. The notion of structural rationality can help to integrate empirical and theoretical findings especially in game-theoretic analysis into a coherent conception of practical rationality instead of dealing with these findings as a threat to or a subversion of rationality (cf. Elster 1979, 1983, 1987).

The notion of structural rationality implies only a moderate theoretical change. It gives up consequentialism as the standard interpretation of decision theoretic rationality. But coherentism is preserved insofar as the overall preferences of a rational person can still be represented by a real-value function (cf. Nida-Rümelin, 1997 chap. VI). In more philosophical terms: starting from a Humean account of motivation, one is forced to acknowledge structural constraints on optimization in order to define practical rationality. And likewise: starting from Kantian apriorism as established by the universalizability test of the *Categorical Imperative,* one has to refer to contingent personal preferences in order to decide what is rational and moral. The notion of structural rationality in that sense covers a middle-ground between orthodox Humeanism and orthodox Kantianism. Humeans should acknowledge that rational persons strive inter alia at inter-temporal consistency (or better: coherence), and this constrains consequentialist optimization, and Kantians should acknowledge that without "contingent" personal preferences (*Neigungen*) the normatively required action (moral if rational in the Kantian sense) would be vastly underdetermined. Structural rationality is incompatible with both Kantian apriorism and Humean consequentialism.

The basic idea of structural rationality is the following: if we decide in concrete situations between specific options, we take into consideration "structural" aspects of our plans of life (to use a Rawlsian term which is easily misinterpreted, as if we choose at a given time once and for all a specific personal plan of life), e.g. we have a preference for participating in a specific social project constituted by many actions over a long period of time by many participants and decide in a concrete situation for such an option which is at least compatible with our further participation or is even part of the overall social activity fostering that project. In cases like these we do not look only at the (causal or probabilistic) consequences of this concrete optional alternatives, instead we decide on the basis of more or less intuitive criteria of fit or compatibility. There is an analogy between the interpersonal and the intrapersonal case. In cooperating with others, we refrain from optimizing subjective values in order to be structurally coherent and reliable for others, and in case of personal projects, we refrain from optimizing parts of our life in order to make our life on the whole coherent. This sketchy characterization is everything but precise and explicit, and we will not try to give a general account of structural rationality here. Instead we pick up some problems of rational decision in game theory which can well be embedded into the notion of structural rationality.

II. GAUTHIER'S DISPOSITIONS

Following Gauthier, the just person is disposed to comply with the requirements of the principle of minimax relative concession in interacting with those other fellows whom he believes to be similarly disposed. There are strong arguments showing that there is no unique solution for rational bargainers as Gauthier assumed, but this criticism is not essential for our argument. Let us assume that Gauthier succeeds and there is one and only one solution for rational bargainers, which is the principle of minimax relative concession. Then we would face again the seemingly paradoxical result that the rational person would decide, "on utility-maximizing grounds, not to make further choices on those grounds" (Gauthier 1986, p.158).

Now the question is: what interpretation is needed to make this claim a sound one? Obviously it would be a sound claim if specific options were available to the agent. Think of the following options: there are ten pills, swallowing one of these pills establishes specific dispositions in the behavior of the agent who took the pill. If among these pills there is one which establishes a disposition to behave in compliance with the principle of minimax relative concession and another pill would establish the disposition to maximize personal utility (to discriminate this kind of maximization, we call it "pointwise maximization"), then the rational utility-maximizing agent would indeed prefer to take the first pill.

The problem with Gauthier's theory is that he does not offer such an interpretation. Instead he gives the impression that, if to behave according to a specific disposition had better results than pointwise maximizing behavior, then rational maximization would require us to act according to this disposition. In this way, Gauthier tries to reconcile two forceful intuitions (1) that rational action is optimizing and (2) that constraints are necessary in order to act rationally (in this case to cooperate). This reconciliation hides the breakdown of pointwise optimization. What Gauthier and many other practical philosophers and decision theorists have discovered is that the optimizing theory of rationality is indeed self-defeating. My proposal here is to face these results and modify the theory with the help of the notion of structural rationality.

In the case of Gauthier, the interpretation is straightforward: the rational agent wants to cooperate. If the principle of minimax relative concession could indeed uniquely establish social cooperation – as Gauthier argues –, then agents wishing to cooperate would indeed comply with this principle. A person complying with this principle would therefore reveal a certain "disposition" in acting and interacting and thus comply with a specific social structure of interaction. But since the person did never choose dispositions – instead she chose concrete actions complying with the principle –, she in fact did not optimize in all these singular cases: which is compatible with Gauthier's idea that the rational person is an overall optimizer of her (social) life. It might well be that the person is doing both in the singular case: optimizing and complying, but this is a mere contingent combination.

The rational person opts for a structural trait of her (inter-)action because she knows that general compliance with this structural trait would optimize her interests or (the weaker claim) because only in complying can she find other persons willing to cooperate – but complying means not to optimize in the singular (general) case.

III. MULTI-PERSON COOPERATION

Let us assume that, depending on the number of cooperators, the individual utility status changes. Let us further assume that the individual utility status of a person is always lowered if this person decides to cooperate instead of not to cooperate, if the rest of cooperators and non-cooperators remain unchanged. Therefore we assume that there is an individual incentive not to cooperate for every person involved. A closer analysis of the multi-person cooperation problem (cf. Schelling 1978, chap. 7) shows that there are some differentiations possible which possibly have an impact on the motivation to cooperate.

For example it could be of some relevance for the motivation of rational agents whether an increase of cooperators would increase or decrease the individual utility status of all the cooperators. In some cooperation problems, there is an optimal number of cooperators (n_3) beyond which there is no cooperative surplus (see illustration below).

It might also influence the motivation of rational persons whether the cooperative surplus suffices to get the utility statuses beyond the level of universal non-cooperation. One might call the number of cooperators necessary for getting a utility status of the cooperators which is higher than the utility status in case of universal non-cooperation the "threshold for self-sustaining cooperation" (n_1). It is not unreasonable to assume that below this threshold the motivation to cooperate is difficult to obtain. If there are more than n_1 potential cooperators, one could expect that they base their cooperation on an agreement to cooperate even if this agreement is not enforced. A person knowing that there are less than n_1 cooperators will probably feel that personal cooperation is senseless.

In general, free-riders profit from the cooperation of others: they are better off if others cooperate. We assume that their utility status rises monotonously with the number of cooperators. Beyond a number of cooperators (n_2) the non-cooperator might be better off than the cooperator in case of universal cooperation.

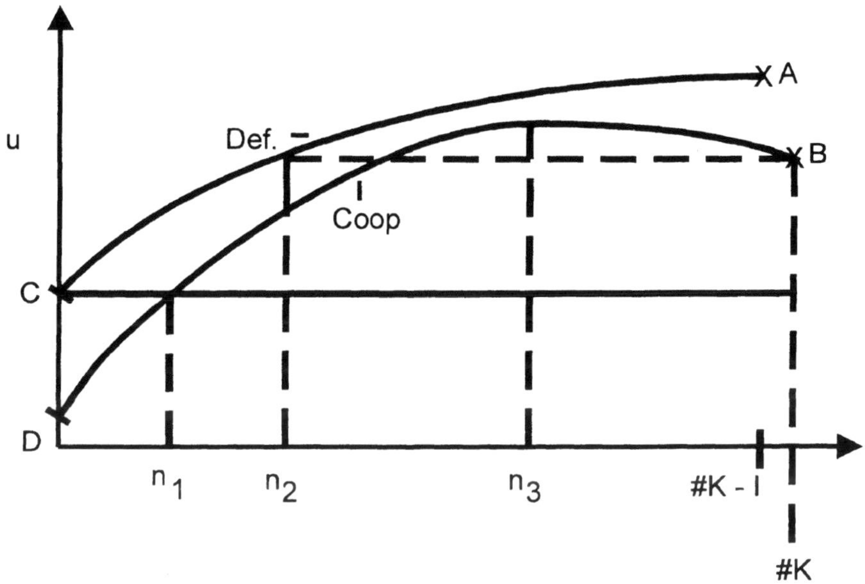

A: Unilateral Non-Cooperation
B: Universal Cooperation
C: Universal Non-Cooperation
D: Unilateral Cooperation
$u(A) > u(B) > u(C) > u(D)$
n_1: threshold for self-sustaining cooperation
n_2: threshold for free-riding improvements to universal cooperation
n_3: co-ordination optimum for cooperators
#K−I: cooperation optimum for non-cooperation
#K: cooperation maximum

Within the standard consequentialist account of rational action all participants would end up in C. But in many everyday situations in which multi-person cooperation is possible, individuals are motivated differently, i.e. under specific conditions they are willing to cooperate even if the consequences of cooperation in terms of their individual utility status are such as described in the illustration. Agents of this kind typically do not cooperate unconditionally. They cooperate because they see their individual action as part of a multi-person collective action. They are in favor of this collective action and therefore do their individual part in this collective action. Since in many cases such a collective (cooperative) action is not constituted by individual actions of the same specific type and in others there might be different possible collective actions which constitute cooperation, there are difficult problems how to make this kind of "belonging" of individual actions as part of a collective action explicit (cf. Tuomela 1995). The situation as illustrated above can therefore be quite complex in detail. A simple version would result e.g. if there were at least n_1 persons being motivated to cooperate under the condition that at least $n_1 - 1$ others are willing to cooperate, too. The individual person then chooses her cooperative action as part of a collective action comprising at least n_1 actors.

IV. METAGAMES AND STRUCTURAL RATIONALITY

The origin of metagame theory goes back to von Neumann and Morgenstern (1953). They expressed the view that in order to analyze two-person zero-sum games in normal form one should analyze the so-called minorant and majorant games, i.e. the games which would result if one of the players chose her strategy in knowledge of the others' choice.

In more general terms, Howard (1971) introduced the notion of a metagame that results for every game in normal form if one of the interacting persons chose her strategy in knowledge of all the others' choices.

In order to construct a metagame one makes use of the fact that to every given arbitrary normal form of a game there corresponds at least one extensive form (assume that the persons choose in some fixed order and that they are not informed about any strategy choices preceding their own, which results in quite big information sets containing all decision nodes of the person).

The extensive form of a metagame for person i results if we develop the extensive form out of the normal form of a game as above with i coming last and then delete i's information set and give her one information set for each of her decision nodes: she chooses last knowing the others' choices. Thus there are n metagames in extensive form resulting out of one n-person game in normal form.

The normal form of the metagame is found if we replace i's strategy set with a set containing all functions from the joint strategy choices of the others to his own strategy choices.

Mixed strategies are excluded by the non-quantitative approach of metagame theory.

Original PD:

		B C	D
	C	3/3	2/2
A	D	4/1	2/2

B – PD:

f_1: C → C & D → C
f_2: C → C & D → D
f_3: C → D & D → C
f_4: C → D & D → D

		A C	D
B	C & C	3/3	1/4
	C & D	3/3	2/2
	D & C	4/1	1/4
	D & D	4/1	2/2

Equilibrium in: D & D/D

On the basis of the simple Prisoners' Dilemma there results a PD-metagame which reveals in addition to universal non-cooperation universal cooperation as an equilibrium point (which is not lost in the further extension of the metagame tree).

Game → i-Metagame
i-Metagame → ji-Metametagame
→ Infinite Metagametree

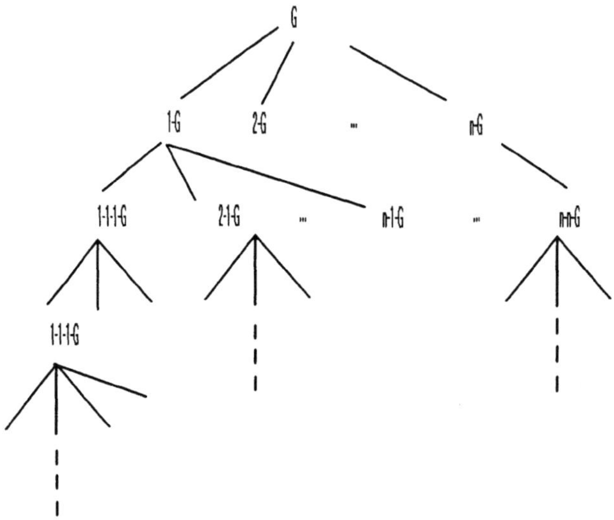

A-B–PD-Metagame

	CC	DD	CD	DC
C C C C	33	14	33	14
D D D D	41	22*	22	41
D D D C	41	22	22	14
D D C D	41	22	33*	41
D D C C	41	22	33	14
D C D D	41	14	22	41
D C D C	41	14	22	14
D C C D	41	14	33	41
D C C C	41	14	33	14
C D D D	33	22	22	41
C D D C	33	22	22	14
C D C D	33	22	33*	41
C D C C	33	22	33	14
C C D D	33	14	22	41

C C D C	33	14	22	14
C C C D	33	14	33	41

Three equilibrium points: 1) DDDD/DD
2) DDCD/CD (pareto-efficient)
3) CDCD/CD (pareto-efficient)

The problem with this metagame "solution" of the Prisoners' Dilemma is that there is no immediate and straightforward interpretation of meta-strategies. The standard conviction in game theory is that the metagame approach should be given up in favor of the supergame (iterated game)-analysis. Indeed some of the most interesting results of metagame theory can be reproduced in supergame analysis. Instead of reformulating the decision problem diachronically (supergames), structural rationality allows for understanding metastrategies as a reconstruction of some motivational process. The above example of the PD-metagame is a good illustration:

(a) CC can be interpreted as: "I cooperate in both cases (case 1: I expect the other person to cooperate, case 2: I expect the other person not to cooperate)."

(b) CD can be interpreted as: "I cooperate if I expect the other person to cooperate and I do not cooperate if I expect the other person not to cooperate."

(c) DDCD can be interpreted as: "I do not cooperate if I expect the other person to be a universal cooperator (CC); I do not cooperate if I expect the other person to be a universal non-cooperator (DD); I do cooperate if I expect the other person to be a conditional cooperator (cooperating only if she expects the other person to cooperate too); I do not cooperate if I expect the other person not to cooperate if she expects me to cooperate, and to cooperate if she expects me not to cooperate."

The new equilibrium point DDCD/CD can now be interpreted synchronically: Neither of the two prisoners has an incentive to change the decision if these decisions are motivated like (c) for prisoner A and (b) for prisoner B.

This motivational structure makes sense only if both prisoners see their individual decision as constitutive part of a (fictitious) structure of interaction, partly reconstructed as metastrategies.

V. RATIONAL INDUCEMENT

In cases of rational inducement there are several equilibrium points with different outcomes for the participants. In such situations a person who is willing to accept a disadvantage can threaten another participant successfully if the other participant is a simple optimizer. Only in situations of complete transparency can one say that the actual non-optimizer succeeds against an optimizer in getting his preferred equilibrium.

Analogously with meta-strategies as discussed above, there is a diachronic interpretation in terms of supergame analysis and a synchronic interpretation in terms of structural rationality.

		B	
		b_1	b_2
A	a_1	3/4	3/3
	a_2	1/1	4/3

Two Equilibria: a_1/b_1 and a_2/b_2

$$\min(a_1) = 3 \qquad \min(b_1) = 1$$
$$\min(a_2) = 1 \qquad \min(b_2) = 3$$

A prefers the equilibrium a_2/b_2; *B* prefers the equilibrium a_1/b_1. If both were consequentialist optizimers, the chosen equilibrium point would be underdetermined. And if there would be no possibility to coordinate, they could end up with any of the four possible strategy combinations. The moment in which one participant informs the other which strategy he will choose, the other is (as a consequentialist actor) bound to accept the "proposed" equilibrium point. The conflict is "solved" by time precedence. But if one of the participants (*A*) is a structurally rational agent, choosing his decisions as part of bigger (structural) strategies, he will possibly opt for a_2 unconditionally. Thus accepting the possible disadvantage of getting 1 instead of 3, if *B* chooses b_1. But if *B* is a consequentialist actor, the underdetermination is resolved because *B* would give in and stay to b_2.

VI. THE UTILITARIAN SOLUTION TO THE COOPERATION PROBLEM

As long as there are interpersonally variant overall action-guiding value-functions, it is to be expected that the cooperation problem persists. In theory though, it is possible to constrain the sets of feasible (interpersonally variant) value functions such that the cooperation problem does not arise. But if we assume – as utilitarianism in general does – that there are two levels of values – the values of the second level being an aggregate of the values of the first –, then we see that under plausible conditions there is only one kind of aggregation; which eliminates the cooperation problem. If we constitute the action guiding moral preferences of the ideally rational and moral person by transfer rates, which adds a portion of the first level-values of other persons, then the only transfer rate which puts an end to the cooperation problem is 1 (100%). If the interpersonal transfer rate is lower, some PD-situations on the first level will survive on the second, and if it is higher, then some non-PD-situations on the first level will transform into a PD-situation on the second. This result can be taken as an argument in favor of utilitarianism, or it can be taken as an argument in favor of a deontological theory of constraints. In the first case one concentrates on withdrawing the cooperation problem and concludes that only the utilitarian constitution of action-guiding moral preferences will have this result and therefore take utilitarianism to be well-founded. In the second case one concludes that persons can cooperate even if cooperation is not individually optimizing.

I am convinced that only the second conclusion can be defended (cf. Nida-Rümelin 1991). But even if one prefers the first conclusion, the case for structural rationality as the adequate normative account would – astonishingly enough – not be lost. Even if we accepted the somewhat radical idea that ideally rational and moral persons share the same action guiding preference relation, there would remain a problem of cooperation which is responsible for the superiority of rule-utilitarianism (if adequately reconstructed and under some specification of application).

The collective irrationality of optimizing agents with interpersonally different utility functions in the generalized PD can be solved by a radical measure. The measure is to eliminate interpersonal differences on the level of action guiding preferences. Utilitarianism is one way to do this. Preference utilitarianism as a special case provides one interpersonally invariant value function by simple aggregation of preferences. The most sophisticated form as presented by Harsanyi, combines decision theoretic Bayesianism with Preference Utilitarianism, it aggregates personal preferences to moral ones. Moral preferences can then be represented by an interpersonally invariant value function.

One might think that in this way the collective irrationality of optimizing behavior has finally ceased (even if some might think that the price for achieving this result is too high). But this is not true. The simple reason is that even rational ethical consequentialists face problems of cooperation. Problems of cooperation transformed into mere coordination problems survive in the ideal world of optimizing utilitarians (having the same interpersonally moral value function); therefore even rational utilitarians ought to change from consequentialist rationality to structural rationality.

Consequentialist utilitarians usually deny this. The most prominent form of this denial is the pretended practical equivalence of act- and rule-utilitarianism. The "proofs" of this thesis were refuted soon (cf. Ezorsky 1968) after they had been presented (cf. Lyons 1965), but the thesis surprisingly survived and is still one of the standard arguments in favor of ethical consequentialism.

Intuitively it is quite simple to understand why the equivalence thesis must be wrong. The argument in favor of the equivalence thesis is that the possible rules should not be taken out of a given list – given by language or conventional rationality – but can be formulated in as complex, detailed and sophisticated a way as you want. Even if there are pragmatic limits for this refinement, in theory all the anti-consequentialist arguments can be met by refining the respective rule. If the conventional rule of truth telling has good utilitarian consequences in general, then one should include the exception in a modification of the rule, and *a limine* you end up with the consequentialist rule to tell the truth if and only if it (this concrete case of telling the truth under these concrete circumstances) has optimal consequences. Rules survive, but only as rules of thumb. Those who think that even in theory (i.e. in the ideal case of perfectly rational and moral agents) one should follow rules also when this rule-following does not have optimal consequences, attribute an intrinsic moral value to rules, and this cannot be accepted within utilitarian approach.

The problem with this argument is that it is based on a parametric, not a strategic (or game-theoretic) view regarding practical rationality. Even utilitarians face the problem of how to coordinate their actions interpersonally. If we look at the matter from a game-theoretic point of view, we easily see that optimizing a collective strategy (i.e. a combination of individual strategies) does not have the same result as optimizing an individual strategy except for rare special cases like the ideal market. The equivalence thesis is therefore wrong. It has to be conceded though that proponents and critics of the equivalence theses did not realize that the equivalence thesis can be read in two different ways: parametrically and strategically. Strategically it is obviously false. Parametrically its falsity is less easily demonstrated and we do not need to do so here.

Harsanyi has proven the non-equivalence of act- and rule-utilitarianism in analyzing some examples of coordinative collective action by game-theoretic tools (cf. Harsanyi 1977, 1983, 1985). Since the argument is straightforward and without doubt, it remains only to interpret the rule-utilitarian device adequately. Rule-utilitarians in Harsanyi's sense choose that action which performed by all other moral and rational (i.e. rule-utilitarian) persons maximizes overall expected preference fulfilment. This means that in the general case the rational and moral person does not optimize the (preference-)utilitarian value function, but that the collective of rule-utilitarian agents jointly optimize the (preference-)utilitarian value function. Thus the rational moral person chooses her action as part of a collective strategy which is optimizing, whereas the individual action is not optimizing. The rational and moral person acts structurally rational.

REFERENCES

Jon Elster, *Ulysses and the Sirens*. Cambridge: Cambridge University Press 1979.

Jon Elster, *Sour Grapes*. Cambridge: Cambridge University Press 1983.

Jon Elster, *Subversion der Rationalität*. Frankfurt a. M./New York: Campus 1987.

Gertrude Ezorsky, "A Defense of Rule Utilitarianism", in: *Journal of Philosophy*, 65, 1968, pp.533-544.

David Gauthier, *Morals by Agreement*. Oxford: Clarendon Press 1986.

J.C. Harsanyi, "Rule Utilitarianism and Decision Theory", in: *Erkenntnis*, 11, 1977, pp.25-53.

J.C. Harsanyi, "Basic Moral Decision and Alternative Concepts of Rationality", in: *Social Theory and Practice*, 9, 1983, pp.231-244.

J.C. Harsanyi, "Does Reason tell us What Moral Code to Follow and, Indeed, to Follow any Moral Code at all?", in: *Ethics*, 96, 1985, pp.42-55.

Nigel Howard, *Paradoxes of Rationality*. Cambridge/Mass.: MIT Press 1971.

David Lyons, *Forms and Limits of Utilitarianism*. New York, Oxford: Clarendon Press 1965.

John von Neumann and Oskar Morgenstern, *Theory of Games and Economic Behavior*. Princeton: Princeton University Press 1944.

Julian Nida-Rümelin, "Practical Reason or Metapreferences? An Undogmatic Defense of Kantian Morality", in: *Theory and Decision*, 30, 1991, pp.133-162.

Julian Nida-Rümelin, *Kritik des Konsequentialismus*. München: Oldenbourg 1993, 1995.

Julian Nida-Rümelin, *Economic Rationality and Practical Reason*. Dordrecht: Kluwer 1997 [in print].

Thomas C. Schelling, *Micromotives and macrobehavior*. New York: Norton 1978.

Reinhard Selten, "The Chain Store Paradox", in: *Theory and Decision*, 9, 1978, pp.127-159.

Amartya Sen, "Rational Fools: A Critique of the Behavioural Functions of Economic Theory", in: *Philosophy and Public Affairs*, 6, 1976/77, pp.317-344.

Raimo Tuomela, *The Importance of Us. A Philosophical Study of Basic Social Motions*. Stanford/Cal.: Stanford University Press 1995.

Georg-August-Universität Göttingen
Philosophisches Seminar
Humboldtallee 19
D-37073 Göttingen
Germany

Robin Pope

Attractions to and Repulsions from Chance

I. Introduction

This paper is concerned with the discussion of the phenomenon sometimes described as "the utility and disutility of chance" both from the descriptive and the prescriptive point of view. Emphasis is not on axioms and formal properties but on the psychological content of decision theoretic constructs.

The author has written on related matters before (Pope 1983, 1984, 1985, 1986, 1988, 1991, 1995, 1997a and 1997b). Necessarily some points raised in earlier papers have to be dealt with here again, but also some new material will be presented. This mainly concerns a sharper distinction between descriptive and prescriptive questions and the problems raised by human motivational tendencies for normative utility theory.

Descriptively it is dubious to what extent the idea of utility is appropriate. Framing and history dependence seem to be important. Therefore the title refers directly to the relevant motivational factors. It will be argued that these motivational forces should also matter for prescriptive decision theory.

II. The Problem

The problem in what is now often called the utility of gambling already has been seen clearly by Blaise Pascal in his *Pensées* as early as 1670.

> Some people live a perfectly pleasant existence in which they continually gamble small amounts. Give one of them every morning the amount he could have won during the day providing he does not gamble, and you will make him miserable.
>
> You may think this is because he is looking for the pleasure of the game rather than the winnings involved. If so, just try letting him play without stakes. He will betray no enthusiasm, only boredom. (English translation by Allais 1979)

The problem is still not resolved in modern decision theory. There are serious difficulties with incorporating the attractions to and repulsions from chance into the usual framework of decision theory. This is clearly expressed by a quotation from Oskar Morgenstern 1974:

> I want to make it absolutely clear that I believe – as von Neumann did – that there may be a pleasure of gambling, of taking chances, a love of assuming risks etc. But what we did say and what I do feel I have to repeat even today after so many efforts have been made by so many learned men, is that the matter is still very elusive. I know of no

W. Leinfellner and E. Köhler (eds.), Game Theory, Experience, Rationality, 95–107.

axiomatic system worth its name that specifically incorporates a specific pleasure or utility of gambling together with a general theory of utility.

The problem that Morgenstern alludes to in this quotation is a pleasure of gambling not taken into account by a convex utility of money. As we shall argue later, the attractions to and repulsions from chance which are important here are not a reflection of outcome utilities but are due to factors such as emotions either occurring in the pre-outcome or arising in the post-outcome period because there was a pre-outcome period. In the pre-outcome period, the period after the decision is taken but before the actual outcome is known, the decision maker may feel a tension caused by his lack of knowledge about how the decision situation will be resolved. The anticipation of this tension before the decision is made may be attractive or repulsive. It and related factors caused by this lack of knowledge are additional factors entering the decision making process apart from but not independent of the satisfaction derived from the outcome.

III. Some Historical Remarks

Table 1 lists a number of important contributions to the problem area. The description of the phenomenon by Blaise Pascal (1670) already has been shown by a quotation in the last section. Daniel Bernoulli (1738) introduced a logarithmic utility of assets evaluated in monetary terms. His work is generally considered to be the start of modern work on utility theory.

However for a long time decision theory was not much further developed, and not much attention was paid to formally modelling the special subject matter of attractions to and repulsions from chance. In 1920 Alfred Marshall discussed the problem. He explicitly spoke about the pleasure of gambling and an adventurous disposition as a motivational force in the behaviour of entrepreneurs. Alfred Marshall connected no value judgment to his description of economic motivation. He thought that an adventurous disposition might sometimes be good and sometimes be bad for an individual and for society as a whole.

Frank P. Ramsey (1926) however explicitly expressed the opinion that love or hatred of excitement is a bad influence on serious decision making. Businessmen should avoid being guided by such emotions. His view has been accepted by many later theorists, e.g. Samuelson (1952b). Jacob Marschak (1938, p.320) citing for support Irving Fisher, Pigou, Hicks, Florance, and Keynes' 1921 *Treatise on Probability*, proposed the view that decisions are guided by a function depending on expected value, variance, skewness and higher moments of a distribution of monetary outcomes. Gerhard Tintner (1942) was the first one to propose a different approach which involves first applying a utility function in order to transform monetary amounts into utilities and then taking expectation, variance and higher moments of the utility distribution as the basis of decision making. One may say that here two kinds of utilities come into the picture. One utility function which is directly applied to monetary amounts, and another one depending on the moments of the distribution of the utilities of the first kind. The directly applied first utility function can be thought of as describing the utility of monetary amounts expected with certainty, and the second utility function then takes account of attitudes toward risk. Tintner himself did not propose this interpretation, but it is clearly implied by the work of Maurice Allais (1952) who took a very similar approach. He elicited judgments about the directly applied utility by questions without any reference to uncertainty. In fact he later (1988) explicitly said that the first applied utility is based on outcomes to be imagined as certain.

Blaise Pascal	1670	Description of the phenomenon
Daniel Bernoulli	1738	Logarithmic utility of assets
Alfred Marshall	1920	Pleasure of gambling adventurous disposition
Frank P. Ramsey	1926	Love or hatred of excitement not good for serious decisions
Jacob Marschak	1938	Summarising Hicks, Keynes, Pigou, Florence, Irving Fisher
Gerhard Tintner	1942	Higher moments of utility
John von Neumann and Oskar Morgenstern	1944 1947	Axiomatisation Utility of chance as a problem
Milton Friedman and Leonard J. Savage	1948	Concave-convex utility of money
D.O. Hebb	1949	Organisation of behaviour
Jacob Marschak	1950	Risk aversion
Maurice Allais	1952	Allais paradox
Leonard J. Savage	1954	Axiomatisation Sure-thing principle
Oskar Morgenstern	1974	Utility of chance as a problem
John Quiggin	1982	Rank dependent utility
Tversky and Kahneman	1992	Generalised prospect theory

Table 1: Historical Calendar Pertinent to the Utility of Chance

In their famous book on games and economic behavior von Neumann and Morgenstern (1944, 1947) presented an axiomatisation of expected utility. They show that under what they saw as intuitively justifiable axioms a decision maker must behave in a way which can be described as maximising the expected value of a utility function determined up to a positive linear transformation. The profession was immediately much impressed by the axiomatic approach. However the content of the axiomatisation was at first not accepted by most theorists who critically reviewed the book by von Neumann and Morgenstern. Critics were opposed to the conclusion that utility is cardinal in the sense that it is determined up to a positive linear transformation in their axiomatic system, and considered it too narrow in omitting higher moments and the utility of gambling. See e.g. William Vickrey (1945); Karl Kaysen (1946), Jacob Marschak (1946) and Paul A. Samuelson (1950).[1]

The situation was dramatically changed by a paper of Friedman and Savage (1948) and a sequel paper (1952). They showed that a decision maker having a concave-convex utility function of monetary assets may at the same time take insurance and buy lottery tickets.

Economic theorists began to think that in this way risk attitudes are sufficiently taken into account by von Neumann-Morgenstern utility. However von Neumann and Morgenstern in their book did not express the opinion that they had successfully dealt with the utility of chance. On the contrary they explicitly said that the utility of chance is a deep problem not covered by their axiomatic system (1947, pp.629-632). A much later quotation to this effect from Oskar Morgenstern (1974) has already been presented in section II.

In 1949 Jacob Marschak turned away from his critical attitude towards expected utility theory and became an adherent of it (1949, 1950, 1951). He, as did others soon after, e.g. Paul A. Samuelson (1952) developed his own axiom system, Marschak (1950) which he considered much more plausible, even if now all these alternative axiomatisations may seem not more than variations of von Neumann and Morgenstern's original work. Von Neumann and Morgenstern explicitly spoke of "utility" in their axioms. This reference to utility, especially to cardinal utility, created difficulties of understanding for the strongly anti-cardinalist younger theorists who had a programme of trying to bury the utility concept, even if von Neumann and Morgenstern's language use can be looked upon as not more than a manner of speaking about ordinal preferences over lotteries.

In his 1949 Jacob Marschak, who had previously coined the term risk aversion for a dislike of variance (1938, pp.312 and 321), was the first to use the term risk aversion in the now commonly accepted sense of concavity of the von Neumann–Morgenstern utility function. It is now a widely held view that attitude about risk is sufficiently covered by convexity or concavity of the utility function. Allais in 1952 strongly opposed this opinion. He presented his famous Allais paradox, a decision task in which most people are inclined to make choices not compatible with expected utility maximisation. Allais considered that decisions are not based on expected utility alone, but also on factors such as variance.

When Allais exposed Leonard J. Savage to the Allais paradox, he fell into the trap and made choices incompatible with expected utility theory. Savage concluded that he had made a mistake which he must correct. Then how do you correct such mistakes? Savage felt the need of relying on a principle in doing this. For this purpose he proposed the sure-thing principle which is often misunderstood as an interpretation of his axiom P2.

However as the author has shown elsewhere (Pope 1991), Savage explicitly referred to the sure-thing principle as an "extralogical" consideration outside his axiomatic system. It has the status of what one might call a clarifying principle. Whenever natural inclinations violate the consistency requirements one wants to achieve, preferences have to be made consistent by correcting natural tendencies as long as one is not willing to abolish any of one's consistency requirements. One of the purposes of this paper is the attempt to point out that there is a need for the exploration and discussion of clarifying principles. Up to now not much has been done in this area.

The joint axiomatisation of expected utility and subjective probability by Savage (1954) is now very popular as the foundation of Bayesian decision theory. The Bayesian point of view is now shared by many economic theorists even if experimental findings (e.g. Tversky and Kahneman 1979) have raised serious doubts. Attempts to account for the experimentally observed anomalies have led to quite a number of so-called non-expected utility theories. Presently the most important of these approaches is rank dependent utility informally proposed by Allais (1952) first formally modelled by John Quiggin (1982).[2]

These theories first rank all monetary outcomes in ascending order, then transform outcomes by a utility function and cumulatative probabilities by a weighting function, and then apply a formula similar to expected utility theory. Tversky and Kahneman's generalised prospect theory (1992) is also based on these ideas.

One piece of literature in Table 1 not yet mentioned up to now, is the very influential book *Organisation of Behavior* by D.O. Hebb (1949). This work is well known in psychology and neurophysiology. But economists, with few exceptions (eg Scitovsky 1976), have not taken notice of the importance of Hebb's ideas. He expressed the opinion that the central nervous system needs constant stimulation and therefore produces behaviour which actively seeks out excitement. Even unpleasant sensations like fear and anxiety are pleasurable if they are experienced with a very low intensity as in the course of viewing a mystery play on television. The motivational forces related to the need of the central nervous system for stimulation also are relevant for those attractions to risk which cannot be covered by expected utility theory.

IV. Time Aspects of Decision Making

As the author has pointed out elsewhere Pope (1983, 1995), it is necessary to distinguish different time spans involved in decision making if one wants to understand the attractions to and repulsions from chance. In the sufficiently simple decision situations on which we concentrate our attention here, this distinction concerns the following time spans:

1 Pre-decision period
2 Pre-outcome period
3 Post-outcome period

The pre-decision period is the time prior to the decision being made. The pre-outcome period begins immediately the decision is made, and extends up to the time in which the outcome becomes known with certainty. It is not important when the outcome actually occurs. This may be earlier or later than when it becomes known. The post-outcome period is the time beginning when the outcome is known.

The significance of this distinction lies in the fact that some of the consequences of making the decision already appear in the pre-outcome period, whereas others only belong to the post-outcome period. During the pre-outcome period, the decision maker may experience pleasant or unpleasant emotions like hope and fear. These emotions may be anticipated in the pre-decision period and thereby influence the choice which is made.

V. The Nature of the Attractions to and Repulsions from Chance

In the last section hope and fear have already been mentioned as examples of emotions in the pre-outcome period whose anticipation is an influence on decision making. Hope and fear pre-suppose the presence of outcomes providing different degrees of satisfaction in the post-decision period. Hope focuses on the chance of a good outcome, and fear on the chance of a bad one. If all outcomes are equally satisfactory, there is no scope for hope or fear. Both emotions arise from the simultaneous presence of good and bad outcomes. In this sense the different outcomes can interact in their effect on pre-decision anticipation of the pre-outcome psychological state. We refer to this as a complementarity between different outcomes.

Other emotions which may arise in the pre-outcome period are wonder, curiosity, suspense and excitement. Unlike hope and fear, these emotions can also be present if all outcomes are ranked equally, but are nevertheless different in some interesting ways. Thus a scientist may wonder which of several theories turns out to be successful in prediction without having a stake in any of them.

Adam Smith (1778) in his inquiry into scientific discoveries in astronomy and other subjects, saw wonder and curiosity as driving forces.

VI. AIM OF PRESCRIPTIVE THEORY

There is no doubt about the fact that descriptively pre-outcome emotions are important for decision making. One may argue as Ramsey (1926) did that at least some of them should not be relevant for prescriptive decision theory. This paper takes the point of view that this is wrong. In order to understand what is the basis for this opinion, it is necessary to spell out what could be considered the aim of prescriptive decision theory. According to the opinion presented here, prescriptive decision theory should help us to be more:

1 Systematic
2 Motivationally comprehensive
3 Able to identify and reduce irrational considerations
4 Able to identify and modify emotions deemed to be substantively irrational
5 Near to natural inclination in combining motivational forces to form a decision.

1. Systematic. Being systematic means that decision making satisfies consistency requirements relating events and outcomes in different situations. Thus transitivity, in the sense that a preference of A over B combined with a preference of B over C implies a preference of A over C, relates decisions in three choice situations. Typically consistency requirements are not directed towards single decisions, but to hypothetical choices in several related decision situations.

2. *Motivational Comprehensiveness.* Apart from substantively irrational motivations no motivational factor should be neglected in decision making. One might be inclined to require that all motivational forces should be duly considered by prescriptive theory. However this would be going too far as will be discussed under heading 4.

3. Ability to Identify and Reduce Irrational Considerations. On reflection, some considerations influencing our decision inclinations may be recognised as misleading. In this connection, we may think of some framing effects and cognitive illusions. It should not matter whether a cup is described as half empty or half full.

4. Ability to Identify and Reduce Considerations deemed to be Substantively Irrational. Some people are strongly motivated by compulsions under which they act against important over-riding goals. In this connection we may think of phenomena like drug addiction, compulsive buying of consumer goods and compulsive gambling. A decision maker having such behavioural tendencies should be helped to become aware of the detrimental consequences and if he wants this, he should be helped to get rid of these compulsions. It does not seem to be adequate to restrict prescriptive decision theory to purely formal requirements. Substantial considerations like those in this and the preceding point need to be included in a decision theory addressed to the improvement of practical decision making.

5. Proximity to Natural Inclination in Combining Motivational Forces to Form a Decision. In utility theory, possible combining procedures are e.g. computation of expected values, or rank dependent probability weightings. Descriptively it is not yet clear how motivational forces are combined to yield a decision. Therefore at the moment it is difficult to achieve the proximity requirement considered here. However in principle, it seems to be important to deviate not more than necessary from natural inclinations.

VII. THE CONSTRUCTION PROBLEM FOR NORMATIVE UTILITY THEORY

Natural preferences rarely satisfy the usual consistency requirement. Therefore either the consistency requirement or the preferences have to be changed. As has been explained already in the section on historical remarks, the correction of preferences requires clarifying principles. In order to provide an example we look more closely at Savage's sure-thing principle. Allais 1952 asked Savage to choose in the Allais paradox. Savage's answers did not conform to expected utility theory. Savage therefore changed his preferences. For this purpose he devised the sure-thing principle.

> *Sure-thing Principle:*
> In the choice between two alternatives, delete common consequences, re-normalise and compare afterwards.

An example may illustrate what is meant by this.

ORIGINAL CHOICE TASK: 're-arranged' to have common consequences

Ticket No.	1, ... , 10	11	12, ... , 100
C	1	1	0
D	5	0	0

MODIFIED CHOICE TASK: after deletions and renormalisation

Ticket No.	1, ... , 10	11
C'	1	1
D'	5	0

In these choice tasks, all of the available tickets, 1 to 100 in the original one, and 1 to 11 in the re-normalised one, have the same probability. The sure-thing principle may easily be confused with Savage's axiom P2, since there a similar connection between two related choice tasks is made. However it is not the consistency requirement expressed by axiom P2 which is expressed by the sure-thing principle. Suppose that in this example the natural inclinations contradict expected utility theory by a strict preference for D over C in the original task, and a strict preference for C' over D' in the re-normalised one. Then the sure-thing principle recommends to correct this violation of expected utility theory by reversing the preference in the original task. One could just as well reverse the preference in the re-normalised task to achieve consistency. This shows that the sure-thing principle is not just a necessary consequence of Axiom P2. Usually violations of consistency requirements can be corrected in many ways. One needs clarifying principles in order to select one of them.

VIII. CONSEQUENCES FOR THE UTILITY OF CHANCE

As has been pointed out in section III, Frank P. Ramsey has been influential in bringing about the view that attractions to and repulsions from chance - which arise from anticipated pre-outcome emotions (section VI) - should not influence serious rational business. Is it really justified to exclude feelings of hope, fear, wonder, excitement and suspense from rational consideration? After all, not only these emotions, but also the satisfaction derived from the outcome, is psychological. Outcomes are desirable or undesirable because of their psychological aspects, not for their objective features.

Therefore there seems to be no reason to exclude emotional reactions from consideration just because they are pre-outcome rather than post-outcome reactions. The requirement of motivational comprehensiveness points in the direction of the attractions to and repulsions from chance being an influence on behaviour which must be reflected in an adequate prescriptive theory. However as we have already said in connection with parts 3 and 4 in our explanation of prescriptive theory, some emotions deemed to be substantively irrational may need to be modified. Thus prescriptive theory should not support behavioural tendencies such as compulsive gambling. However a desire to get excitement and hope out of a moderate amount of gambling does not seem to be more objectionable than a desire for entertainment by theatre plays. Both may be valued because of their potential for excitement, and both may be similarly expensive or inexpensive. Similar comments apply to the attractions to chance in research, export and other business decisions. Therefore the author takes the point of view that apart from extreme cases, the influences of attractions to chance should not be excluded from consideration by prescriptive decision theory.

Obviously a similar argument can be made for repulsions from chance not covered by a concave utility function. Of course also here, exaggerated and unrealistic fears like agoraphobia may be deemed to be irrational and therefore fall under those influences which should be overcome or at least mitigated by rational decision making.

It is not clear to the author how prescriptive theory should deal with such tendencies in need of therapy, but the author feels strongly that such tendencies in need of therapy cannot simply be swept away by looking at decision theory merely concerned with abstract consistency requirements.

The author also notes that tendencies in need of therapy are not limited to attractions to and repulsions from chance, though this impression is often created in the decision literature since about 1952 and the rediscovery of Frank P. Ramsey's view.[3] This post Ramsey literature focuses on instances of the attractions to and repulsions from chance that are possibly or definitely detrimental to the person. This post Ramsey literature lacks balancing examples of neutral or positive instances of these attractions as were discussed earlier by Alfred Marshall (1920) with his instances in which the adventurous disposition resulted in handsome business dividends from discovery and exploitation,[4] and in the example supplied here of getting excitement from theatre plays. This post Ramsey decision literature may also be described as unbalanced in that its focus is on the compulsive gambler losing his fortune in the casino through his attraction to chance. This is not counterbalanced by examples of those other compulsions under which people lose their fortune through attractions unrelated to chance (and therefore permitted under the standard decision literature notion of rationality). These include compulsive buying, compulsive eating, compulsive drinking, and so forth.

IX. The Illusion of Certainty

Many theorists, e.g. Savage and Allais, base their interpretation of utility on the comparison of outcomes considered to be certain. However actual decision making often attaches special importance to certainty in the evaluation. For such persons, the evaluation of outcomes is not independent of whether they are certain or uncertain. Neglecting such tendencies may exclude some behavioural tendencies which are not irrational *per se*. One may say that somebody guided by a clarifying principle involving comparisons between certain outcomes as a basis for judgments on uncertain prospects may thereby create an illusion of certainty which unnecessarily distorts that person's natural behavioural tendencies.

Ramsey maintained that expected utility theory cannot be applied unless the utility consequences of the outcomes are independent of the probability of those outcomes. In this connection he spoke of tastes independent of beliefs (1950, p.177). The author thinks that an interdependence of tastes and beliefs should be reflected in prescriptive decision theory where it is present. Where this is the case, clarifying principles are needed which not unduly emphasise comparisons between certain and uncertain outcomes. Savage's sure-thing principle also is based on a presupposition of tastes independent of probabilities. If this independence is lacking it may not be justified to eliminate common consequences and to re-normalise since the probabilities attached to the remaining outcomes are changed by this operation.

NOTES

* I thank Reinhard Selten for valued suggestions and discussions.

1. On continuing embarassment, difficulties and perceived needs for rationalisations concerning the cardinal utility index of von Neumann and Morgenstern's axiomatised expected utility procedure, see e.g. William Baumol (1958) and Daniel Ellsberg (1954).
2. Other developments in rank dependent models include Lopes 1984 and 1995, Segal 1984 and Yaari 1987. See also Quiggin 1993.
3. Ramsey's 1926 lecture containing this view was published in 1950 and soon after brought to the attention of Paul A. Samuelson, Ken Arrow and others of the Rand circle by Keith Dalkey.
4. Edwin Canaan (1926) also reflects a balance of positive and negative instances in that he describes the attractions to chance positively in terms of "an adventurous disposition", and repulsions from chance negatively as "timidity".

REFERENCES

M. Allais, "Fondements d'une Theorie Positive des Choix Comportant un Risque" ('Foundations of a Theory of Random Choice'), Colloque International de Paris sur les 'Fondements et Applications de la Theorie du Risque en Econométrie', Paris, 12-17 Mai, 1952. Econométrie, Colloques Internationaux du Centre National de la Recherche Scientifique, 40, 1952, pp.127-140, translated in Allais 1979.

M. Allais, "The Foundations of a Positive Theory of Choice Involving Risk and a Criticism of the Postulates and Axioms of the American School", in: M. Allais/O. Hagen (eds.), *Expected Utility Hypotheses and the Allais Paradox Contemporary Discussions of Decisions under Uncertainty with Allais' Rejoinder*. Dordrecht: Reidel 1979, pp.27-145.

M. Allais, "The General Theory of Random Choices in Relation to the Invariant Cardinal Utility Function and the Specific Probability Function. The (U, Θ) Model: A General Overview", in: Bertrand Munier (ed.), *Risk, Decision and Rationality*. Dordrecht: Reidel 1988, pp.231-289.

William J. Baumol, "The Cardinal Utility which is Ordinal", in: *Economic Journal*, 68 (Dec.), 1958, pp.665-672.

D. Bernoulli, "Specimen Theoriae Novae de Mensura Sortis", an English translation by Dr Louise Sommer is "Exposition of a New Theory on the Measurement of Risk", in: *Econometrica*, 22, 1954 (1738), pp.23-36.

E. Canaan, "Profit", in: Henry Higgs (ed.), *Palgrave's Dictionary of Political Economy, New Edition* . New York: Kelley 1963.

Daniel Ellsberg, "Classic and Current Notions of 'Measurable Utility'", in: *Economic Journal*, 64 (Sept.), 1954, pp.528-556.

M. Friedman/L.J. Savage, "Utility Analysis of Choices Involving Risk", in: *Journal of Political Economy*, 56, 1948, pp.279-304.

M. Friedman/L.J. Savage, "The Expected Utility Hypothesis and the Measurability of Utility", in: *Journal of Political Economy*, LX(6), 1952, pp.463-474.

Donald O. Hebb, *The Organisation of Behavior*. New York: Wiley 1949.

Daniel Kahneman/Amos Tversky, "Prospect Theory: An Analysis of Decision Under Risk", in: *Econometrica*, 47 (2), 1979, pp.263-291.

C. Kaysen, "A Revolution in Economic Theory?", in: *Review of Economic Studies*, XIV, 1946, pp.1-15.

J. Keynes, *A Treatise on Probability*. London: Macmillan 1921.

Lola L. Lopes, "Risk and Distributional Inequality", in: *Journal of Experimental Psychology: Human Perception and Performance*, 10, 1984, pp.465-485.

Lola L. Lopes, "On Modelling Risky Choice: Why Reasons Matter", in: J.P. Caverni/M. Bar-Hillel/F.H. Barron/H. Jungermann (eds.), *Contributions to Decision-Making I*, Amsterdam: Elsevier 1995, pp.29-50.

J. Marschak, "Money and the Theory of Assets", in: *Econometrica*, 6, 1938, pp.311-325.

J. Marschak, "Neumann's and Morgenstern's New Approach to Static Economics", in: *Journal of Political Economy*, LIV (2), 1946, pp.97-115.

J. Marschak, "Measurable Utility and the Theory of Assets", in: *Econometrica*, 17, 1949, pp.63-64.

J. Marschak, "Rational Behavior, Uncertain Prospects, and Measurable Utility", in: *Econometrica*, 18, 1950, pp.111-141.

J. Marschak, "Why 'Should' Statisticians and Businessmen Maximize 'Moral Expectation'?", Proceedings of the Second Berkeley Symposium on Mathematical Statistics and Probability, University of California Press, Berkeley, 1951, pp.493-506.

A. Marshall, *Principles of Economics*. London: Macmillan 1920 and 1956.

Oskar Morgenstern, "Some Reflections on Utility" in: Maurice Allais/Ole Hagen (eds.), *Expected Utility and the Allais Paradox*, Dordrecht: Reidel 1974 and 1979, pp.175-183.

Blaise Pascal, *Pensees*. Paris: Hachette 1845 (1670).

R. Pope, "The Pre-Outcome Period and the Utility of Gambling", in: B. Stigum/F. Wenstøp (eds.), *Foundations of Utility and Risk Theory with Applications*. Dordrecht: Reidel 1983, pp.137-177.

R. Pope, "The Delusion of Certainty in Savage's Sure-Thing Principle", in: *Journal of Economic Psychology*, 12 (2), June, 1991, pp.201-241.

R. Pope, "Towards a More Precise Decision Framework: A Separation of the Negative Utility of Chance from Diminishing Marginal Utility and the Preference for Safety", in: *Theory and Decision*, 39, 1995, pp.241-265.

R. Pope, "Debates on the Utility of Chance: A Look Back to Move Forward", in: *Zeitschrift für Wissenschaftsforschung*, 1997, pp.1-41.

R. Pope, "Reconciliation with the Utility of Chance by Elaborated Outcomes Destroys the Axiomatic Basis of Expected Utility Theory", paper to be presented to the Eigth International Conference on the Foundations and Applications of Utility, Risk and Decision Theory, Faculties Universitaires Catholiques de Mons, July, 1997b.

John Quiggin, "A Theory of Anticipated Utility", in: *Journal of Economic Behavior and Organisation*, 3, 1982, pp.323-43.

John Quiggin, *Generalised Expected Utility Theory: The Rank Dependent Model*. Boston: Kluwer 1993.

F. Ramsey, "Truth and Probability" in: R. Braithwaite (ed.), *The Foundations of Mathematics and Other Logical Essays*. New York: Humanities Press 1950 (1926), pp.156-184.

Paul A. Samuelson, "Probability and the Attempts to Measure Utility", in: *Economic Review*, 1 (3), 1950, pp.117-123.

Paul A. Samuelson, "Probability, Utility, and the Independence Axiom", in: *Econometrica*, 20, 1952a, pp.670-678.

Paul A. Samuelson, "Utility, Preference and Probability", paper submitted to the International Symposium on Risk, Paris, May; Symposia, Colloques Internationaux XL, Econometrie, 1953, Centre National de la Recherche Scientifique, Paris, 1952b, pp.141-150.

L. Savage, *Foundations of Statistics*. New York: Wiley 1954.

Uzi Segal, "Nonlinear Decision Weights with the Independence Axiom", University of California Los Angeles, Working Paper 353, November 1984.

Tibor Scitovsky, *The Joyless Economy: An Inquiry into Human Satisfaction and Consumer dissatisfaction*. Oxford: Oxford University Press 1976.

Adam Smith, *The Principles which Lead and Direct Philosophical Enquiries: Illustrated by the History of Astronomy*. 1788 (reproduced: Indianapolis: Liberty Press 1982).

G. Tintner, G., "A Contribution to the Non-Static Theory of Choice", in: *Quarterly Journal of Economics*, 61, 1942, pp.274-306.

Amos Tversky/Daniel Kahneman, "Advances in Prospect Theory: Cumulative Representation of Uncertainty", in: *Journal of Risk and Uncertainty*, 5, 1992, pp.297-323.

W. Vickrey, "Measuring Marginal Utility by Reactions to Risk", in: *Econometrica*, 13, 1945, pp.319-333.

J. von Neumann/O. Morgenstern, *Theory of Games and Economic Behaviour*. Princeton: Princeton University Press 1944 (second ed. 1947).

Menahem E. Yaari, "The Dual Theory of Choice Under Risk", in: *Econometrica*, January, 55, 1987, pp.95-115.

Work Postal Address:
Department of Mathematical Economics
Bielefeld University
Postfach 10031
D-33501 Bielefeld
Germany

MATTHIAS HILD
RICHARD JEFFREY
MATHIAS RISSE

AGREEING TO DISAGREE: HARSANYI AND AUMANN

I. INTRODUCTION

In "Agreeing to Disagree" [1], Robert Aumann proves that a group of agents who once agreed about the probability of some proposition for which their current probabilities are common knowledge must still agree, even if those probabilities reflect disparate observations. Perhaps one saw that a card was red and another saw that it was a heart, so that as far as that goes, their common prior probability of 1/52 for its being the Queen of hearts would change in the one case to 1/26, and in the other to 1/13. But if those are indeed their current probabilities, it cannot be the case that both know them, and both know that both know them, etc., etc.

Aumann suggests that his result may undermine what he calls "The Harsanyi Doctrine" (p. 1238), i.e., the view that rational agents—e.g., players in a many-person game—are best thought of as starting with a common prior probability distribution over the field propositions of interest to them, with differences in their subsequent observations accounting for the divergent posterior probabilities on the basis of which they make their moves in the game. But that view of the matter seems to be based on exaggerated expectations of transparency to each other of different agents' states of mind, once these are made public. We discuss this point in Sec. III below.

In Aumann's framework new probabilistic states of mind can only arise by conditioning old ones on new knowledge. In such a framework, current probabilities must derive from what is in effect knowledge, i.e., true full belief. But in Sec. II below we derive Aumann's result from common knowledge of the shared value of a probability, however arrived at. We work with possible worlds in which the agents' probabilities and their evolution are matters of fact, represented within the model.

Independence of particular update rules is a central feature of the new framework. But of course we need some constraint on how agents update their probabilities. For this we use Goldstein's requirement that current expectations of future expectations equal current expectations. This is the workhorse for our proof of the Generalized 'No Agreement' Theorem.

II. THE THEOREM GENERALIZED

Notation. $1, \ldots, N$ are the agents, Ω is a non–empty set of "possible worlds", and $\mathcal{A}$ is a σ–field over Ω, each element of which specifies a complete history—past,

W. Leinfellner and E. Köhler (eds.), Game Theory, Experience, Rationality, 109–115.

present, and future. So far, our framework is like Aumann's; but now the change comes. In our framework agents belong to worlds, and as time goes by, their probabilities concerning their own and other agents' probabilities evolve along with their probabilities concerning other matters—e.g., my probability for your probability for my probability for a Republican President in the year 2000.

We model this via a time index t taking values in some linearly ordered set T, and probability measures $pr_{i\omega t}$ representing ideally precise probabilistic states of mind of agents i in worlds ω at times t. In the present framework there might be different common priors in different worlds ω, ω': If the prior time is $t = 1$, perhaps the probability measures $pr_{i\omega 1}$ are the same for all i, and so are $pr_{i\omega' 1}$, but $pr_{i\omega 1}(A) \neq pr_{i\omega' 1}(A)$ for some A.

Now Def. B^t defines *Belief* as 100% probability (i.e., certainty, full belief), and Def. K^t defines what we shall call "knowledge" as true full belief:

(Def. B^t) $$B_i^t A = \{\omega : pr_{i\omega t}(A) = 1\}$$

(Def. K^t) $$K_i^t A = A \cap B_i^t A$$

In terms of knowledge we then define mutual and common knowledge by Def. MK^t and Def. CK^t respectively:

(Def. MK^t) $$M_0^t A = A$$
$$M_{n+1}^t A = \bigcap_{i=1}^{N} K_i^t M_n^t A$$

(Def. CK^t) $$\kappa^t A = \bigcap_{n=0}^{\infty} M_n^t A$$

In the generalized theorem, $t = 1$ and $t = 2$ are times at which agents have their old and new probabilities for some hypothesis $H \in \mathcal{A}$. In world ω those probabilities are $pr_{i\omega 1}(H)$ and $pr_{i\omega 2}(H)$—which we now write simply as $P_\omega(H)$ and $Q_\omega(H)$, respectively, with the subscript i understood, and with the work of the time subscripts 1 and 2 done by writing P and Q:

(Shorthand) $$P_\omega \text{ for } pr_{i\omega 1}, \quad Q_\omega \text{ for } pr_{i\omega 2}$$

Thus, with $t = 2$, Def. B^t is written:

(Def. B^2) $$B_i^2 A = \{\omega : Q_\omega(A) = 1\}$$

In the generalized theorem the proposition C^2 specifies $q_1, \ldots, q_N$ as the agents' new probabilities for H:

(Def. C^2) $$C^2 = \bigcap_{i=1}^{N} \{\omega : Q_\omega(H) = q_i\}$$

The crucial hypothesis of the generalized theorem is Goldstein's principle (G) that *old probabilities = old expectations of new probabilities*. Here the integrand

$Q(A)$ is a random variable, a P_ω-measurable function of ω which takes real values $Q_{\omega'}(A)$ as ω' ranges over Ω :

(G) $$P_\omega(A) = \int_\Omega Q(A)\, dP_\omega$$

The second hypothesis says that whenever A is in $\mathcal{A}$, so are the N propositions saying—perhaps, falsely—that at time 2 the several agents are sure that A is true. This guarantees that $\mathcal{A}$ is closed under all the operations K_i^2, M_n^2, and κ^2.

(More Shorthand) B for B_i^2, κ for κ^2, C for C^2

To prove the theorem we use two lemmas:

Lemma 1 *While something is common knowledge, everyone is sure it is:* $\kappa A \subseteq B\kappa A$.

Proof. By Def. CK^2, MK^2, and K^2. ■

Lemma 2 *If (G) holds, then* $\int_{\Omega-\kappa C} Q(\kappa C)\, dP_\omega = 0$

Proof. $P_\omega(\kappa C) = \int_\Omega Q(\kappa C)\, dP_\omega = \int_{\kappa C} Q(\kappa C)\, dP_\omega + \int_{\Omega-\kappa C} Q(\kappa C)\, dP_\omega$ by (G). By lemma 1 the first term of this sum $= \int_{\kappa C} 1\, dP_\omega = P_\omega(\kappa C)$, so the second term = 0. ■

Theorem 3 (Aumann's 'No Agreement' Theorem Generalized)
Hypotheses: (G) holds, $B : \mathcal{A} \to \mathcal{A}$, P_ω *is the same for all* i, $P_\omega(\kappa C) > 0$.
Conclusion: $P_\omega(H|\kappa C) = q_1 = \ldots = q_N$.

Proof. $P_\omega(H|\kappa C) = \dfrac{\int_\Omega Q(H \cap \kappa C)\, dP_\omega}{\int_\Omega Q(\kappa C)\, dP_\omega}$ by (G),

$$= \frac{\int_{\kappa C} Q(H \cap \kappa C)\, dP_\omega + \int_{\Omega-\kappa C} Q(H \cap \kappa C)\, dP_\omega}{\int_{\kappa C} Q(\kappa C)\, dP_\omega + \int_{\Omega-\kappa C} Q(\kappa C)\, dP_\omega},$$

$= \dfrac{\int_{\kappa C} Q(H \cap \kappa C)\, dP_\omega}{\int_{\kappa C} Q(\kappa C)\, dP_\omega}$ by lemma 2 since $Q(A \cap \kappa C) \leq Q(\kappa C)$,

$= \dfrac{\int_{\kappa C} Q(H)\, dP_\omega}{\int_{\kappa C} 1\, dP_\omega}$ by lemma 1,

$= \dfrac{\int_{\kappa C} q_i\, dP_\omega}{\int_{\kappa C} dP_\omega}$ by Def. C^2, $= q_i$, i.e. the same for $i = 1, \ldots, N$. ■

III. UNCOMMON KNOWLEDGE

In common usage, 'common knowledge' refers to what everybody knows, i.e., technically, first degree mutual knowledge. Common knowledge in the technical sense is another matter, harder to come by. Non-linguistic animals may have modest degrees of mutual knowledge, but in practice, the higher degrees seem to require language, as common knowledge seems to do in principle.

Consider, e.g., the puzzle of the three dirty-faced children: Alice, Bob, Claire. Each sees that the other two have dirty faces, so each knows that *not all their faces are clean*, i.e. mutual knowledge of degree 1. And of degree 2, as well: Each sees that the other two have dirty faces, so each sees that each of the other two sees that the other one's face is dirty; and of course, in seeing a dirty face, each knows that he or she knows that not all their faces are clean. But third degree mutual knowledge that not all their faces are clean is lacking, as can be seen by considering what Alice thinks possible. As far as she knows, it is possible that her face is clean:

$$\langle \text{Alice} \rangle (\text{Alice has a clean face})$$

(In general, the notation $\langle i \rangle$ is to be read: *As far as i knows, it is possible that.*) And as far as Alice knows, it is possible that as far as Bob knows, both of their faces are clean—for as far as she knows, her face may be clean, in which case Bob would see that, and still think his face might be clean:

$$\langle \text{Alice} \rangle \langle \text{Bob} \rangle (\text{Alice and Bob have clean faces})$$

And, indeed, as far as Alice knows, Bob, thinking his face may be clean, may see that her face is clean, and think that as far as Clare knows, all faces are clean:

$$\langle \text{Alice} \rangle \langle \text{Bob} \rangle \langle \text{Claire} \rangle (\text{All their faces are clean}).$$

(Indeed, any permutation of the names in this statement will leave it true.) Then the fact that *not* all their faces are clean is not mutual knowledge of degree 3, and, so cannot be common knowledge.

Now suppose their mother tells them: "Not all your faces are clean!" They believe her, and in fact it is common knowledge among them that they do. (The story is told about a mother and her children in order to make all this seem a matter of course.) Their mother has stated a physical fact that each already knows, but because of their common knowledge of their trust in her, their individual knowledge that she has made that statement informs them of mental facts they had not known—in particular, that *all know that all know that all know* the fact their mother enunciated. And this crumb of third degree mutual knowledge, combined with successive revelations of their ignorance, can tell all three children that their faces are dirty:

(1) Mother says: "Any who know your faces to be dirty, raise your hands!" No hands go up. (2) Mother asks a second time, and again no hands go up. (3) Mother asks a third time, and now all their hands go up.

What happened? Their mother's announcement that not all their faces were clean removed the possibility that $\langle \text{Alice} \rangle \langle \text{Bob} \rangle \langle \text{Claire} \rangle (\text{All faces are clean})$, and removed the possibilities corresponding to permutations of the three names in that statement. The proposition she enunciated was something they already knew; indeed, it was second degree mutual knowledge among them. But the fact that their mother enunciated it made it common knowledge among them. And the important point is that it made it mutual knowledge among them of degree three, which was exactly enough to let the second non-showing of hands show them that all their faces must be dirty; the additional degrees of mutual knowledge that make up common knowledge were not needed to solve their problem.

What their mother told them was a fragment of what they could all see for themselves. She told them that at least one of their faces was dirty, and each could each see that two of their faces were dirty. But each saw two different faces; they had three different bodies of evidence, of which each separately implied what their mother told them, and any two together implied that all their faces were dirty. Their mother structured the game so that they could not simply pool their evidence, and could not discover from what she told them that they all had dirty faces, but could discover it from the fact of her telling them what they could see for themselves.

This sort of reasoning was deployed by Geanokoplos and Polemarchakis [3] to analyze the process by which successive announcements by $N = 2$ agents of their current probabilities for truth of A *as the announcements were made* eventually ends any disagreement in their probabilities for A—if it is common knowledge that each takes these successisve revelations into account by conditioning on the disjunction of the propositions that the other could have conditioned upon to get the announced probability for A. In that process, too, it may be that no minds are changed until several announcements have been made, at which point the next announcement reveals total agreement ensuing from the previous announcement; and the rationale may be rather abstruse.

Evidently there are circumstances in which common knowledge can be hard to come by. Common knowledge need not be what every fool knows—as it may be in special cases, e.g., in the case of *self-evident* propositions, where[1]

$$A \text{ is self-evident} \Leftrightarrow \text{for all } i,\ \mathcal{P}_i(\omega) \subseteq A \text{ for all } \omega \in A$$

A self-evident event is a true self-evident proposition. For such events, it is evident to all that all have the same, conclusive evidence. But such conclusiveness is possible for other events as well, e.g., for the event of the mother's declaring that not all faces were clean. Before that event took place, the fact that not all their faces were clean was not self-evident, for although all could see that it was true, all could also see that all had different visual evidence, and could puzzle out the fact that what they could plainly see was not mutual knowledge of degree three.[2]

We take these considerations to weigh against Aumann's view of his theorem as undermining the Harsanyi "Doctrine" that rational people will have the same prior:

> John Harsanyi (1968) has argued eloquently that differences in subjective probabilities should be traced exclusively to differences in information—that there is no rational basis for people who have always been fed precisely the same information to maintain different subjective probabilities. This, of course, is equivalent to the assumption of equal priors. The result of this paper might be considered evidence against this view, as there are in fact people who respect each other's opinions and nevertheless disagree heartily about subjective probabilities. (p.1237–8)

Now as Harsanyi points out, many games *are* aptly modelled by assuming a simple (1) *common prior*, shared by all players. On the other hand, as we have been pointing out, (2) *common knowledge of a shared posterior probability* may be very hard to come by; Geanokoplos and Polemarchakis's program for eliminating disagreement in such a way as to reach common knowledge of a shared posterior probability is not the sort of thing we easily understand and implement.[3] And, finally, as Aumann points out, we may well respect each other's opinions but (3) *disagree about posterior probabilities.* It is (1), (2), (3) *together* that are refuted by Aumann's theorem; one or more must go. We submit that in the game context, (2) is far the least plausible of the three, far the likeliest candidate for rejection.

On the other hand, in the broader epistemological context in which Carnap (1950) defended it, the common prior assumption looks less plausible than the assumption of common knowledge of a shared posterior probability. Certainly it is the more alien of the two, inapplicable to humans. Carnap's inductive logic would be more appropriate for artificial intelligences that are "born" fully able to produce *Protokollsätze* in response to experience as it comes. But in a collaborative search for truth, such intelligences could get common knowledge of a common posterior probability simply by pooling their (non-indexical!) *Protokollsätze* and conditioning their common prior on all of them. (Non-indexical: Each intelligence could write its *Protokollsätze* on slips of paper to be placed in a common bin, whence they could all be retrieved and formed into a single long conjunction which would make the same sense irrespective of who produced them, and when.)[4]

NOTES

1. We adapt the terminology of Osborne and Rubinstein, p. 73.
2. Lewis's [7], (II, §1) often cited concept of common knowledge is often misrepresented as mutual knowledge of all finite degrees, our κ. In fact, his 'common knowledge' refers to states of affairs (sc., *bases* for common knowledge) in which, as he points out, modest degrees of mutual knowledge are expectable, with κ holding under idealized assumptions of a sort commonly made in the game theory literature. His discussion is of interest in the present context.
3. Furthermore, it presupposes updating by conditioning on new certainties, and so is not generally applicable in the broader framework of Sec. II here. See [6], sec. 2, for some alternatives to simple conditioning.
4. Neurath [8] seems to have had something like this in mind.

REFERENCES

[1] R.J. Aumann, "Agreeing to Disagree", in: *Annals of Statistics*, 4, 1976, pp.1236–1239.

[2] R. Carnap, *Logical Foundations of Probability*. Chicago: University of Chicago Press 1950 (2nd ed. 1962).

[3] J.D. Geanakoplos and H.M. Polemarchakis, "We Can't Disagree Forever", in: *Journal of Economic Theory*, 28, 1982, pp.192–200.

[4] M. Goldstein, "The Prevision of Prevision", in: *Journal of the American Statistical Association*, 78, 1983, pp.817–819.

[5] J. Harsanyi, "Games of Incomplete Information Played by Bayesian Players", Parts I-III, in: *Management Science*, 14, 1967–1968, pp.159–182, 320–334, 486–502.

[6] M. Hild, R. Jeffrey and M. Risse, "Aumann's 'No Agreement' Theorem Generalized", in: C. Bicchieri, R. Jeffrey, and B. Skyrms, (eds.), *The Logic of Strategy*. Oxford: Oxford University Press 1997.

[7] D. Lewis, *Convention*, Cambridge, Mass.: Harvard University Press 1969.

[8] O. Neurath, "Protokollsätze", in: *Erkenntnis*, 3, 1932, pp. 204-214. Translated in his *Philosophical Papers 1913-1946*, Dordrecht: Reidel 1983.

[9] M. Osborne and A. Rubinstein, *A Course in Game Theory*. Cambridge, Mass.: The MIT Press 1994.

Matthias Hild
Balliol College, Oxford
UK

Richard Jeffrey
Department of Philosophy, Princeton University
Princeton, NJ
USA

Mathias Risse
Department of Philosophy, Princeton University
Princeton, NJ
USA

EDWARD F. MCCLENNEN

RETHINKING RATIONAL COOPERATION

I. INTRODUCTION

I want to extend here a line of reasoning that I pursued in *Rationality and Dynamic Choice* (1990). In that book I argued that the standard Bayesian model of expected-utility reasoning needs to be revised to accommodate a capacity, on the part of rational decision-makers, to effectively coordinate with their own future selves – to be guided by plans that they have deliberately adopted. I also suggested that an analogous line of reasoning might be employed to show that rational agents could engage in rule-governed coordination of their choices with one another to a greater extent than the standard theory would seem to admit. It is this suggestion that I now want to explore more fully in the present paper. It is not that the standard theory denies the possibility of such coordination. On that account, however, coordination can only be achieved by the adoption of an effective enforcement system. Since, however, any enforcement scheme requires the expenditure of scarce resources, the standard theory seems committed to the view that rational agents can achieve only a second best outcome. It is worth considering whether thoroughly rational persons who know each other to be such might not do better than this. It is also the case that the process whereby individuals come to agreement on the terms of their coordination schemes is conceived as driven essentially by non-cooperative considerations, specifically the relative bargaining power or threat advantage of the participants. Bargaining outcomes based on this principle, however, tend to generate mutually disadvantageous conflict, and thus there are greater costs yet to be borne by "rational" interacting agents. I shall have to leave the exploration of this second problem to another occasion, however.

II. MUTUAL GAINS AND LOSSES

Starting with Adam Smith, and running like a bright thread throughout virtually all the subsequent theoretical literature on political economy, one can mark a preoccupation with the conditions under which individuals can transact with one another to their mutual advantage. In the more formal literature, this concern culminates, in the middle of twentieth century in the first fundamental theorem of welfare economics, according to which individuals can, under conditions of perfect competition achieve an outcome that is Pareto-optimal and characteristically strictly Pareto-efficient relative to the outcome of no transactions, that is, each does better as a result of such transactions. (For brevity's sake I will, in what follows, use the phrase "the Pareto condition" to refer to the conjunction

W. Leinfellner and E. Köhler (eds.), Game Theory, Experience, Rationality, 117–129.

of these two conditions, and use the qualifiers, "optimal"or "efficient" only when it is important to the argument.)

The theme of Pareto-efficient changes in institutional structures is also central to Coase's important work, both on the theory of the firm (1937) and the problem of social cost (1960), to Posner's economic analysis of law (1986), the public choice tradition initiated by Buchanan and Tullock (1962), and Axelrod's work on iterated prisoners' dilemma games (1984) and the evolution of co-operation (Axelrod and Dion, 1988), and Ullman-Margalit's study of the emergence of norms (1978: Chs II, III). The Pareto condition also figures centrally in virtually all axiomatic bargaining and social choice models.

To be sure, these explorations have often been accompanied by the suggestion that arrangements satisfying the Pareto condition and changes that are Pareto-efficient are perhaps best secured indirectly rather than by any deliberate attempt by participating agents. This is, of course, the idea that is so strikingly captured by Smith's "invisible hand" metaphor: the wealth of nations is a byproduct of each person's pursuing his or her own personal interests. It also finds powerful reincarnation in some evolutionary accounts of the emergence of institutional structures (see Alchian, 1950; Hayek, 1967; and Sugden, 1989).

Finally, where satisfaction of the Pareto condition is not assured (e.g., public goods problems, externalities, and single-shot prisoners' dilemma games), much effort is invested in the search for mechanisms that will overcome this problem. The recent "folk" theorems on indefinitely repeated games are a case in point (see, e.g., Fudenberg and Tirole, 1992: Ch. 5). The objective is to show that Pareto-efficient outcomes can be secured by backing up a coordination scheme with an appropriate set of sanctions.

The good news about the emergence of structures that are Pareto-efficient, however, has been tempered by the bad news (1) that the historical processes of institutional development tend to be path-dependent in ways that work against adaptive efficiency, and (2) that virtually all institutional arrangements are subject to manipulation by special interests, to the short term advantage of some, but often to the long-term disadvantage of all. The former theme is central to the work of Arthur (1994) and also to the new institutional theory associated with North (1990). The latter theme is central to the literature on "rent-seeking" (see Buchanan, Tollison and Tullock, 1981), to North's work (1990), and Knight's recent study of social conflict (1992). Taken as a whole, in fact, recent work in political economy has identified deep pressures on human interaction that tend to prevent the satisfaction of the Pareto condition.

In one sense, there should be nothing surprising about these pessimistic conclusions. After all, the historical record hardly supports any other picture. What is surprising, however, is that many these findings are based on models of *ideally* rational beings who have common knowledge of each other's rationality. On many of the accounts offered, failure of the Pareto condition does not flow from assuming that some of the parties involved are less than fully rational, or that parties have to make decisions under conditions of limited knowledge. Free-riding, rent-seeking, and conflict over the distribution of goods can be due to choices made by persons who are hyper-rational and fully informed. How does it happen that such agents manage to do so poorly?

III. MODELING COOPERATIVE INTERACTION

Although my ultimate concern is with models of on-going interactions which involve a mixture of conflict and cooperation, it will help to begin with a review of the way in which the theory of rational choice has been articulated for simple one-stage cooperative games, and, in particular, for a special class of such games, namely, *pure-coordination* games. These are games in which (1) outco mes are ranked identically by all participants, so that there is a complete convergence of interests, (2) there is more than one combination of choices that satisfies the Pareto condition (relative to what they can expect if no attempt is made to coordinate their choices), and (3) the parties are not able to freely communicate with one another.

As Schelling (1960) and Lewis (1969) clearly recognized, it is the combina tion of (2) and (3) above that generates a coordination problem. If there is common knowledge of the structure of the game and each participant's evaluation of outcomes, and (2) is false – that is, there is a unique Pareto outcome – then there is an obvious choice for each. Each should choose so as to ensure the realization of that outcome. Thus, there is no coordination problem. Correspondingly, if (3) is false – that is, the parties can fully communicate with one another, then it should be possible for them to agree on which Pareto outcome to realize, and proceed to execute that agreement. Again, there is no coordination problem.

Since we are concerned here with games in which persons deliberately attempt to coordinate their choices with one another, it is natural to think of participants as responsive to Pareto considerations. Schelling, however, starts with a much more general account, and then particularizes this framework in a manner that ends up marginalizing the Pareto condition. What he offers is a model of the processes whereby convergence on a specific combination of strategies can take place, and, in particular, the process by which persons represent to themselves the strategic dimension of the situation – that is to say, the interdependency between their choices and their expectations of how others will choose. Moreover, and most importantly, this theme of interdependence of expectations and choices is laid out initially in abstraction from any specific criterion of choosing rationally. He merely speaks of a situation in which one can choose "wisely" in the face of an expectation as to how the other player will choose (Schelling, 1960, pp.86-7). As it turns out, however, Schelling proceeds to give content to the notion of a "wise" or rational choice by appealing to a series of interdependent situations for which he assumes (without any direct or explicit argument!) the appropriate criterion of rational, interdependent choice is the equilibrium requirement. That is, he assumes that under conditions of common knowledge the strategy choices of each player must be an expected-utility maximizing response to set of choices of the other players. Lewis (1969, p.27ff) proceeds in much the same fashion. He initially characterizes the coordination problem as one involving interdependency of choice and expectation, in abstraction from a specific criterion of rationality, and, once again without any explicit defense, he then moves directly to the equilibrium condition. For Schelling and Lewis, then, and for the many who have followed them, the task has been to model the process whereby rational agents converge on some particular equilibrium of strategy choices.

To be sure, preoccupation with equilibrium pairs does not have to mean abandonment of concern for Pareto outcomes. In such games all Pareto combinations of strategies are also equilibrium combinations. To see this, suppose that the combination is optimal but not in equilibrium. Then there is some participant who could, by choosing differently, increase the utility of the outcome, on the assumption that everyone else continues to play for the original combination. Since, however, the parties agree on the utilities of the outcomes, this must mean that *everyone* would prefer to coordinate on this combination rather than the original one. Thus, contrary to the hypothesis, the original combination did not yield an optimal outcome. (The converse, of course, is not necessarily true: a strategy pair can be in equilibrium but its outcome need not be optimal.) But, given the commitment to the equilibrium requirement, then, what has predictably emerged is an account in which the Pareto condition typically serves as a secondary requirement on a solution, that is, as merely one basis for selecting from among multiple *equilibria*.

The equilibrium approach, of course, had already been employed in the analysis of the other limit case - the zero-sum (pure conflict) game. With the two limit cases linked in this manner it was, of course, inevitable that the same conceptual machinery would be used to bridge the game types in between. Yet Schelling himself suggests that the game of pure coordination introduces something distinctive, and something that is bound to be overlooked by those who preoccupy themselves with games of pure conflict. This is the need for the parties to *coordinate* their choices. In a game of pure conflict of interest, what one hopes for is precisely that one's choice is *not* correctly anticipated by the other player. The appropriate model is hide and seek, avoidance and pursuit. Any anticipation of the choice of the other that works to one's own advantage must work to the disadvantage of the other. But in the pure coordination game, the task is to meet up, to correctly anticipate each other's choices and to capitalize on this in a manner that works to mutual advantage. To be sure, Schelling does pivot his discussion of psychological salience on the imperative of meeting up. The whole analysis of saliency, however, is still subordinated to the equilibrium requirement.

IV. WHAT DRIVES THE ANALYSIS

Why have Schelling, Lewis, and others chosen to develop an account of pure coordination games within the framework of the standard equilibrium account of rational interdependent choice? The answer is to be found at a very deep level, in the very way in which interactive choice has been conceptualized from the very outset. Originally, the problem for a general theory of games was to provide the solution to an ideal type of interdependent choice situation, characterized by the following conditions:

(1) *Mutual Rationality*. Both players are rational.

(2) *Common Knowledge*. There is common knowledge of (a) the rationality of both players, (b) the strategy structure of the game for each player, and (c) the preferences that each has with respect to outcomes.

(3) *Consequentialism*. Each player's choice among available strategies is consequentially oriented, i.e., choice is strictly a function of the preferences that player has with respect to the outcomes associated with the various strategies.

(4) *Maximization*. Each player's choice behavior with regard to any abstractly conceived space of outcomes (including probabilistically characterized outcomes) can be captured by supposing the player maximizes a von Neumann-Morgenstern utility function, defined over that same space of outcomes.

The solution proposed was originally based, in turn, on a thought experiment or "indirect" argument, which proceeds along the following lines. First one imagines that a theory of games has been worked out and that both players know the theory. One then tries to conclude something about the content of the theory itself by tracing out the implications of the assumption that there is a theory which all the players know (Von Neumann and Morgenstern, 1953, pp.146-8). The most frequently cited version of such an indirect argument is the one presented by Luce and Raiffa (1957, p.63). They begin by assuming that if there exists an absolutely convincing theory of the zero-sum, two-person game, then, under conditions (1) and (2), mutual rationality and common knowledge, each player knows what the theory tells each of them to do. But given this, they argue, it is plausible to suppose that if the theory picks out specific strategies r_i and c_j, for row and column player respectively, mere knowledge that this is what the theory prescribes should not cause either to choose a different strategy. How could such knowledge cause a given player to choose differently? Implicitly invoking conditions (3) and (4), Luce and Raiffa suppose that a rational row player would, given an expectation that the other player will choose c_j, change from r_i to some other strategy, if the *outcome* associated with that other strategy has a higher utility for row player than the *outcome* associated with r_i. There would be no change only in the case that r_i was a utility maximizing reply to c_j. When r_i and c_j are such that knowledge of the theory would not lead either player to make such a change, then (r_i, c_j) is said to be in *equilibrium*. On this way of reasoning, then, the pair of strategies singled out as the solution to a game must satisfy the condition of being in equilibrium.

Implicit in all of this is an assumption that if conditions (1) through (4) hold, and there is a well-defined theory about how rational players should choose, then this (perhaps together with certain additional assumptions) will enable each player to frame expectations about the behavior of the other (also rational) player, expectations that are sufficiently determinate to enable the player in question to treat his or her own decision problem as a simple maximization problem. On this way of thinking, then, in the ideal case each player is presumed to be in a position to take the behavior of the other (rational) player as a given, i.e., a well-defined parameter.

Finally, as indicated, the appeal to this "indirect" argument was first made in connection with the analysis of zero-sum two-person games. But nothing in the argument presupposes that agents have conflicting preferences with regard to outcomes. Thus the argument is just as applicable to a game of pure coordination as to any other game.

V. LOOKING MORE CLOSELY AT THE ARGUMENT

Is it really so clear that a rational player should always view a interdependent decision situation as posing a problem of maximizing expected-utility under conditions of parametric choice? Let us grant the intuitive appeal of conditions (3) and (4). It still needs to be recalled that the objection of suboptimality that can be posed to the standard theory is a consequentialist objection. Agents who reason and choose in the standard fashion end up doing less well, in terms of preferred consequences, than those who are able to reason in an alternative manner. This suggests that what the argument turns upon is not just maximiza tion and consequentialism, as such, but how maximization and consequentialism are to be conceptually *anchored*, i.e., from what standpoint or perspective they are to be applied. What is assumed, in effect, is that in deliberating about the choice of means to preferred outcomes, one should abstract from the context of the interactive problem itself, and consider how one would be prepared to evaluate such options, were it the case that one faced a problem of individual decision-making against nature. The intuitive notion is simply that it is one and the same whether the outcome of a choice of an action is conditioned by choices that another agent makes or by natural events. Stated somewhat more formally, what is implicit in the standard argument is an appeal to something like the following condition:

(5) *Separability*. Let G be any game, and let D be the problem that as a player in G one would face, were the outcomes of the strategies available to one in G conditioned, not by the choices of another player, but by some "natural" turn of events in the world, so that one faces in D a classic problem of individual decision-making under conditions of risk and/or uncertainty; and suppose that one's expectation with regard to the conditioning events corresponds to one's expectations with regard to how the other player will choose in G: then one's preference ordering over the options in G must coincide with the preference ordering one would have over the options in D.

For a fuller discussion of this organizing assumption, and its importance, see McClennen (1990, 1992, and 1997).

VI. RETHINKING RATIONAL COOPERATION

What I now want to argue is that there is something odd about conceptualizing a pure-coordination game as one in which each agent is preoccupied with ensuring that the choice made is a utility maximizing response to the choice that the other player makes. The intelligibility of that approach cannot be denied in the case of zero-sum games of pursuit and avoidance, where the gains to one are matched by losses to the other. But that is not the situation in the game of pure cooperation. There the players face a problem of how to *co*ordinate their choices – how to "meet up" at an outcome that is maximally preferred by each. This suggests, as I have already noted, that it is the Pareto criterion (rather than the equilibrium condition) on which solution theory should focus.

As already noted, of course, in a game of pure coordination every Pareto outcome is reached by a set of strategies that are in equilibrium. But it does not follow that the process of drawing the expectations of the various players to a focus is appropriately modeled in terms of the standard equilibrium theory. An analogy here should be helpful. It is a (logical) fact about two-person zero-sum games in which mixed strategies are permitted that every equilibrium combina tion of strategies is also a maximin pair of strategies. But it does not follow that maximining is the appropriate way to think about such games. Von Neumann and Morgenstern did in fact focus on that alternative solution concept, but in subse quent work the maximin approach rapidly gave way to the equilibrium approach. The point is simply that the concept of trying to coordinate upon a Pareto combination of strategies can itself be understood to furnish an alternative account of the process of convergence. Intuitively, the problem that the two agents face in a game of pure-coordination is that each is able to identify a number of combinations of strategies as generating a maximally preferred outcome, but, by definition, neither agent is in a position to unilaterally secure the outcome that is maximally preferred by that agent. The preferred outcome can be achieved only by that agent *co*ordinating his or her choice with the other agent. Happily, however, there is no conflict of interest in this case: their problem is only one of meeting up. And meeting up means simply converging on a strategy combination whose outcome is maximally preferred by *both*. Consider, then, a model in which each agent is, in addition to (1) through (4), governed not by (5) but by the following principle:

> (6) *The Pareto Principle.* Rational agents who know each other to be such will, *ceteris paribus*, confine their choice to strategies that can, in combina tion with the choices of the other agent(s) generate outcomes that satisfy the Pareto condition.

A *ceteris paribus* clause is needed here since when players face a communication barrier, each may judge that they improve their chances of avoiding what they mutually agree would be an even worse outcome, if they focus on a second best outcome.

Focusing on the set of Pareto outcomes leaves in place all of what Schelling and Lewis have to say about the role of psychological salience. Within the framework of the pure-coordination game, where there is a barrier to communication, there is a need for some way of sorting through the alternative Pareto combinations. And it also leaves in place Schelling's and Lewis's notion of a process by which expectations are drawn to a focus. Suppose that there is common knowledge of each other's rationality, and the strategy and payoff structure of the game, and that the problem in a pure-coordination game is to effectively coordinate choices so as to achieve what is mutually perceived to be a best outcome. Suppose, moreover, that there is a strategy combination (r_i, c_j) that is salient among the multiple combinations that satisfy the Pareto condition. To say that (r_i, c_j) is salient is to say that Row expects Column to choose c_j, and that Column expects Row to choose r_i. And given such expectation on the part of each, and an understanding that the task is to coordinate on a combination that satisfies the Pareto condition, Row now has a reason for choosing r_i, and Column has a reason for choosing c_j. Moreover, each can replicate the reasoning of the other. It is because (r_i, c_j) is salient that each expects the other to play for this combination. But then each also expects that the other expects that they will play for this combination, etc., etc.

Just as on the standard account, coordinated choice on this model involves conditional strategies: one's best choice depends upon what one expects the other player to do. Similarly, there is a strategic dimension to coordinated choice: by choosing a certain strategy one may hope to induce the other player to respond in a manner that contributes to the furthering of these coincident interests. More generally, the best choice depends in part upon how one expects that others will respond to one's own choice. But it is not part of the logic of this alternative model that one expects others to unilaterally choose to maximize utility in response, any more than that one thinks that the rationality of one's own choice turns on its maximizing utility in response to what one expects the others to do.

Lewis makes the following interesting and relevant remark about equilibrium outcomes in the course of his development of a theory of coordination:

> This is not to say that an equilibrium combination must produce an outcome that is best for even one of the agents (though if there is a combination that is best for everyone, that combination must be an equilibrium). In an equilibrium, it is entirely possible that some of or all the agents would have been better off if some or all had acted differently. (1969, p.8).

Although this is not his intention, this can surely be read as a brief against making satisfaction of the equilibrium condition a *necessary* condition of a rational solution. Suppose an outcome is such that all would have been better off if all or some had acted differently. Even if that outcome did result from an equilibrium combination of strategies, it is not at all clear why this constitutes a successful resolution of the coordination problem. It is true that if the outcome results from an equilibrium combination, then each has done the best they can, given what the other has done. And, no doubt, their expectations are concordant. But what does this have to do with successful coordination?

If decision and game theorists have tended to think about this kind of coordination problem in the wrong way, so have many moral philosophers. The philosophical literature on rule governed behavior is discouraging in this regard. It is often argued that rule-governed behavior is rule *bound* behavior and, as such, is consequentially indefensible. The suggestion seems to be that what is defensible from the point of view of consequences is a flexible policy of making exceptions to following the rules be adopted – whenever allowing for such exceptions contributes to the realization of the very ends for which the rules themselves were adopted. On this account, rules are to be understood as merely maxims – "rules of thumb". Such an account, however, completely ignores the logic of coordination problems as set forth above. For two notable exceptions to this tendency among philosophers, however, see Rawls (1953) and Bratman (1987).

The alternative account of cooperation proposed here carries with it one very important implication. Rational beings will not approach pure coordination problems as if they call for reasoning in accordance with the separability principle and the demands of parametric reasoning. To be sure, since in pure coordination games Pareto outcomes are invariably also equilibrium outcomes, their choice behavior will not, in fact, be inconsistent with such principles; but those principles, in effect, will play no role in the reasoning of such persons.

VII. EXTENDING THE ARGUMENT TO MIXED GAMES

The graph of a game of pure conflict of interest consists in a set of points that lie on a straight line with negative slope, i.e., it has only NW-SE dimensionality.

The graph of a game of pure coordination consists in a set of points that lie in a straight line with positive slope, i.e., it has only NE-SW dimensionality. Every game that lies in between these two extremes, including the prisoners' dilemma game, is a mixed-motive game, which involves, then, both a conflictual and a cooperative dimension. The potential for cooperation is to be found in its NE-SW dimensionality, the potential for conflict in its NW-SE dimensionality.

The potential for mutual gain or loss does not correspond, of course, to any particular player's potential for gain or loss. One may be able to gain by being able to effect a NW-SE dimensional shift in outcome (in which case, one's gain is matched by a loss on the part of the other player). Within this framework, moreover, there is no simplistic identification of cooperative, as distinct from conflictual, interaction with whether or not the gains from coordination are shared equally or not. Coordination, in one form or other, could result in any number of different outcomes, each of which would improve the position of each player, relative to some specified "status quo" situation, but each one of which has different distributive implications.

The full significance of this indeterminacy emerges only within the context of an exploration of the other problem facing rational players who know each other to be such: the bargaining or negotiation problem of how the gains from coordination are to be distributed. As I indicated at the outset, I must defer discussion of this issue. But we can abstract from this dimension of a mixed motive game, by considering games in which in so far as agents manage to coordinate their choices, they do so on distributive terms that are exogenously fixed in some fashion or other, e.g., by the structure of the interactive situation. What I want to suggest is that the alternative model of coordination presented above plausibly extends to this case as well. That is, in this case, no less than the case of a pure coordination game, ideally rational persons, who have com mon knowledge of their rationality and the game they are playing, will be disposed to cooperate to ensure a Pareto outcome. Within the context of mixed games, such a model of cooperation has the important implication that rational agents who know each other to be such will not confine their attention to equilibrium combinations of choices. That is, if parametric or equilibrium reasoning leads to a suboptimal outcome, and there is a unique combination of coordinated choices which satisfies the Pareto condition, say, relative to that equilibrium outcome (as in the case of the prisoners' dilemma game), then they will coordinate on it.

This leads in turn to another natural extension of the model. Suppose that there is not a unique alternative satisfying the Pareto condition, but rather a number of alternatives between which the agents are indifferent. We may suppose here, in keeping with Schelling's own analysis, that successful coordination will turn on considerations of psychological saliency. But suppose now that the agents can communicate with one another, and agree on a coordination scheme. The proposed model of rational coordination implies that this scheme will be regulative of the choices that each then makes, despite the fact that it does not prescribe an equilibrium set of choices for the participants. This defines one class of games to which the folk theorems for indefinitely iterated games are addressed. For these games, our alternative model of coordination yields precisely the conclusion that is reached by those theorems: rational players will not have to settle for a Pareto inferior outcome.

VIII. PLAN OR RULE GOVERNED CHOICE

With this last extension to a case, in which players can tacitly or explicitly agree upon a coordination scheme, we arrive at a model in which sense can begin to be made of the notion of rule-guided choice (McClennen, 1997). What we have just described are persons who are capable of tacitly or explicitly identifying a coordination scheme and then choosing in a rule-guided manner, even when that choice is not supported, or mandated, by standard (expected) utility maximizing considerations. The central question, for one who is capable of such a way of thinking about coordination is *not*: What available alternative is mandated by expected utility reasoning? It is, rather: What course of action is called for here, given some coordination scheme on which we have tacitly or explicitly agreed? This is not to say that such a person is slavishly committed to rules. The commitment to a given rule will characteristically be a conditional commitment. That is, it will be a defeasible commitment, one that can be set aside by reference to any one of a number of considerations. All that we need to suppose here is that such a person will not be disposed to free-ride on the cooperative dispositions of others.

It is precisely this sort of choice-behavior that the standard account of rationality cannot accommodate. Parametric reasoners cannot make even this very modest sort of "commitment" to act in accordance with a coordination rule. For such a reasoner, what was agreed upon at some point in the past will characteristically be irrelevant. The only relevant question will be whether the expected utility of acting in conformity with that plan or rule is greater than the future benefits of deviating. Such reasoners will, of course, be able to implement coordination schemes. But since each will be disposed to "free-ride" on the coordination commitments of the others, they will have to arrange to have the coordination scheme backed up by some sort of enforcement device.

To be sure, those who accept the equilibrium account will object that this is not quite true. An obvious functional substitute for social sanctions or formal enforcement system is a set of shared cultural beliefs regarding the "appropria teness" or "correctness" of acting in accordance with the rules in question. But the appeal to shared cultural values is an appeal to what is not a matter of a deliberate, and voluntary, agreement between persons. It involves an appeal to a "non-rational" factor. My concern here is with rational, deliberate choice.

IX. THE CASE FOR THE REVISED MODEL OF RATIONALITY AND AGAINST SEPARABILITY

Consider now the separability principle that has been shown to drive the argument for the equilibrium condition and hence the standard theory of cooperation. It is open to the defender of the standard theory to insist that the presuppositions of that theory capture in a particularly compelling way deep features of what it means to be rational. What I have suggested in reply is that they have a plausible case to make with regard to maximization and consequentialism, but that (1) their argument goes through only by appeal to the additional assumption of separability, and (2) there are *consequential* considerations that militate against the acceptance of the separability principle. This is the thrust of the findings briefly summarized in Section 2. If compliance with mutually advantageous coordination schemes by "rational" agents requires that in each situation in which the coordination rule prescribes a certain form of behavior, there be a corresponding sanction that "enforces" the rule, and without which, it would not necessarily be rational for the agent in question to comply, the costs of cooperative interaction become significantly greater. Those who can act voluntarily in accordance with rules will realize savings with respect to enforcement and surveillance systems, which is to say that such a way of organizing their joint activities will be Pareto-efficient relative to the adoption of a system of enforcement and surveillance. But then it follows that there is a *consequentialist* argument for not approaching every interdependent choice situation from the separable perspective developed by the standard theory. A rule-governed approach to certain interactive situations is one that will better serve the goals or interests of the participants than one that is not so governed.

To be sure, the standard approach also appeals to consequential considerations. The issue, then, is whether consequences must be assessed from the perspective that virtually all decision and game theorists have adopted, or whether there is an alternative perspective on assessment that is not only abstractly possible but which can motivate choice. One can begin by noting that, since those who master the business of rule or plan guided ways of transacting with one another will tend to flourish relative to those who must use elaborate enforcement devices, one can expect, at the very least, that purely "accidental" or "imperfectly rational" disposition can lead, over the long haul, to the emergence of stable "voluntary" schemes of coordination. But what I have sought, in this paper, is a more ambitious conclusion than this. Since those who are capable of plan or rule governed behavior do better in terms of furthering whatever interests they have than those who do not, this surely should have some direct bearing on the rationality of various alternatives, at least for those who count themselves as consequentialists. For a consequentialist, the bottom line must be which way of approaching such games yields the more preferred outcomes. Within the Coasian framework, then, according to which human beings are disposed to seek to realize the gains that cooperation makes possible, the following stronger conclusion seems quite plausible: rational persons who come to appreciate the costs of enforcement devices will be motivated to deliberately seek out others who are also aware of the gains to be realized by voluntarily choosing in accordance with rules, and transact with them in a plan- or rule- guided manner.

REFERENCES

A.A. Alchian, "Uncertainty, Evolution and Economic Theory". in: *Journal of Political Economy*, 58, 1950, pp.211-21.

W.B. Arthur, *Increasing Returns and Path Dependence in the Economy*. Ann Arbour: The University of Michigan Press 1994.

R. Axelrod, *The Evolution of Cooperation*. New York: Basic Books 1984.

R. Axelrod/D. Dion, "The further evolution of cooperation", in: *Science*, 242, 1988, pp.1385-90.

M.E. Bratman, *Intention, Plans, and Practical Reason*. Cambridge, Mass.: Harvard University Press 1987.

J.M. Buchanan/R.D. Tollison/G. Tullock, *Toward a Theory of the Rent-Seeking Society*. College Station: Texas A & M University Press 1981.

J.M. Buchanan/G. Tullock, *The Calculus of Consent*. Ann Arbor: University of Michigan Press 1962.

R.H. Coase, "The nature of the firm", in: *Economica*, 4, 1937, pp.386-405.

R.H. Coase, "The problem of social cost", in: *Journal of Law and Economics*, 3, 1960, pp.1-44.

D. Fudenberg/J. Tirole, *Game Theory*. Cambridge, Mass.: The MIT Press 1992.

F.A. Hayek, "Notes on the evolution of systems of rules of conduct", In: *Studies in Philosophy, Politics, and Economics*. Chicago: University of Chicago Press 1967.

D. Lewis, *Convention*. Cambridge, Mass: Harvard University Press 1969.

J. Knight, *Institutions and Social Conflict*. Cambridge: Cambridge University Press 1992.

R.D. Luce/H. Raiffa, *Games and Decisions*. New York: Wiley & Sons 1957.

E.F. McClennen, *Rationality and Dynamic Choice. Foundational Explorations*. Cambridge: Cambridge University Press 1990.

E.F. McClennen, "The theory of rationality for ideal games", in: *Philosophical Studies*, 65, 1992, pp.193-215.

E.F. McClennen, "Rationality and Rules", in: Peter Danielson (ed.), *Modeling Rational and Moral Agents*. Oxford: Oxford University Press 1997, pp.11-39.

D.C. North, *Institutions, Institutional Change and Economic Performance*. Cambridge: Cambridge University Press 1990.

R.A. Posner, *Economic Analysis of Law*. Boston, Mass: Little, Brown 1986.

J. Rawls, "Two concepts of rules", in: *The Philosophical Review*, 64, 1955, pp.3-32.

T.C. Schelling, *The Strategy of Conflict*. Cambridge, Mass.: Harvard University Press, 1960.

R. Sugden, "Spontaneous order", in: *Journal of Economic Perspectives*, 3, 1989, pp.85-97.

E. Ullman-Margalit, *The Emergence of Norms*. Oxford: Oxford University Press 1978.

J. von Neumann/O. Morgenstern, *Theory of Games and Economic Behavior* (Third Edition) Princeton: Princeton University Press 1953.

Department of Philosophy
Bowling Green State University
Bowling Green, OH 43403
USA

KARL SIGMUND
MAARTEN BOERLIJST
MARTIN A. NOWAK

AUTOMATA AND INNER STATES FOR REPEATED GAMES

The repeated Prisoner's Dilemma is often used as a game-theoretic paradigm to analyse the evolution of reciprocity, of mutual aid, and of trade. Conceivably, it can also help in understanding the emergence of inner motivational states, like contrition or outrage, which are essential for fuelling the ethics of communal life. A highly internalised sense of fairness and the readiness for moralistic aggression, the ability to be provoked and the feeling of guilt are important and apparently ubiquitous aspects of human socialisation. Like Ridley (1996) and Frank (1988) we believe that emotional commitment is a major factor in the economics of every-day life.

Let us consider a two-player game where both players have the same two strategies and the same payoff matrix. We denote the first strategy by **C** (for 'cooperate') and the second by **D** (for 'defect') and write the payoff matrix as

$$\begin{pmatrix} R & S \\ T & P \end{pmatrix}$$

Such games include the Prisoner's Dilemma, where $T > R > P > S$, and the Chicken game, where $T > R > S > P$. (In the Prisoner's Dilemma case, R stands for the *reward* for mutual cooperation, P is the *penalty* for mutual defection, T is the *temptation* payoff for unilaterally defecting and S the *sucker* payoff for being exploited. The strategy **D** dominates **C**, so that rational players will defect and thus earn only P instead of R. One also assumes $2R > T + S$.)

Let us assume now that the game is repeated with a constant probability w. The number of rounds is a random variable with expected value $(1-w)^{-1}$. The total payoff is given by $\sum A_n w^n$, with A_n as payoff in the n-th round and w^n the probability for the occurence of an n-th round. In the limiting case $w = 1$ (the infinitely iterated game) one uses as payoff the limit in the mean, i.e. $(A_1 + ... + A_n)/n$ (provided it exists). For the PD game and w sufficiently large, there exists now no strategy which is best against all comers (see Axelrod, 1984). For $w > (T-R)/(T-P)$, for instance, the best reply against $Always\mathbf{C}$ is to always defect, whereas against $Grim$ (the strategy that cooperates up to the first time it has been exploited, and then always defects) it is best to always cooperate.

The so-called folk theorem for repeated games implies that there exist infinitely many Nash-equilibria (see e.g. Fudenberg and Tirole 1992 or Binmore 1994) – in particular, every feasible pair of payoff-values larger than the maximin value can be attained by Nash-equilibria. On the other hand, it is easy to see that there exist no strict Nash equilibria, and in fact no evolutionarily stable strategies in the strict sense (see Binmore and Samuelson 1982; with a different notion of evolutionary stability, see Lorberbaum 1994). A strategy E is said to be evolutionarily

W. Leinfellner and E. Köhler (eds.), Game Theory, Experience, Rationality, 131–139.

stable if in addition to the Nash-equilibrium condition $A(E', E) \leq A(E, E)$ one has $A(E', E') < A(E, E')$ for all strategies $E \neq E'$ (here $A(E', E)$ is the payoff for a player using strategy E' if the co-player uses E). This implies that a population of E-players cannot be invaded by a minority of E'-players under the effect of natural selection. (See Maynard Smith, 1982, and Hofbauer and Sigmund, 1988).

It is easy to see, for example, that the strategy *Tit For Tat* (or *TFT*) is not evolutionarily stable (although this has been claimed occasionally). *TFT* is the strategy which plays **C** in the first round and from then on repeats whatever the co-player did in the previous round. This strategy did very well in a series of computer tournaments (see Axelrod 1984). But a player using $Always$**C** does as well, both against *TFT* and against its own kind, as *TFT*. Intuitively speaking, in a population of *TFT*-players, invading $Always$**C**-players can spread by neutral drift, ultimately allowing defectors to cash in.

Another weakness of *TFT* is that it suffers from occasional errors against its like, as has been pointed out by Selten and Hammerstein (1982) already. If, by some mistake, a *TFT*-player plays **D** against another *TFT*-player, this leads to a long vendetta of alternating unilateral defections which can only be stopped by a further mistake: such a second mistake, however, can lead just as well to a regime of simultaneous defections. This lowers the average payoff in a *TFT*-population considerably and allows more generous strategies to invade.

In Nowak and Sigmund (1993, 1995) evolutionary chronicles have been studied by computer simulations of large populations of players using strategies defined by the propensities p_R, p_S, p_T and p_P to play **C** after having experienced the payoff R, S, T resp. P in the previous round (and, for $w < 1$, the probability to play **C** in the first round). We note in passing that Boerlijst et al. (1997) have shown that in case $w = 1$, there exist for every value π between P and R uncountably many strategies of this type with the property that every co-player obtains π as payoff, no matter which strategy he is using. In particular, for every pair of values π and ψ between P and R, this yields Nash-equilibria composed of pairs of such 'equaliser' strategies, and provides a variant of the folk theorem.

The evolutionary simulations lead frequently to cooperative regimes dominated by the *Pavlov* strategy, the strategy with $p_R = p_P = 1$ and $p_S = p_T = 0$. This strategy, which is based on the win-stay, lose-shift principle of repeating the former move if and only if it leads to a high payoff, i.e. to R or to T, has the property of being error-correcting. If, in a game between two *Pavlov*-players, one erroneously plays **D**, then in the next round both play **D** and after that resume mutual cooperation.

Sugden (1988) has proposed another, more sophisticated strategy called *Contrite TFT* (*cTFT*). It is based on the notion of *standing*, which is associated to each player and can be g (good) or b (bad). In each round, the player acts (i.e. opts for **C** or **D**) and obtains a new standing which depends on his action and on the previous standing of both players. The rules for updating the standing are the following: if the other player has been in good standing, or if both have been in bad standing, one receives a good standing if one has cooperated, and a bad standing otherwise. If one has been in good standing and the other player in bad standing, one receives a good standing no matter what one is doing.

Thus if one cooperates in a given round, one will always obtain a good standing: but if one defects, one will be in good standing only if one has been in good standing and the opponent in bad standing in the previous round.

In a given round, a player can be in three possible states: Cg, Dg and Db: the first means that he has cooperated (which automatically entails good standing), the second that he has defected with good reason (i.e. while 'provoked'), the third that he has wantonly defected and feels the pangs of conscience. The state *of the game* in a given round is made up of the states of the first and the second player. There are nine such combinations: (Cg, Cg), (Cg, Dg), (Cg, Db), (Dg, Cg), (Dg, Db), (Db, Cg), (Db, Dg), (Db, Db) and (Dg, Dg). It is easy to check that this last state can never be reached: we therefore omit is, and number the remaining eight states in this order.

cTFT is the strategy which cooperates except if it is in good standing and the other player is not. This means that the player defects when provoked, but not otherwise. If he defects by mistake, he knows that he lost his good standing, and meekly accepts punishment, i.e. keeps cooperating even if the other player uses **D** on him.

In other words, the strategy *cTFT* begins with a cooperative move, and cooperates except if provoked (or by mistake). If two *cTFT*-players engage in a repeated Prisoner's Dilemma, and the first player defects by mistake, then he loses his good standing. In the next round, he will cooperate, whereas the other *cTFT*-player will defect without losing his good standing. From then on both players will be in good standing and resume their mutual cooperation in the following round.

In Boerlijst et al (1997) we have studied evolutionary chronicles for strategies defined by their propensities $(q_1, ..., q_8)$ to play **C** when in one of the eight states previously enumbered. This large class contains the (p_R, p_S, p_T, p_P)-strategies in the form $(p_R, p_S, p_S, p_T, p_P, p_T, p_P, p_P)$. The strategy *TFT* is for instance given by $(1, 0, 0, 1, 0, 1, 0, 0)$, the strategy *Pavlov* is $(1, 0, 0, 0, 1, 0, 1, 1)$ and the rule for *cTFT* is $(1, 1, 0, 1, 0, 1, 1, 1)$.

These chronicles show that *cTFT* works very well indeed. We may distinguish two situations. If the temptation T is large (for instance, $T = 5.5$, $R = 3, P = 1, S = 0$) then the simulations end up in about 80 percent of all runs with a population dominated by *cTFT*, and in most other runs with a population dominated by $Remorse$. $Remorse$ is the strategy $(1, 0, 0, 0, 0, 1, 1, 1)$ which cooperates only if in bad standing, or if both players had cooperated in the previous round.

If on the other hand the temptation to defect is small (for instance $T = 3.5$, and the other values as above) then 70 percent of all runs end up with *cTFT* or some similar strategy, 20 percent with *Pavlov* and the rest with a strategy called *Weakling*, which plays **C** if and only if it is in bad standing, and which obtains against itself a sub-Pareto optimal payoff (namely $(R + P)/2$ for $w = 1$) by simultaneously cooperating and defecting in every other round. It is surprising that this suboptimal trapping is only observed in situations with low temptation to defect.

In contrast to *Pavlov*, which does very poorly in a population of $Always$**D** players (it plays **C** in every second round), *cTFT* and $Remorse$ are as good at invading an $always$**D**-population as are the strongly retaliatory strategies *TFT* or $Grim$.

This is one of the reasons for the success of *cTFT*; it does not need a strategy which catalyses cooperation, as *Pavlov* does.

In order to better understand why the outcomes described in the previous simulations are stable, we can use a minor variant of the concept of limit-ESS introduced by Selten (see Selten, 1984, and Leimar, 1997). Let us first note that the description of a repeated game by its extensive form suffers from the fact that there is no uniform upper bound for the length of the branches of the game-tree. If we consider only finite state automata, this problem is overcome, since we have only to investigate a finite number of nodes. This yields a 'closure' of the representation of the repeated game.

For simplicity let us consider first only strategies depending on the moves of the previous round, like *TFT* or *Pavlov*. In the repeated game, there occur only four pairs of states, namely $(\mathbf{C}, \mathbf{C})$, $(\mathbf{C}, \mathbf{D})$,$(\mathbf{D}, \mathbf{C})$ and $(\mathbf{D}, \mathbf{D})$. The history of every repeated game corresponds to a path connecting these four vertices. Let us consider the fate of a player invading a population of *TFT* players, for instance (see Fig. 1).

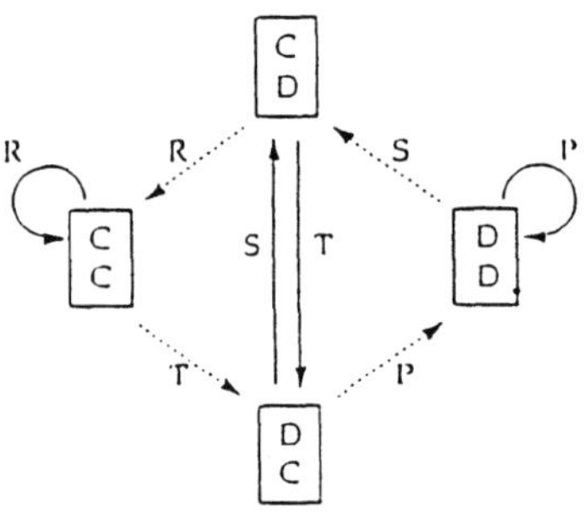

Figure 1

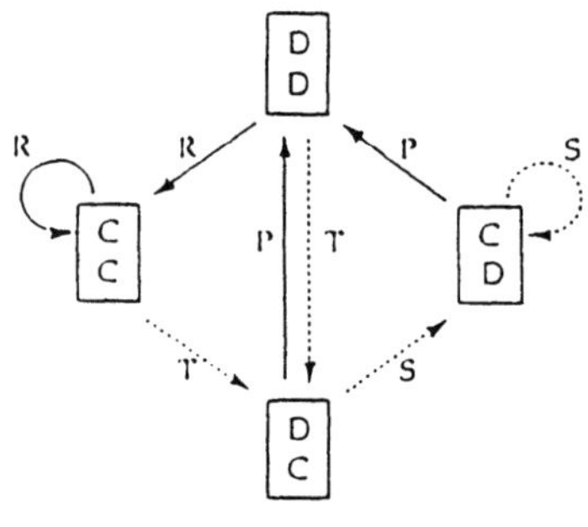

Figure 2

In every state, the move of the co-player, who belongs to the resident population, is specified by the *TFT* rule. The invader has two alternatives. He can use the move prescribed by the *TFT*-rule, which we describe in the graph by a full arrow, or he can use the other option, described by the broken arrow. One sees immediately that it is better, in the state $(\mathbf{C}, \mathbf{D})$, not to use *TFT*, since this would yield as outcome for the next two rounds only $T + wS$; it would be better to cooperate and thus to reach $(\mathbf{C}, \mathbf{C})$ since this yields (for the next two rounds) $R{+}wR$. This latter option is better provided w is sufficiently large. We note that in a game between two *TFT* players the node $(\mathbf{C}, \mathbf{D})$ can only be reached by mistake. According to Selten's theory of perfect equilibria, such states cannot be neglected.

Let us consider, in contrast, a player invading a *Pavlov*-population (see Fig. 2). In this case, one sees that as long as $T + wP < R + wR$, it is best to do, in every node, as the residents do. If the converse inequality holds, it is better to deviate in the states $(\mathbf{C}, \mathbf{C})$ and $(\mathbf{D}, \mathbf{D})$. We say that for $T + wP < R + wR$, *Pavlov* is a limit ESS (see Wu and Axelrod, 1995).

More generally, we say that a strategy E implemented by a finite-state automaton is limit-ESS if for every pair of states which can be reached in the normal course of a game between two E-players or after mis-implementations, it is better to follow the E-rule than the alternative (i.e. the full arrow yields always the path with the highest payoff). In Fig. 3 we check that $Remorse$ is a limit-ESS if $T + wP > R + wR$ (just the opposite as with *Pavlov*) whereas Fig. 4 shows that *cTFT* is always a limit ESS. Olof Leimar (1997) has developed an extensive theory of limit-ESS for state space strategies, and has shown that for automata with three or four states, there exist thousands of such strategies which (like *Pavlov*, $Remorse$ or *cTFT*) are Pareto-optimal for large w.

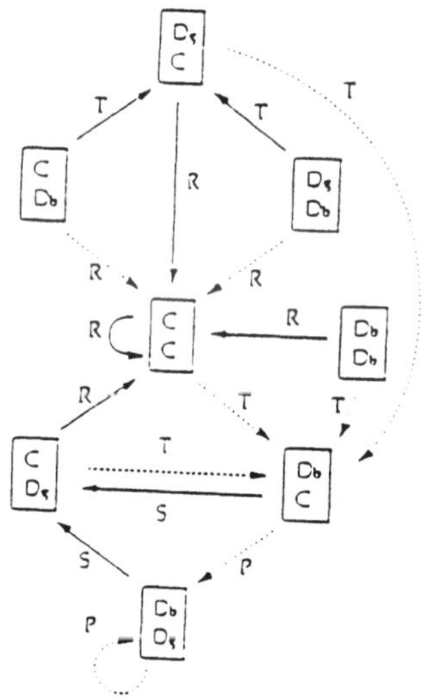

Figure 3

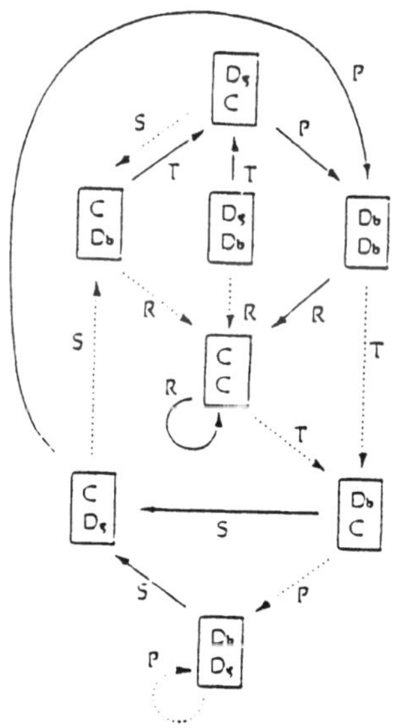

Figure 4

On the other hand, while *cTFT* is immune to errors in the implementation of a move, it is not immune to errors in the perception of a move. If, in a match between two *cTFT* players, one player mistakenly believes that the other is in bad standing, this leads to a sequence of mutual backbiting, just as with *TFT*. In contrast to this, if in a game between two *Pavlov*-players one player mis-interprets a move (by his co-player or by himself), then the situation is quickly redressed and mutual cooperation resumed after two rounds.

Errors in perception – rather than implementation – have been studied in Nowak et al. (1995), among others. In order to see that they are really different from errors in implementation, let us return to the class of (p_R, p_S, p_T, p_P)-strategies and assume that the error probability is ϵ. If this is an error in implementing a move (i.e. in choosing **C** or **D**) then p_R turns into $(1-\epsilon)p_R+\epsilon(1-p_R)$ etc. so that the correction term is $\epsilon(1-2p_R, 1-2p_S, 1-2p_T, 1-2p_P)$. If the error affects the perception of the co-player's move (i.e. if it confuses an R with an S or a T with a P) then the correction term is $\epsilon(p_S - p_R, p_R - p_S, p_P - p_T, p_T - p_P)$. If the error affects the perception of the own move (i.e. if it confuses an R with a T or an S with a P) then the correction term is $\epsilon(p_T - p_R, p_P - p_S, p_R - p_T, p_S - p_P)$. For $w = 1$ and the limit $\epsilon \to 0$, errors in implementation yield as payoff $(2P + 2S + T)/5$ for a $(1, 0, 0, 0)$-player against a $(0, 0, 1, 0)$-player, whereas errors in perceiving the opponent's move yield as payoff $(S + T)/2$.

As soon as we allow mistakes in perception, we lose the concept of a public state (like the *standing*). In the elementary set-up we are considering, there is no referee to tell the players who is in which state. Every player can only monitor his own state. This state need not be restricted to his own previous move, or to his own standing.

For instance, we can realise the *cTFT* strategy by an automaton with three states $1, 2, 3$. State 2 is (b, g) (my standing is bad, the co-players good); state 3 is (g, b); and state 1 is (g, g) or (b, b). The *cTFT* rule then is to play **C** when in state 1 or 2, and **D** when in state 3. Outcome R (mutual cooperation) leads always to state 1, outcome S leads from states 1 or 3 to 3 and from state 2 to 1, outcome T leads from 1 and 2 to 2 and from 3 to 1, and outcome P leaves all states unchanged. If two *cTFT*-player make no mistakes in interpretation, then they will either both be in state 1, or one player in 2 and the co-player in 3 (see Leimar, 1997). But misperceptions can change this symmetry, and lead to vendettas.

The *trembling hand* doctrine of Selten assumes that the player is completely lucid. He can make a wrong move, but he is immediately aware of his gaffe. Such a player is still too rational for the kind of evolutionary games we have in mind. A player is also liable to errors in interpreting the outcome R, S, T or P of the previous round, and in updating his state as a function of this outcome.

This suggests considering strategies implemented by automata of the following kind. Each consists of a finite set Ω of inner states and a (possibly stochastic) action rule with the probability p_ω for the player to choose the move **C** when in state $\omega \in \Omega$. In addition, there exists a (possibly stochastic) transition rule ρ which specifies the state in the next round, depending on the current state and the outcome R, S, T or P of the current round. If we allow mistakes in implementing a move, p_ω will not attain the extremal (deterministic) values 1 oder 0, and each outcome R, S, T or P can occur with positive probability. If we assume mistakes in the interpretation, this implies that the transition rule $\rho\colon \Omega \to \Omega$ is also stochastic. To specify the strategy completely, we also need the initial state (i.e. a probability distribution q on Ω).

The actual repeated game between two automata (Ω, p, ρ, q) and (Ω', p', ρ', q') is described by a path visiting the vertices of the state space (Ω, Ω') of the game. The transitions within this state space are described by a Markov chain. When in state (ω, ω'), the outcome of the next round, i.e. R, S etc. is given by the probabilities $p_\omega p'_{\omega'}$, $p_\omega(1-p'_{\omega'})$ etc. and this outcome leads with the probabilities defined by ρ and ρ' to the follow-up states. Let us denote the transition probability from the state (ω, ω') to the state (α, α') by $P(\omega, \omega'; \alpha, \alpha')$.

If one starts in the state (ω, ω'), the first player obtains a total payoff whose expected value is $E(\omega, \omega')$. It is easy to compute these values by using the fact that in the next round (which occurs with probability w) the whole process starts as if from scratch (it is a renewal process). Hence

$$\begin{aligned} E(\omega, \omega') &= Rp_\omega p'_{\omega'} + Sp_\omega(1-p'_{\omega'}) + T(1-p_\omega)p'_{\omega'} + P(1-p_\omega)(1-p'_{\omega'}) \\ &+ w\sum_{\alpha,\alpha'} E(\alpha, \alpha')P(\omega, \omega'; \alpha, \alpha') \,. \end{aligned}$$

This yields a finite system of linear equations for the $E(\omega, \omega')$. We only have to multiply these values with the probabilities $q_\omega q'_{\omega'}$ for the initial states (ω, ω'), and obtain by summation the expected total payoff for the automaton (Ω, p, ρ, q) against (Ω', p', ρ', q').

We can formulate the notion of limit-ESS in this context too. For an automaton to be a limit-ESS, we must require that in every state which can be reached by the automaton playing against itself (either in the normal run of the game or after a mistake), both the prescribed move and the prescribed transition are the unique optimal moves. This implies immediately that such rules have to be deterministic – for otherwise, two alternatives would fare equally well. It seems improbable that there exist Pareto-optimal strategies which are in this sense limit-ESS. It does not suffice that the strategy is immune against errors in perception and implementation (as for instance *Pavlov* if $T + wP < R + wR$). If one *Pavlov*-player is for instance in state $(\mathbf{C}, \mathbf{D})$ (and believes himself to have been exploited) whereas the other *Pavlov*-player is in state $(\mathbf{C}, \mathbf{C})$, then both will be in the next round in state $(\mathbf{D}, \mathbf{C})$, in the following round in state $(\mathbf{D}, \mathbf{D})$, and only from then on in $(\mathbf{C}, \mathbf{C})$. It would obviously have been better if the first player had chosen **C** right away. But this would have been the wrong move if the co-player had shared the same view of the situation with the first player.

Evolutionary simulations in the world of such automata should be quite interesting. Even in the very restricted case of two-state automata, this could well produce new perspectives, since most simulations so far did not take errors of perception into account. But the real challenge lies in automata with more states. If strategies like *cTFT* emerge, this would be a strong hint that inner states like guilt or wrath serve an economic purpose, and can be viewed as instruments to make our societies tick. One way to design simulations able to introduce new structures among automata – new states, for instance – has been shown by Lindgren (1993), and mimicks gene duplications. One mutation could duplicate a state ω: the resulting two states would act in the same way, so that no new behaviour emerges. Such a mutation is 'silent' and selectively neutral. In a second step, other mutations could modify the action rules and the transition rules in the two states in different ways. These mutations would now be able to differentiate the two states, and hence to build new, more complex automata.

Lindgren (1991) used this principle to investigate strategies dependent on the moves of the last two, three, four rounds etc. *cTFT* and *Remorse* are strategies of a different nature. They only depend on the state of the previous round, but this state, now, does not depend only on the actions **C** or **D** of the two players, but on the standing – good or bad – after a defection. The rules for determining this standing seem quite natural: we can identify with a player who feels bad after having committed a defection erroneously, or who feels provoked by the unilateral defection of the co-player after a string of mutual cooperation. The rules embody a certain notion of 'fairness' which seems to be rather common. If it should indeed turn out that this notion is a human universal, we would have to explain how it emerged.

It seems highly plausible that there exists a wide variety of workable 'taggings' or states which yield interesting ESS's. The question is whether an evolution based on mutation and selection would tend to lead to one form of states rather than another. This could ultimately shed light on why humans developed a sense of fairness, feelings of guilt, and highly effective social norms. So far, however, this is science fiction.

REFERENCES

R. Axelrod, *The Evolution of Cooperation.* (Reprinted 1989) Harmondsworth: Penguin 1984.

D. Abreu/A. Rubinstein, "The structure of Nash equilibria in repeated games with finite automata", in: *Econometrica*, 56, 1988, pp.1259-1282.

K.G. Binmore, *Game Theory and the Social Contract, vol. I*, Cambridge, Mass.: MIT Press 1994.

K.G. Binmore/L. Samuelson, "Evolutionary stability in repeated games played by finite automata", in: *Journal of Economic Theory*, 57, 1992, pp.278-305.

M. Boerlijst/M.A. Nowak/K. Sigmund, "The Logic of Contrition", in: *Journ. Theor. Biol.*, to appear 1997.

M. Boerlijst/M.A. Nowak/K. Sigmund, "Equal Pay for the Prisoners", in: *AMS Monthly*, to appear 1997.

R. Boyd, "Mistakes Allow Evolutionary Stability in the Repeated Prisoner's Dilemma Game", in: *J. Theor. Biol.*, 136, 1989, pp.47-56.

R.H. Frank, *Passion within Reason.* New York: Norton 1988.

D. Fudenberg/E. Maskin, "Evolution and Cooperation in Noisy Repeated Games", in: *American Economic Review*, 80, 1990, pp.274-279.

D. Fudenberg/J. Tirole, *Game Theory.* Cambridge, Mass.: MIT Press 1992.

J.C. Harrington/R. Axelrod, "Tit For Tat with Contrition is an Evolutionarily Stable Strategy in the Iterated Prisoner's Dilemma with Noise". Mimeo 1995.

J. Hofbauer/K. Sigmund, K., *The Theory of Evolution and Dynamical Systems.* Cambridge: Cambridge University Press 1988.

O. Leimar, "Repeated Games: a State Space Approach", in: *Journ. Theor. Biol.*, to appear 1997.

K. Lindgren, "Evolutionary Phenomena in Simple Dynamics", in: C.G. Langton et al. (Eds.), *Artificial Life II*, Santa Fe Institute for Studies in the Sciences of Complexity, Vol. X, 1991, pp.295-312.

J. Lorberbaum, "No Strategy is Evolutionarily Stable in the Repeated Prisoner's Dilemma", in: *Journ. Theor. biol.*, 168, 1994, pp.117-130.

J. Maynard Smith, *Evolution and the theory of games*. Cambridge: Cambridge University Press 1982.

M. Nowak/K. Sigmund, "Win-Stay, Lose-Shift Outperforms Tit-for-Tat", in: *Nature*, 364, 1993, pp.56-58.

M. Nowak/K. Sigmund, "Automata, Repeated Games, and Noise", in: *J. Math. Biol.*, 33, 1995, pp.703-732.

M. Ridley, *The Origin of Virtue*. Harmondsworth: Viking Press 1996.

A. Rubinstein, "Finite Automata Play the Repeated Prisoner's Dilemma", in: *Journal of Economic Theory*, 39, 1986, pp.83-96.

R. Selten, "Re-Examination of the Perfectness Concept for Equilibrium Points in Extensive Games", in: *International Journal of Game Theory*, 4, 1975, pp.25-55.

R. Selten/P. Hammerstein, "Gaps in Harley's Argument on Evolutionarily Stable Learning Rules and in the Logic of Tit for Tat", in: *Behavioural and Brain Sciences*, 7, 1984, pp.115-116.

K. Sigmund, *Games of Life*, Harmondsworth: Penguin 1995.

R. Sugden, *The Evolution of Rights, Co-Operation and Welfare*. Oxford: Blackwell 1986.

J. Wu/R. Axelrod, "How to Cope With Noise in the Iterated Prisoner's Dilemma", in: *Journal of Conflict Resolution*, 39, 1995, pp.183-189.

Karl Sigmund
Institut für Mathematik, Universität Wien
Strudlhofgasse 4, A-1090 Vienna, Austria

Maarten Boerlijst
Department of Zoology, University of Oxford
South Parks Road, Oxford OX1 3PS, U.K.

Martin A. Nowak
Department of Zoology, University of Oxford
South Parks Road, Oxford OX1 3PS, U.K.

LUTZ BEINSEN
ULRIKE LEOPOLD-WILDBURGER

TOWARDS BOUNDED RATIONALITY WITHIN RATIONAL EXPECTATIONS – SOME COMMENTS FROM AN ECONOMIC POINT OF VIEW

I. INTRODUCTION

Rationality has been a principle widely assumed in economics from early on. In contrast, rationality in the formation of economic expectations is rather new. Since the term "rational expectations" has meanwhile become a kind of slogan for diverse issues in economics as in related fields, there is some danger of authors' not always being aware of the true meaning of this technical term. Sometimes it is used, in a context where expectation *formation* about uncertain events is essential, as if rational expectations solved this fundamental problem.

In reconsidering the 1961 rational expectations article of John F. Muth against the background of the Rational Expectations Revolution which started about a decade later, and which may still last for many years, one may ask how the initial article and the subsequent revolution fit together, and, eventually, what the consequences for expectation formation are. Furthermore, the question remains what kind of rationality is actually meant, or how "rational" rational expectations really are.

In section II we present a brief rehearsal of Muth's starting point. In section III we discuss the meaning of the term "rational expectations" outside and inside Rational Expectations theory. In section IV we try to make clear in brief why the hypothesis of rational expectations was (and is) so important. In section V we explain why we consider empirical and experimental research adequate in this field. In section VI we make some principal statements on the formation of (rational) expectations. In section VII we report on experiments in expectation formation. Finally, in section VIII we try to draw some conclusions and to provide an outlook on further problems.

II. REREADING JOHN F. MUTH

The pre-history of rational expectations concerns adaptive and (to some degree) extrapolative expectations. With adaptive expectations the forecast is continuously adapted by an error adjustment mechanism. This will in general reduce but not eliminate the forecast error. So the question arises of why people do not learn (enough) from their past mistakes and why they do not use their knowledge of

W. Leinfellner and E. Köhler (eds.), Game Theory, Experience, Rationality, 141–152.

economic theory about the interdependent relations between economic variables.

The answer of John F. Muth (1961, p.315ff) was that "In order to explain fairly simply how expectations are formed we advance the hypothesis that they are essentially the same as the predictions of the relevant economic theory." and that "The way expectations are formed depends specifically on the structure of the relevant system describing the economy." Although he is obviously concerned with the way expectations are *formed*, this idea proves, at least partly, to have been lost or even deliberately abandoned later on (Lucas/Prescott, 1971, p.660), not by Muth himself, but by important followers.

Every theorist certainly will agree with Muth that economic forecasts should have a reasonable basis in economic theory. But basing a forecast on a certain theory does not by itself explain the formation of the expectation in an understandable way. One single theory may lead to several equations for the empirical estimation, if only for the different possible lags. Theories in general are not very precise about lags. Equations differing only in lags may have quite different stability properties. The latter are very often the most important properties of the result. This means that the right lag structure has to be found by empirical estimation. The latter is, indeed, an important issue in the empirically oriented line of Rational Expectations theory.

And theories are often not precise as far as the exact functional form of the interdependence between the variables is concerned. Even if we all believe that imports (M) depend positively on income (Y) - why is the income-dependent import function always so simple ($M=mY$; m constant)? Or, if this example is not strong enough, we should imagine the many forms of utility functions, most of them telling the same principal "truth" of diminishing marginal utility. The preparation of utility functions for empirical estimation also goes different ways. This, too, points to the fact that for rationality sometimes theory has to be completed by empirical investigation.

The conclusions concerning the formation of (rational) expectations depend to a considerable extent on whether the empirical estimation is or is not a necessary element. The answer is yes, as we see above. Doubtlessly, this is in agreement with Muth (1961). Of course, if we content ourselves with conditional qualitative forecasts there is less need for empirical work. But since the Rational Expectations theory provides the conclusion of policy ineffectiveness, qualitative prediction is not sufficient: People make economic policy ineffective by anticipating correctly, say, the percentage of a monetary expansion and by increasing all kinds of prices in all kinds of contracts by the same percentage at the same moment. So there can be no real (i.e. quantity) effects on demand and employment. This behaviour would certainly presuppose an exact quantitative estimation. That means: empirical investigation is a necessary part of and has an important role in the *formation* of rational expectations.

III. WHAT DOES "RATIONAL EXPECTATION" MEAN OUTSIDE AND INSIDE RATIONAL EXPECTATIONS THEORY?

Outside Rational Expectations theory rational expectations can be understood as the result of a rational forecast. A rational economic forecast is some empirical prediction procedure, mostly an econometric estimation based on accepted economic theory. A precondition of a forecast is the collection of relevant information. A theory that aims at explaining the formation of (rational) expectations has to explain (i) which theory, or more precisely, which economic model to apply, (ii) which empirical method to use and (iii) what sort of information to collect. If information were free and its collection not time-consuming then rationality would presuppose full information. Since both just mentioned conditions are generally not fulfilled, rationality necessarily assumes incomplete information. If we assume rising marginal costs and decreasing marginal value in the process of information, then the collection of information is to be stopped when marginal cost equals marginal value. Here the problem arises that we can never be quite sure of the decreasing marginal value of information since it is the intrinsic character of information that it is unknown and cannot be judged in advance. So there is no definite end to the collection of information. This rather points in the direction of bounded rationality.

In Muth's 1961 article there seems to be no doubt that theory-based empirical application is the very objective of Rational Expectations theory. But when his seminal idea of 1961 was widely taken up a decade later, especially in the New Classical Macroeconomics (NCM), the kind of reasoning changed considerably. Today we may well ask whether we have learnt anything about the *formation* of rational expectations. And the answer, at least as far as the pure theory is concerned, may be no, although we have learnt many other things from this debate.

Rather early on, economists realized that Rational Expectation is an equilibrium concept (introduced by Lucas/Prescott, 1971, specifically p.664, footnote 9). Meanwhile a whole class of models belongs to the Rational Expectations type, most of them within the paradigm of NCM. Models of this type are characterised (besides market clearing on account of perfect price flexibility) by the property that forecasts of the agents are not worse than the forecasts of the economic advisors of the government (or of the "economist", to follow Sargent, 1988, p.76). In other words: the agents have the "true" model of the economy. The last sentence is the normal premise in these models today. It is a premise, indeed, not a result. This becomes even clearer when one considers a rational expectation simply as an - on balance - *correct* expectation (Turnovsky 1996, p.6), which is already very close to perfect foresight, except for the - not too big - stochastic error term. It has neither to do with the way expectations are formed nor with empirical estimation. Consequently, we observe an obvious contradiction with section II. To reconcile the two views of Rational Expectations, we propose to distinguish two branches which both have their serious merits. The first has become a self-contained approach to macroeconomic theory within the NCM-paradigm. It is in use without any reference to empirical facts and has to be considered as a development in pure theory. The second branch is theory-based empirical research following more closely the original lines of John Muth in so far as it seeks to combine theoretical and empirical research. Here, the interest in the way expectations are formed is still upheld.

We did not overlook that there is a rather rich literature in econometrics on learning to form rational expectations, partly as rational learning, partly as boundedly rational learning (Savin 1988, p.83). Although we recommend here experimental methods as especially interesting in this field, we agree with the above development, since it is concerned with the *formation* of expectations. We must insist, however, that Rational Expectations in the pure theoretical sense of NCM is not compatible with learning. The reason is that, for economic policy to be ineffective (as maintained in NCM), the correct equilibrium must be perfectly and instantaneously foreseen.

IV. WHY IS THE RATIONAL EXPECTATIONS HYPOTHESIS SO IMPORTANT?

In a sense, adaptive expectations are always lagging behind events. The forecast error adjustment is sluggish, which means that agents do not learn fast enough from their previous mistakes. This behaviour may be stabilising (as in the case of slow adaptation of investment to increasing demand) but can, however, also be misleading in that it gives a stability illusion if expectations are actually not adaptive but rational. With Rational Expectations, however, agents go straight ahead to forecast and expect the correct future values of all relevant variables. Since present decisions are based on the expectations, future values determine current values and these in turn influence expectations. This is a considerable change in thinking. By this approach we learnt quite a lot about the significance of correct anticipations of the future in economics. We learnt what it means if agents anticipate the future equilibria correctly. Many previous results of macro-economics were abandoned. We learnt, e.g., that under conditions where economic policy is effective with adaptive expectations, it is ineffective with rational expectations. We also learnt from the policy ineffectiveness debate that economic policy is not altogether ineffective, but on what conditions it depends. The policy ineffectiveness conclusion is valid primarily for monetary policy and certain forms of cyclical public expenditure and tax policy (the latter being no great surprise).

If expectations are formed as the correct future equilibrium values (the first branch), it looks, at first glance, as if expectations are no longer bound by past values. Therefore, stability may not be as natural as under adaptive expectations. Whereas physical quantities (like the stock of capital or the flow of real national product) change only slowly, some kinds of prices (like share prices and exchange rates) and other valuations may prove to be "jump variables"

(Turnovsky, p. 7). In general, we think of price expectations as being more volatile than quantity expectations, a view which was rather enforced by the running debate. That it has not led to a Rational Expectations macroeconomic instability issue, may be due to the fact that Rational Expectations always appear in the NCM as a twin of the market clearing assumption. So, jumps can only be equilibrium jumps and therefore do not attract much attention.

If, on the other hand, expectations are formed by theory-based empirical estimations (the second branch) using statistical methods, the estimates will in general be dependent on past values of the relevant variables. Even if we find an optimal predictor for every time horizon, the data set of the time series underlying the estimation stems from former years, if not decades. In following the original lines of John Muth, though, it does not seem very realistic to become completely independent of the past. Nevertheless, this is still a sensible way to obtain quantitative results.

The conclusion is different if we look at experimental methods. Here we cannot proceed to precise quantitative forecasts of economic variables. But we are, in fact, completely independent of the past. What we can do in experimental economics is to draw conclusions about the general behaviour of the agents towards different economic situations, on their kind of decision-making, and on their way of forming expectations. In addition, the way is open to quantitative statistical evaluation methods of the experimental findings. Furthermore, whereas in econometric estimation we can never be quite sure whether the proposed "leading indicators" of expectation formation are the same as those actually used by the agents, we can be much more certain in experimental investigation because the informational basis is provided by the experimenter.

V. WHY EMPIRICAL AND EXPERIMENTAL WORK?

"It is better to make many empirically supported *ad hoc* assumptions than to rely on a few unrealistic principles of great generality and elegance." (Selten 1989). Empirical and experimental work show that human behaviour is *ad hoc*: Different principles are applied to decisions of the same sort.

Empirical research, and especially formal reports of laboratory experiments are rather new ideas in economics: Chamberlin (1948), Allais (1953), Flood (1958), Stone (1958), Sauermann and Selten (1959), Siegel and Fouraker (1960) were, in a broad sense, the first articles about experimental economics.

Speaking about "laboratory" experimentation means that the economic environment is completely under the control of the experimenter, whose access to the experimental subjects is relatively unimpeded. The distinction between "laboratory" and "field" experiments is sensible and necessary, it relies mainly on the control of the environment and access to the subjects.

Looking at experimentation in general we can say: In a certain sense economists and psychologists have common intellectual roots. Both have some of their origins in the philosophy of the "British empiricist school" which was started in the seventeenth century. Bentham's utilitarianism from the beginning of the nineteenth century led to an important area of research in psychology and economics. During the last century the schools have obviously drifted further apart but a sharp boundary does not exist. Furthermore, there is the field of economic psychology as a connection. This idea goes back to the French social psychologist Gabriel Tarde in 1902. A particular school of thought within economics, mainly the late nineteenth century Austrian school, created the idea of "psychological economics", which, however, did not take root directly, and various interactions between psychologists and economists during the early twentieth century were neither successful nor fruitful. The concept is found occasionally in writings by leading economists (Hayek 1978) and historians of economic theory (Coats 1976). In general, we find experimental assumptions behind economics mainly in the discussion of the meaning of utility maximization or rationality postulates.

VI. EXPECTATION FORMATION

In reexamining John Muth (1961, p.315), we find his ideas on the limits of a systematic theory of fluctuations in markets or in the economy. This is caused by the lack of explanation of the way expectations are formed. "To make dynamic economic models complete, various expectation formulas have been used. There is, however, little evidence to suggest that the presumed relations bear a resemblance to the way the economy works". These statements still have current relevance.

The explanation of the way in which economic agents form their expectations about the future is still a matter of controversy in macroeconomics. The simplest ways of modelling expectations in economic theory are either to assume that conditions prevailing today will be maintained in all subsequent periods or to accept the hypothesis of adaptive expectations.

Muth's concept is to suppose that the forecasts of "experts" agree with the corresponding mathematical expectation values calculated from a true model of the economic system. In his articles on "rational expectations" his connotation is in some sense misleading and could be better expressed by the term "forecasts".

We need quantitative theories of bounded rationality, but as rational as possible, supported by experimental evidence which can be used in economic modelling as an alternative to exaggerated rationality assumptions.

For a long time economists have shown increasing interest in the analysis of expectation formation. Following the approach of Hayek (1948), a lot of macroeconomic theories have been developed which are at least partly based on variations in expectations. However, the explanation of the way in which economic agents form subjectively expected utility results in different fields of research. We know, however, that "ideal" rationality of expectations is very often violated – as observed by experimentation and by empirical work in general.

One attempted explanation is that individual forecasts or predictions – and consequently the average opinions too – depend in very subtle ways on knowledge and information and do not follow the simple schemes which are often used by theorists.

Numerous empirical investigations have often shown that the hypothesis of extrapolative forecasts has to be rejected, because there are frequently systematical underestimations of changes and outrange effects whenever a time-series is supposed to reach a line of demarcation, or a sort of ceiling or floor.

Numerous empirical investigations in the last two decades which used forecasts of experts do not support the Rational Expectations hypothesis. Especially the Livingston-series had been analysed very frequently with different conclusions but mostly with results that reject the Rational Expectations hypothesis.

VII. PREDICTION EXPERIMENTS

Experimental investigations on the formation of expectations yield similar results. The first experiments by economists were conducted independently of each other in the early sixties by Fisher (1962) and Becker (1967). The main characteristic of these experiments is that the participants have to give a one-span or multi-span forecast of a time-series, and utilisable information available to them about this time-series restricts itself to the foregoing realizations plotted on a chart; i.e. nothing is said about the real meaning of the series. Some further experiments differ especially in the award structure. (Schmalensee (1976), Bolle (1988), Mason (1987), Becker and Leopold (1996)).

The assumption that subjects see figures of a series in a chart but don't know anything else about the events which they have to predict can be justified on experimental reasonings as a first step which has to be modified to get a more realistic picture. A lot of further information about the meaning of the series and factors which have an influence is given to the participants in the experiment of Garner (1982). Although the results of his pilot study are not very encouraging, it is inevitable that one should proceed on this level of complexity.

This paper presents some results of a forecasting experiment which was conducted by Becker and Leopold (1996) over several years as an interesting example of experimentation in economics. In this analysis, the data are compared with predictions that conform to the Rational Expectations hypothesis.

The "experts" in the experiment were subjects which had to predict a time-series. Nothing was said about the meaning of the series. The participants had to predict the values of a time series for the next following period. The renderings of this time series are presented to the respondents on a chart. Worksheets were used and after each prediction the participants were told the true value. The subjects had to draw this new data-point in the chart and to connect the points to get a closed line. Then they had to predict the next value.

The information available to them was restricted to a chart which shows the prior renderings of the series (basic version). In two further versions the subjects see in their chart an additional time-series (extended versions 1 and 2) which serve - in the terminology of business cycle theory - as a coincidental indicator.

These indicators are known for period t when the participants had to make a prediction for this period. The subjects were told in advance that they have to predict as precisely as possible. In this way they got predictions based on poor information (basic version), good information (extended version 1) and partially misleading information (extended version 2).

The objectives were: to get some insights into the individual processes of expectation formation under different information sets, to test the Rational Expectations hypothesis, and to seek better explanations for the average forecasts.

Some of the results which are of general interest are: individual predictions differ significantly from each other, depend very strongly on the information given, show remarkable prediction errors, and are not rational expectations in the sense of Muth. Average predictions have comparatively small prediction errors, show - as a more or less statistical phenomenon - a tendency towards rational expectations, and can be best explained by regression equations based on the information at hand.

It was possible to distinguish the difference between a bad and a good prediction of a subject, or in other words: It can be shown that subjects who have predictions nearer to the mean of the group have better results than the other subjects. This already indicates the idea that several aspects exert a more strongly weighted influence upon the collective mean than upon the individual predictions. Because the connection is a strong one, the distance to the group mean can be accepted as a measure of success for the individual performances.

From a macroeconomic point of view, however, the explanation of average forecasts is more important than the explanation of the individual behaviour. Therefore, the differences between average forecasts and the rational expectations in the three versions of the experiment are compared. The mean errors are comparatively small and suggest employing a certain modified version of the Rational Expectations hypothesis suggested in the paper. This method seems to be at least better than making use of naive expectation schemes. Even if better theories explaining the forecasts might exist, the use of the Rational Expectation Hypothesis seems to be a good advice in this situation. The experiment used there has confirmed in a certain way the relevance of Rational Expectations at least by a modification using gestalt-psychological ideas.

VIII. OUTLOOK AND CONCLUSIONS

There are arguments in principle against human ability of full rationality, discussed e.g. by Loasby (1996, p.3), who states that "there is no general equilibrium of economic agents all equipped with the correct model of the universe (...)". Impossibility problems in a context of self-reference which also apply to Rational Expectations are discussed by Knudsen (1993). Conlisk (1996, p.686), besides citing a considerable amount of literature in favour of bounded rationality, argues that "By its most common definition, economics concerns scarcity. Because human reasoning ability is scarce, one could as well argue that economists are by definition *required* to study bounded rationality." He, as many others, considers the arguments for full rationality as "too extreme to be convincing", but concedes "many useful insights" (ibid.) concerning the treatment of rationality. And, from a methodological perspective, his view that every economic theory must be "open to empirical challenge" (ibid.) is certainly to be shared.

The Rational Expectations hypothesis, despite its undoubted influence on a significant change in economic thinking, is continually put in question in such a way that cannot be overlooked. But it is questioned in a way that its main new insights, although not the extreme ones, may be preserved. Quite a few approaches to criticism, too many to be mentioned here, are emerging from the modern macroeconomic debate (see e.g. Hargreaves Heap, 1992). None of them referred to adaptive or extrapolative expectations; most of them made interesting qualifications to rational expectations. They more or less modified both New Classical as well as New Keynesian Macroeconomics.

One of the most serious lines of attack concerns the *interdependence* of expectations: if all agents expect the same rate of inflation for all intermediate and final goods and services, then this will be the correct expectation whatever its value may be. Here it is the self-fulfilling properties of price (or also of quantity) expectations; in other cases there are game-theoretic approaches which lead to multiple equilibria. With respect to the latter it is difficult to say what "rational expectation" might mean.

Another criticism refers to the fact that Rational Expectations as well as most other theories was developed as a theory of the closed economy. So, in an open economy, what does the "true model" mean? It is not just a closed economy completed by some balance of payments figures, as presented very often in the national accountancy. Instead, it assumes full knowledge of present and future values of all relevant variables and of all future concepts and measures of economic policy in the home country and in the countries of the main present and future trading partners. This is a difficult task (which, to our knowledge, has not yet found adequate consideration in the literature). Its solution can hardly be achieved by the agents.

Whereas Muth (ibid., p.316) argued that "Information is scarce, and the economic system generally does not waste it." the new trade-and-growth theory treats information as a (partly) public good (see e.g. Grossman/Helpman, 1993). To solve this ambivalence we propose considering *new* information as a scarce and *old* (i.e. already existing) information as a public good. New information is costly and will be acquired only in accordance to an individual optimisation judgement (which has no logical solution; see above). Old information is free and will be used to a great although not unlimited extent, i.e. with respect to the time-consuming nature of its collection and evaluation. The public goods property of information is recognised in the literature (e.g. Allen, 1990) but has not yet been worked out on a broader scale. Old information as a public good may but should not be wasted. Anyway, we expect people to work with scarce information that is new and plentiful information that is old. But not everybody works with the same information, because of different preferences underlying the information demand functions. From here it is a long way to rational expectations, but a short one to bounded rationality.

Asking what remains and what does not remain of Rational Expectations theory, we observe that at least two very important things are preserved: (i) the principal foundation of forecasts and expectations on economic theory; (ii) the orientation of forecasts and expectations on future values of the economic variables instead of on their past values. In stable equilibrium the future equilibrium values can, indeed, be assumed to give the correct tendency of change. They need not be in any way more bounded to past values than the future equilibrium itself. What was not preserved is mainly the pretended perfectness of expectations; a failure which is mainly due to the difficulty of forming such expectations.

Today the Rational Expectations hypothesis in its purely theoretical version is very much alive but nonetheless it is under attack on account from many sides of its extreme rationality postulate. On the other hand, since this theory opened fundamental new insights, nobody seems to be willing to go back to the *status quo ante*. After having learnt our lesson from the extreme pure-theory version of Rational Expectations there are enough reasons to proceed to preserve and apply the new insights as much as is possible under more limited forms of rationality.

REFERENCES

Maurice Allais, "Le comportement de l'homme rationel devant le risque: critique des postulats et axiomes de l'école americaine", in: *Econometrica,* 21, 1953, pp.506-546.

Beth Allen, "Information as an Economic Commodity", in: *American Economic Association Papers and Proceedings*, 80, 2, 1990, pp.268-273.

Otwin Becker, "Experimentelle Untersuchung der Erwartungsbildung für eine Zeitreihe", in: Heinz Sauermann (ed.): *Beiträge zur Experimentellen Wirtschaftsforschung*. Tübingen: Mohr 1967, pp.226-254.

Otwin Becker/Friedl Bolle, *Expectation in Economics: Rational or not? Evidence from Experiments*, Discussion Paper # 39, Europa-University Viadrina, Frankfurt/Oder 1995.

Otwin Becker/Ulrike Leopold-Wildburger, "The Bounds and Likelihood Procedure – A Simulation Study Concerning the Efficiency of Visual Forecasting Techniques", in: *Central European Journal for Operations Research and Economics,* 4, 2-3, 1996, pp.223-229.

Friedl Bolle, "Testing for Rational Expectations in Experimental Predictions", in: Shlomo Maital, *Applied Behavioural Economics*, 1, Brighton 1988, pp.319-334.

Edward H. Chamberlin, "An Experimental Imperfect Market", in: *Journal of Political Economy*, 56, 1948, pp.95-108.

Alfred William Coats, "Economics and Psychology: the Death and Resurrection of a Research Programme" in: S.J.Latsis (ed.). *Method and Appraisal in Economics*, Cambridge (Mass.): Cambridge University Press 1976, pp.42-64.

John Conlisk, "Why Bounded Rationality?", in: *Journal of Economic Literature*, 34, 2, 1996, pp. 669-700.

Franklin M. Fisher, *A Priori Information and Time Series Analysis*, Amsterdam: North Holland 1962.

Merill M. Flood, "Some Experimental Games", in: *Management Science*, 5, 1958, pp.5-26.

Gene M. Grossman/Elhanan Helpman, *Innovation and Growth in the Global Economy*, Cambridge (Mass.): MIT-Press 1993.

Friedrich A. Hayek, *New Studies in Philosophy, Politics, Economics and the History of Ideas*. Chicago, IL and London: The University of Chicago Press and Routledge & Kegan Paul Ltd. 1978/85.

Shaun Hargreaves Heap, *The New Keynesian Macroeconomics. Time, Belief and Social Interdependence*, Hants (UK), Vermont (US): Edward Elgar 1992.

Christian Knudsen, "Equilibrium, Perfect Rationality and the Problem of Self-Reference in Economics", in: Uskali Mäki/Bo Gustafsson/Christian Knudsen (eds.), *Rationality, Institutions and Economic Methodology*, London, New York: Routledge 1993.

Ulrike Leopold-Wildburger, "Induction as a Connection between Philosophy, Psychology and Economics", in: *Grazer Philosophische Studien – Internationale Zeitschrift fuer Philosophie*, 48, 1994, pp.114-128.

Brian J. Loasby, "Knowledge, Institutions and Evolution", *Graz Schumpeter Lectures*, 1996, forthcoming.

Robert E. Lucas, Jr., "An Equilibrium Model of the Business Cycle", in: *Journal of Political Economy*, 83, 6, 1975.

Robert E. Lucas, Jr./Edward C. Prescott, "Investment under Uncertainty", in: *Econometrica,* 39, 5, 1971, pp. 659-681.

Timothy Mason, *Expectation Formation in a Controlled Laboratory Environment*, Bloomington,(Ind.) 1988.

John F. Muth, Rational Expectations and the Theory of Price Movements, in: *Econometrica,* 29, 3, 1961, pp.315-335.

Thomas J. Sargent/Neil Wallace, "Rational Expectations and the Theory of Economic Policy", in: *Journal of Monetary Economics* 2, 1976, pp.169-183.

Thomas J. Sargent, "Rational Expectations" in: John Eatwell/Murray Milgate/ Peter Newman (eds.), *The New Palgrave Dictionary of Economics*, London: Macmillan, New York: Stockton, Tokyo: Maruzen 1987, repr. 1988, Vol. 4, pp. 76-79.

Heinz Sauermann/Reinhard Selten, "Ein Oligopolexperiment", in: *Zeitschrift für die Gesamte Staatswissenschaft*, 115, 1959, pp.427-471.

N. Eugene Savin, "Rational Expectations: Econometric Implications", in: John Eatwell/Murray Milgate/Peter Newman (eds.), *The New Palgrave Dictionary of Economics*, London: Macmillan, New York: Stockton, Tokyo: Maruzen 1987, repr. 1988, Vol. 4, pp. 79-85.

Joseph A.Schumpeter, *History of Economic Analysis*, New York NY. Oxford University Press, 1954.

Richard Schmalensee, "An Experimental Study of Expectation Formation", in: *Econometrica*, 44, 1976, pp.17-41.

Reinhard Selten, *Evolution, Learning and Economic Behavior*, Nancy L.Schwartz Memorial Lecture, Working Paper. Kellogg Graduate School of Management. Northwestern University, 1989.

Alan Shapiro, "On the Implications of Behavioral Decision Making Theory to Economics", in: Alan J., Heather W. (eds.), *Economic Psychology*, Amsterdam, 1986, pp.621-644.

Sidney Siegel/Lawrence E. Fouraker, *Bargaining and Group Decision Making: Experiments In Belated Monopoly,* New York: McGraw-Hill 1960.

Jeremy Stone, "An Experiment in Bargaining Games", in: *Econometrica*, 26, 1958, pp.286-297.

Gabriel Tarde, *La psychologie économique*. Paris. Alcan (2 Vol.), 1902.

Stephen J. Turnovsky, *Trends and Fads in Macroeconomic Dynamics*, Paper presented at the Institute for Advanced Studies, Vienna 1996.

Lutz Beinsen
Institut für Volkswirtschaftslehre, Universität Graz
Universitätsstraße 15
A-8010 Graz
Austria

Ulrike Leopold-Wildburger
Institut für Statistik, Ökonometrie und OR, Universität Graz
Universitätsstraße 15
A-8010 Graz
Austria

THOMAS RUSSELL

MACROECONOMICS AND BEHAVIORAL FINANCE: A TALE OF TWO DISCIPLINES

I. INTRODUCTION

In the last fifty years there have been two revolutions in macroeconomics. The first, the Keynesian revolution, gave birth to macroeconomics as a sub-discipline distinct from microeconomics. By combining broad behavioral relationships, (the consumption function, the investment function, the liquidity preference function, etc.) into a single system, Keynesian macroeconomics demonstrated the existence of output levels at which involuntary unemployment was an equilibrium state.

To restore the economy to full employment in this case required a program of public spending and/or tax cuts to raise aggregate demand to its full employment level. Such expansionary fiscal policy could easily lead to budget deficits, but concern with the level of budget deficits was viewed as a sign of small-mindedness, attributable to a lack of understanding of how the economy worked.

The second revolution, the Rational Expectations revolution, in one swoop undid most of the Keynesian innovations. Macroeconomics was restored to its microeconomic foundations by insisting that all aggregate behavioral functions be derived from first principles of constrained maximization. In addition, expectations, a crucial component of all demand and supply functions, were required to be equal to the forecasts of the macroeconometric model itself. From now on there was to be no distinction between the thought processes of the observer and the observed.

For the Rational Expectations school, long term involuntary unemployment did not occur. Any government misguided enough to try to follow a systematic policy of demand expansion would achieve no lasting effect on real output. Since budget deficits did nothing to reduce unemployment and since on other grounds government spending was viewed as less productive than private spending, a balanced budget became the new policy goal.

It is easy to see what caused the first of these revolutions. The mass unemployment of the 1930s was difficult to reconcile with the then standard doctrine that supply creates its own demand, and a theory identifying unemployment with unsold goods due to lack of demand had obvious immediate appeal. It also did no harm that the new theory called on governments to take immediate action to hire back the unemployed. In a crisis there is a bias towards action.

The cause of the second revolution is less clear. To be sure high inflation rates and the oil shock induced unemployment of the 1970s put a significant dent in the Keynesian promise of full employment with stable prices forever. The failures of the Keynesian revolution, however, did not lead automatically to the success of Rational Expectations. Indeed, the first shot in this second revolution,

W. Leinfellner and E. Köhler (eds.), Game Theory, Experience, Rationality, 153–159.

Muth's paper [1961] which introduced the notion of rational expectations, went completely unheeded in the macroeconomic debates of the 1960s.

Moreover, though the Rational Expectations school met with some empirical success, even now the econometric record is both thin and mixed. What seems to have carried the day for Rational Expectations was not so much its superiority at explaining data as much as its successful insistence that this was the only approach to data allowable to economic science. In particular, and in contradistinction to the ad hoc assumptions of the early Keynesians, the Rational Expectations revolution was founded on the view that individually rational decision making was the hallmark of economic science and that to proceed sans this assumption was a sign of a bad scientist.

More than anything, then, the Rational Expectations revolution was a revolution in viewpoint on which model of man was most appropriate for the analysis of macroeconomic fluctuations. Keynesian man with his propensities to do this and his propensities to do that must be replaced with Muthian man, a creature whose behavior was always in exact accord with the theorems of advanced microeconomics.

Since the Rational Expectations revolution was a revolution in thinking about the psychology of individual judgment, where in all of this was the influence of the behavioral psychologists who specialized in the study of individual decision making under uncertainty? Notable, one would have to say, by their absence. This is all the more surprising since behavioral psychology was enjoying considerable success during the years of ascendancy of the Rational Expectations school. The work of Kahneman and Tversky, for example, shows the strength of this alternative approach, see, for example, Kahneman et al [1982]. A reading of the macroeconomic literature of the 70s and 80s, however, gives no clue that in other disciplines scientists were over and again rejecting the hypotheses which lay at the foundation of the Rational Expectations school.

This situation in macroeconomics stands in sharp contrast to the situation in the neighboring sub-discipline of financial economics. Here, albeit not without a struggle, data based anomalies forced the acceptability (if not the acceptance) of the premise that not all individual behavior is rational. In this paper we argue that the recent success of behavioral methods in financial economics has lessons for the future of macroeconomics.

II. Behavioral Finance: An Unlikely Discipline

There are at least two reasons why, of all the branches of economics, one expects financial economics to be the least amenable to analysis based on irrational behavioral effects[1].

1. The goal of agents who operate in financial markets is reasonably clear. Individuals buy stocks to make positive returns, the higher the returns the better. For that reason we expect that at a minimum financial markets will

offer no arbitrage opportunities. No rational agent would pass up the chance to buy an asset at one price and simultaneously sell it at a higher price.
2. Even if some agents fail to act rationally, one expects that the marginal actor i.e. the agent whose behavior sets the market price, will have the highest level of financial skills. Since a fool and his money are soon parted, the financial institutions with staying power are likely to be those whose analysts systematically have high returns, and those analysts are likely to stay close to the canons of rationality.

For these reasons among others, financial economics was dominated for some time by models which predicted that assets would yield normal returns conditional on all the available information. This 'efficient market hypothesis' includes the Rational Expectations hypothesis since 'all the available information' includes the models of the economy relevant to forecasting future yields.

Stated with this generality, the efficient market hypothesis is difficult to reject. Normal returns must allow some premium for risk, and any ex post pattern of returns can be said to be normal if we define risk to make it so. For example, the data clearly reject the Capital Asset Pricing Model (CAPM) in which risk is defined as market covariance or 'beta', see Fama and French [Forthcoming], but this does not preclude the possibility that some other risk factors in conjunction with beta can explain cross section patterns of returns.

For that reason, the most interesting attacks on the 'hard core'[2] of the rationality hypothesis in finance have come from a different source. In the U.S., financial markets provide prices for portfolios of assets (called closed end mutual funds) at the same time as they price the underlying assets in these portfolios. Rationality dictates that these funds will sell for close to their net asset value (i.e. the value of the assets in the fund.) The facts, however, are otherwise and over time these funds have sold at both discounts and premia to underlying asset value.

Several possible rationality based explanations have been offered for the premia and discounts observed in these funds, but none of these can explain

> why funds get started ... (and) ... why discounts fluctuate over time. Perhaps most important, each of these explanations deals with the puzzle of closed end funds selling at discounts, and fails to explain why sometimes funds sell at premia, particularly when they are started. (Lee/Shleifer/Thaler [1991] p. 82.)

To explain this, investigators have dropped the assumption of rationality and replaced it with the assumption that there are traders in financial markets whose behavior is less than fully rational, what Thaler and I elsewhere have called 'quasi rational', Russell and Thaler [1985]. These traders respond to noise, see Black [1986], and noise traders generate a new source of risk which drives asset returns away from their fundamental values, see De Long et al [1991]. The presence of irrational forces in financial markets can be detected indirectly by testing to see if premia in closed end mutual funds are correlated with signs of positive sentiment in other sectors of the financial market. Lee et al [op.cit.]

have confirmed that positive levels of sentiment on initial public offerings (I.P.O.s) are positively correlated with premia in closed end funds, a result which steps outside the canons of rational behavior.

'Behavioral Finance', an approach to the analysis of financial markets which drops the assumption of rationality and replaces it with assumptions on behavior consistent with those found by psychologists, is now a well established scientific field. If the assumption of rationality must be dropped in financial economics, an area where monetary gain is the dominant goal, questions must be raised concerning other areas of economics where the preconditions for rational behavior are not so well established. Because of its importance for policy, one obvious area is the field of macroeconomics.

III. Behavioral Macroeconomics

Keynes founded macroeconomics on a common sense view of human nature:

> We should not conclude that everything depends on waves of irrational psychology...
> We are merely reminding ourselves that human decisions affecting the future, whether personal or political or economic, cannot depend on strict mathematical expectation, since the basis for making such calculations does not exist: and it is innate urge to activity which makes the wheels go round, our rational selves choosing between the alternatives as best we are able, calculating where we can, but often falling back for our motive on whim or sentiment or chance. ([1936] pp.162-163)

As Lucas [1981] has noted, however, the post Keynes' history of macroeconomics can be seen as a succession of attempts to give each of the aggregate components of the Keynesian model a rational basis. The seeds of the Rational Expectations revolution were already sown by Keynes own followers. In the light of developments in financial economics, however, it is interesting to speculate what macroeonomics would look like if it were to return to the behavioral foundations first proposed by Keynes.

In this paper we can do no more than touch on some general points.

1. As we have noted, agents enter financial market with the single goal of increasing their wealth. In the economy's goods and labor markets, however, there may be other motivating factors. For example, in labor markets wages may reflect considerations of fairness, see Kahneman, Knetch, and Thaler [1986]. This would have wide ranging implications for the determination of the level of unemployment.

Similarly well known agent principal issues, see e.g. Jensen and Meckling [1983], call into question the precise objective function which determines the level of corporate investment. For example, Jensen [1993] finds that managers at General Motors wasted over $100 billion of stockholders wealth in the 1980s by investing free cash in ventures whose chief purpose was to preserve the manager's realm of control. This can only happen when corporate control mechanisms fail, but when this happens, the level of private investment will be determined in part by the whims of corporate managers. This suggests that the level of investment may be determined by forces closer to the 'animal spirits' of Joan Robinson and the early Keynesians than to the rational present value calculations later proposed by Jorgenson.

2. Even if we could take motivation as given, it seems obvious that individuals differ in their ability to compute optimal plans. For example, to compute optimal current consumption, it is necessary to use a delicate stochastic dynamic programming algorithm which uses statistical information on the probability of death and which recognises the option value of delayed retirement. It seems plausible that many consumers, finding such rational calculations well beyond them, will determine current consumption by a simple rule of thumb which assumes retirement at 65 because that is the social custom. Moreover, unlike financial markets, if the agents in goods and labor markets fail to act rationally, there is no cadre of rational agents whose behavior will drive prices in the rational direction.

This difficulty in computing optimal behavior may even extend to the Government agents charged with managing the economy. Mayer [1990] has analysed the observed actions of the Federal Open Market Committee in behavioral terms. Obviously if the Fed were to act irrationally no rational 'Shadow Fed' could undo its actions.

The development of a behavioral macroeconomics can be expected to lead to a better understanding of the sectoral components of the macro model. Important as this is, of even greater importance is the light which a behavioral macroeconomics will shed on the central policy debate of whether or not government action can have any long run real effects.

For the purposes of this debate, an understanding of the behavioral foundations of the supply decisions of individuals and firms is more important than a reworking of the demand side. As Grandmont [1992] has recently shown, under reasonable demographic assumptions aggregate demand functions can generate a unique stable economic equilibrium no matter what the shape of the underlying individual demand functions. In particular, as he notes, there is no need to assume that individual demand is rational. Moreover the rationalisation of the various demand sectors of the Keynesian model did not change in any way the fundamental Keynesian interventionist policy conclusion. The Rational Expectations revolution was a revolution on the supply side.

On the other hand, if money illusion is prevalent, and if corporations in the face of a price level fall are reluctant to cut money wages on grounds of equity, a policy of government stimulus, even if anticipated, will have real effects. It is a completely legitimate objection to this scenario that it is not consistent with the facts. However, it is not a valid objection that it is not consistent with rationality.

IV. Conclusion

The high water mark of the Rational Expectations revolution has probably now passed. Nevertheless, the effect of the school's insistence on rationality as a basis for macroeconomics lingers. For example, a recent textbook (which includes a summary of a paper by Shafir, Diamond, and Tversky [Forthcoming] demonstrating that individuals indeed do suffer from money illusion, the height of non-rationality) states

> Fairly quickly the idea that rational expectations was the right working assumption gained wide acceptance. This did not happen because all macroeconomists believe that people, firms, and participants in financial markets always form expectations rationally. But rational expectations appears to be a natural benchmark, at least until economists have made more progress in understanding whether and how actual expectations systematically differ from rational expectations. (Blanchard [1997], p.617)

Rationality is no longer a natural benchmark in financial economics.[3] The data simply does not support such a simplified view of human nature. The progress made by financial economists in understanding market prices in behavioral terms has many lessons for macroeconomics. These lessons will be developed in future work.

Notes

1. In their pioneering work in behavioral finance De Bondt and Thaler [1994] have also noted the a priori likelihood that financial markets will be rational.
2. In the Lakatos [1970] sense. See Russell [1997] for further discussion of Lakatos' concepts in the context of developments in finance and macroeconomics.
3. Indeed a recent review of the financial literature by Haugen [1995] is entitled The New Finance: The Case Against Efficient Markets.

References

F. Black, "Noise", in: *Journal of Finance*, 41, 1989, pp.529-543.

O.J. Blanchard, *Macroeconomics*. Prentice Hall: N.J. 1997.

W. De Bondt/R. Thaler, "Financial Decision Making in Markets and Firms: A Behavioral Perspective", NBER Working Paper No. 4777. Cambridge 1994.

J.B. De Long/A. Shleifer/L.H. Summers/R.J. Waldmann, "Noise Trader Risk in Financial Markets", in: *Journal of Political Economy*, 98, 1990a, pp.703-738.

E.F. Fama/K.R. French, "The CAPM is Wanted, Dead or Alive", in: *Journal of Finance* (forthcoming).

J.M. Grandmont, "Transformations of the Commodity Space, Behavioral Heterogeneity, and the Aggregation Problem", in: *Journal of Economic Theory* 57 (1), 1992, pp.1-35 .

R.A. Haugen, *The New Finance: The Case Against Efficient Markets*. Englewood Cliffs: Prentice Hall: 1995.

M.C. Jensen/W. Meckling, "Theory of the Firm: Managerial Behavior, Agency Costs, and Ownership Structure", in: *Journal of Financial Economics*, 3, 1976, pp.305-360.

M.C. Jensen, "The Modern Industrial Revolution, Exit and the Failure of Internal Control systems", in: *Journal of Finance*, 48 (3), 1993, pp.831-880.

D. Kahneman/J.L. Knetsch/R. Thaler, "Fairness as a Constraint on Profit Seeking: Entitlements in the Market", in: *American Economic Review*, 76 (4), 1986, pp.728-738.

D. Kahneman/P. Slovic/A. Tversky, *Judgement Under Uncertainty: Heuristics and Biases*. Cambridge: Cambridge University Press 1982.

J.M. Keynes J.M., *General Theory of Employment, Interest, and Money*. London: Harcourt Brace 1936.

I. Lakatos, "Falsification and the 'Methodology of Scientific Research Programs", in: I. Lakatos/A. Musgrave (Eds.), *Criticism and the Growth of Knowledge*. Cambridge: Cambridge University Press 1970, pp.91-195.

Charles Lee/A. Shleifer/R.H. Thaler, "Investor Sentiment and the Closed End Mutual Fund puzzle", in: *Journal of Finance*, 46, 1991, pp.75-109.

R.E. Lucas, *Studies in Business Cycle Theory*. Cambridge: M.I.T. Press 1981.

T. Mayer, "Minimizing Regret: Cognitive Dissonance as an Explanation of F.O.M.C. Behavior", in: T. Mayer (Ed.), *The Political Economy of American Monetary Policy*. Cambridge: Cambridge University Press 1990.

J.F. Muth, "Rational Expectations and the Theory of Price Movements", in: *Econometrica*, 29, 1961, pp.315-35

T. Russell/R. Thaler, "The Relevance of Quasi-Rationality in Competitive Markets", in: *American Economic Review*, 75, 1985, pp.1071-1082.

T. Russell, "The Rationality Hypothesis in Economics: From Wall Street to Main Street", in: *Journal of Economic Methodology*, (Forthcoming 1997).

E. Shafir/P. Diamond/A. Tversky, "Money Illusion", in: *Quarterly Journal of Economics*, forthcoming.

Department of Economics
Santa Clara University
Santa Clara, CA 95053
USA

BRIAN SKYRMS

BAYESIAN SUBJUNCTIVE CONDITIONALS FOR GAMES AND DECISIONS

I. INTRODUCTION

The theory of rational decision has always been implicitly involved with subjunctive and counterfactual conditionals. *"If I were to do A, this would happen; if I were to do B that would happen."* When I have done A, I use the counterfactual: *"If I had done B, the outcome would have been worse."* Counterfactuals are handled so smoothly in decision theory and game theory that they are hardly ever explicitly discussed except in cases where they cause problems. For many philosophers, Newcomb's paradox was such a critical case (See Gibbard and Harper (1978)). In game theory, questions about beliefs off the equilibrium path in extensive form games and related issues of backward and forward induction provoked explicit discussion of counterfactuals (See Aumann (1995), Bicchieri (1988), Binmore (1987), Harsanyi and Selten (1988), Reny (1993), Stalnaker (1994), (forthcoming)). Where subjunctive conditionals cause problems, we need an explicit theory. What theory should we use? There is the selection function semantics of Stalnaker (1968) based on the idea of "most similar possible world in which the antecedent is true", and related theories of Woodruff (1969), Lewis (1973), Pollock (1976), Nute (1980), which relax some of Stalnaker's assumptions.

The Stalnaker selection function account, however, has some difficulty in dealing with chance. Suppose a coin is not flipped in the real world. Is the closest possible world in which it was flipped one in which it came up heads or one in which it came up tails? There are difficulties in dealing with randomized strategies on this approach. I do not want to suggest that such difficulties are insuperable. One might, for instance, introduce deterministic "hidden variables". But I would like to pursue a different approach which seems to me to treat these issues in a more natural way. I would like to use the Bayesian theory of subjunctive conditionals that I introduced in Skyrms (1984) which has as its key notion a family of partitions. The Bayesian theory has the advantages of (1) greater generality and (2) primitives which fit naturally into decision theory and game theory.

W. Leinfellner and E. Köhler (eds.), Game Theory, Experience, Rationality, 161–172.

II. THE BAYESIAN THEORY OF SUBJUNCTIVE CONDITIONALS

The leading idea of the Bayesian theory of conditionals is that the objective value of a conditional in a possible situation is the *conditional chance* of its consequent given its antecedent. In analogy with a standard Bayesian construction of "chance" as probability conditional on a partition (or sigma-algebra) we construe "conditional chance" as probability conditional on a family of partitions.

Suppose a ball is to be drawn from a well-mixed urn containing 100 red and black balls, but you are uncertain about the composition of the urn. You take the composition of the urn to determine the *chance* of drawing a red ball in the natural way. To get a reasonable degree of belief that the ball drawn will be red, you average the chances using as weights your degrees-of-belief in the possible distributions of colors. Conversely, you can take *Chance of red* to be the random variable on your degree-of-belief probability space, defined as probability of red conditional on the partition of possible compositions of the urn.

Now suppose that you are to draw twice from the urn with replacement. The *conditional chance* of red on draw 2 given red on draw 1 is likewise a random variable, defined on your degree-of-belief probability space as the conditional probability of red on draw 2 given red on draw 1, conditional on the same partition. Assuming the urn is well-mixed after replacement, the draws are independent in the chance probabilities. For example, let b99 be the element of the partition specifying 99 red balls and one black ball. At any point in the probability space in b99, the three random variables: chance of red on draw 1, chance of red on draw 2, conditional chance of red on draw 2 given red on draw 1, all have the same value of .99.

Since the degree-of-belief probabilities are gotten by expecting over the possible chances, the two draws are not independent in the degree-of-belief probabilities. A draw of red on the first ball leads by Bayes theorem to a greater degree of belief on the chance hypotheses which favor red, and so the degree-of-belief conditional probability of red on two, conditional on red on one is greater than the unconditional degree-of-belief probability of red on two. The effect is particularly striking if the prior degrees of belief are concentrated at or near the extremes. For example, suppose that you are almost certain that the balls are either all red or all white. Mixing of chance probabilities has created a spurious correlation in the degree-of-belief probabilities. The spurious correlation disappears when we condition on the chance partition. I belabor the point because it will be of some importance later in the application of the theory to decisions and games.

Let us look at an example where conditional chance plays a less trivial role. Suppose that a coin is loaded with iron so as to bias it either towards heads or towards tails and a magnet is either on or off so that the chances of heads are as given in table 1.

Table 1

Chance (H)	BH	BT
OFF	2/3	1/3
ON	5/6	1/6

The coin is to be flipped. To evaluate the chance of heads conditional on the magnet being ON, it is natural to take the partition [BH, BT]. Thus at every point in BH, the chance of Heads conditional on ON is 5/6. This is true for points in BH & OFF as well as those in BH & ON. Likewise, at every point in BT, the chance of Heads conditional on ON is 1/6.

But if we instead take the condition to be BH, the appropriate partition to use is [OFF, ON]. The condition partially specifies the factors affecting the chance of heads, and the partition is suitably chosen to complete the specification. If the condition is tautological, then the appropriate partition must be fine enough to specify the chance of heads all by itself: [BH&ON, BH&OFF, BT&ON, BT&OFF]. If the condition is specific enough to specify the chance of heads all by itself, then the appropriate partition is degenerate, having as its only member the whole space. A general theory of conditional chance must use not a single partition but a family of partitions.

The basic objects of the theory are (1) a probability space, representing an agent's degrees of belief and (2) a family of partitions. Non-empty measurable sets of (1) will be called consistent propositions. We assume that the probability measure is regular, in the sense that each consistent proposition has positive probability. (Uncountable probability spaces will require infinitesimal probabilities.) A family of partitions, π, is a function which maps a consistent proposition, p, onto a partition, $\pi(p)$, each of whose members is consistent with p. By regularity, if $b \in \pi(p)$, then $pr(b \cap p) > 0$. If the domain of a family of partitions consists of all the consistent propositions in the language, then it is a total family of partitions, otherwise a partial family of partitions.

Then we can define *conditional chance*, for a condition in the domain of the family, as:

$$\text{For } w \in b \in \pi(p),\quad \text{Chance}(q|p) \text{ at } w = pr(p \cap q \cap b) \,/\, pr(p \cap b)$$

We take the *objective value of a conditional* whose antecedent is in the domain of a family to be the conditional chance of its consequent given its antecedent:

$$V(p \rightarrow q) = \text{Chance}(q|p)$$

The decision maker may be unsure of the *objective value* of a conditional, and in this case will take as its *assertability value* the epistemic expectation of its objective value, $E[\text{Chance}(q|p)]$.

The example of the loaded coin can be generalized to a class of cases which are especially transparent (Skyrms, 1988). Suppose that the chance of q can naturally be thought of as being determined by values of a list of factors, x_1, x_2, ... , x_n. These factors can be thought of as parameters of the chance set-up for q. A natural condition relative to such a list of factors is a proposition specifying the values of some subset of the list of factors. A natural conditional for such a family will have a natural condition as its antecedent and q as its consequent (the null set of factors being specified by a tautology). The natural family of partitions relative to such a list of factors maps each natural condition onto the partition generated by specification of all the factors on the list not specified in the condition.

Where the foregoing model fits, and for the special conditionals that are dealt with, the account given is straightforward and compelling. It may perhaps be viewed as a probabilistic realization of the parametric ideas in Selten and Leopold (1982). We note that natural families generated by a list of factors are *monotonic* in the following sense:

For p, q in the domain of π,
If $p \subseteq q$ then $\pi(p)$ is a coarsening of $\pi(q)$.

This monotonicity is a consequence of the division of labor between the antecedent of the conditional and the corresponding partition.

Natural families buy their naturalness at the price of partiality. For example, the natural family generated by the two factors in the biased coin case does not specify conditional chances for the disjunctive condition: BH $\cup$ OFF. And there does not seem to be any compellingly natural answer to this question on the basis of the information given.

Suppose, we want a theory for all conditionals with consistent antecedents, and for this purpose insist on a total family of partitions. Monotonicity is too strong to be a plausible a requirement for such a total family, but I proposed (1984) a weaker requirement which I called *Omonotonicity*:

A family of partitions is Omonotonic if for consistent p, q, if p is a subset of q, then every member of $\pi(q)$ whose intersection with p is non-empty is a subset of some member of $\pi(p)$.

A further condition on families of conditionals is motivated by the desire that the operative notions of conditional chance and unconditional chance fit together well. Given the construal of chance as probability conditional on an appropriate partition, conditionals with chance consequents:

If p, then $\text{Chance}(q) = a$.

are covered by the preceding theory. The consequent, $\text{Chance}(q)=a$, is given a truth value at every point by the construction of the random variable, Chance(q), as probability conditional on a partition. We want Chance(q)$=a$ to be equivalent to the conditional: If tautology, then Chance(q)$=a$. So we shall choose the partition used to define unconditional chance as $\pi(T)$, the partition that the family assigns to the tautology. Now we guarantee that the antecedent of a conditional together with the appropriate partition determines the chance of the consequent in line with the chance partition, $\pi(T)$, by adding to Omonotonicity, the condition *MESH*:

> For every consistent p and every element, b, of the partition $\pi(p)$, there is an element c, of the chance partition, $\pi(T)$ such that $b \cap p \subseteq c$. [Skyrms (1984)]

Woodruff (1996) shows that Omonotonicity and MESH together are equivalent to *Priority of Chances*:

> If $w \in b \in \pi(p)$ and $w \in c \in \pi(T)$, then $p \cap b = p \cap c$

That is to say that the partition assigned to p divides up p in exactly the same way that the chance partition does.

A special case will be of interest in connection with the Stalnaker selection function account of conditionals to be considered in the next section.

> A family of partitions is *deterministic* if for all p and for all $b \in \pi(p)$, $b \cap p$ contains exactly one point, w.

III. RELATION TO SELECTION FUNCTION ACCOUNTS

A *selection function*, f(w,p) maps all pairs of points (or "possible worlds") w, and consistent propositions, p, onto points such that:

> (i) $f(w, p) \in p$
> (ii) If $w \in p$, then $f(w, p) = w$

A selection function is *Stalnaker* if, in addition:

> (iii) If $f(w, p) \in q$ and $f(w, q) \in p$, then $f(w, p) = f(w, q)$

[Stalnaker himself includes inconsistent propositions in the domain of the selection function and includes an "impossible world" which is selected when p is inconsistent. All propositions are then counted as true in the impossible world. However, for the purposes of this paper, it is simpler and more natural to restrict the domain of the selection function.]

A deterministic family of partitions induces a selection function if we let:

$f(w, p)$ = the unique $w' \in b \cap p$ (where $w \in b \in \pi(p)$)

It follows immediately from the definition and determinism that f is well defined and satisfies conditions (i) and (ii) for a selection function. Let SEL[π] be the selection function induced by the family of partitions, π.

Conversely, a selection function induces a deterministic family of partitions, if we let:

$\pi(p)$ = the set of inverse images of points in p under $f(*, p)$

The set of inverse images of any function is a partition. Each element of the partition $\pi(p)$ has non-empty intersection with p by (i) and (ii): $f(w, p) \in p$ by (i) and then $f(f(w, p), p) = f(w, p)$ by (ii). Determinism follows because $f(*, p)$ is a function. Let PAR [f] be the deterministic family of partitions induced by a selection function, f.

It is evident from their definitions that PAR and SEL have unique inverses and that each is the inverse of the other. There is a one-to-one correspondence between selection functions and deterministic families of partitions.

Furthermore, it is shown in Skyrms (1984) that a selection function is *Stalnaker* if and only if the corresponding family of partitions is *Deterministic* and *Omonotonic*. First, Consider the selection function, f, induced by a deterministic, omonotonic family of partitions. Let $f(w_1, p) = w_2$ and $f(w_1, q) = w_3$ and $w_2 \in q$ and $w_3 \in p$. Then by (i) w_2 and w_3 are both in $p \cap q$. By Omonotonicity the member, b, of $\pi(p)$ which has as members w_1 and w_2 and the member, c, of $\pi(q)$ which has as members w_1 and w_3 are subsets of some member, d, of $\pi(p \cap q)$. By Determinism, $w_1 = w_3$. Thus, f is Stalnaker.

Conversely, let f be Stalnaker. Suppose that $p \subseteq q$, $b \in \pi(q)$ and $w \in b \subseteq p$. The set b is the inverse image of w under $f(*, q)$. We write $b = f_q^{-1}(w)$. For Omonotonicity, it is to be shown that b is a subset of some member of $\pi(p)$. The required member is $f_p^{-1}(w)$. For if $f_q^{-1}(w)$ were not a subset of $f_p^{-1}(w)$ there would have to be a point, x, in the former but not in the latter. But then $f_q(x)$, which by hypothesis is w, would be in p and thus in q since $p \subseteq q$. Then by Stalnaker (iii), $f_p(x) = f_q(x) = w$, so x is in $f_p^{-1}(w)$ after all.

Stalnaker takes the conditional $p \rightarrow q$ to be true at point w if $f(w, p) \in q$, otherwise false. So for a Deterministic Omonotonic family of conditionals, the *Objective Value* of a conditional at a point w, according to the Bayesian account, agrees with the truth value according to the corresponding Stalnaker selection function, taking truth = 1 and falsity = 0. And thus, under these conditions the *Assertability Value* of the conditional on the Bayesian Account agrees with its subjective probability of truth on the Stalnaker account.

These identities break down in the general case. In particular, conditional chances only behave like truth values in the case of Determinism. The deterministic case can, therefore, be somewhat misleading. Non-deterministic families of partitions induce *set-selection functions*, if we let:

$F(w, p) = b \cap p$ (where $w \in b \in \pi(p)$)

Such set selection functions had already been considered as generalizations of Stalnaker selection functions by Woodruff (1969), Lewis (1973), and Chellas (1975), Pollock (1976), Nute (1980). The connections between these accounts and the Bayesian account are discussed in Stalnaker (1994) and Woodruff (1996). They are not quite so straightforward as the connection with Stalnaker in the deterministic case. In particular, the set selection functions may not correspond to partitions.

IV. DECISIONS

In this section we consider static decision problems, where a single individual at a single time seeks to optimize a fixed, but possibly unknown, state of nature. The decision maker's degrees of belief over everything relevant are formalized by a probability space. The decision maker chooses among alternative acts, *A*, which form a partition of this space. The decision maker attaches value to consequences, *C*, of these acts, which form another partition. The acts influence the chances of consequences, and together with underlying states of the world, *K*, determine the chances of consequences. These underlying states also form a partition of the space. Intuitively, the decision maker is interested in subjunctive conditionals of the form:

If I were to do act A, then consequence C would ensue.

The appropriate background partition is the partition of states of the world.

Let us see how this fits into a decision theoretic framework. The act has tendencies, or propensities to bring about consequences. These are captured in the conditional chances of consequence given act. The objective value of an act to a decision maker is the expectation of the value of the consequences where the weights of the expectation are these conditional chances:

$$\Sigma_j \text{ CHANCE}(C_j \mid A)\ V(C_j)$$

The value of these conditional chances may, however, depend on features of the world about which the decision maker is uncertain. Thus the subjective value of the act will be a subjective expectation of its objective value:

$$\Sigma_i \Pr(K_i)\ \Sigma_j \text{ CHANCE}(C_j \mid A)\ V(C_j)$$

(where the *K*s are the elements of the appropriate partition for conditional chance.) It is important that expected value have the form of this double expectation, and that probabilities of states be unconditional, in order to avoid fallacies of spurious correlation. I will return to this point shortly. Note that in the special case where the state together with the act determine the consequence, the conditional chances are 0 or 1, and the objective value of the act can be construed as a random variable on the space of states of the world. This special case of determinism gives an expected value that looks more like that of Savage (1954).

The relevance of the Bayesian theory of subjunctive conditionals is now apparent. The subjective expected value of act *A* can be rewritten in terms of that theory as:

$$\Sigma_j \text{ Assertability Value } (A \rightarrow C_j)\ V(C_j)$$

That is to say that the assertability value of a subjunctive conditional is just what a rational agent needs to calculate expected value. *Subjunctive conditionals are decision-making conditionals.*

The significance of the subjunctive lies in its connection with conditional chance and its role in avoiding the fallacy of basing one's decisions on spurious correlations. We tend to think of our acts as uncorrelated with anything which precedes them because they are an exercise of "free will". Any serious investigation of the matter will show that no notion of free will that we can plausibly be thought to exercise can support such an extravagant claim. Our spouses and our friends can often predict actions which we freely choose.

In some such cases an act may be a symptom or sign of a state of the world which is relevant to determining its value. Going to the doctor may be a sign that one is really ill. Eating a lot of salt is a symptom of Addison's disease, and it is helpful in the management of that disease. Suppose that in such cases, one calculated expected utility using the degree-of-belief conditional probabilities as weights: $\Sigma_j\, Pr(C_j|A)\ V(C_j)$, as in Jeffrey (1965). Then one would make the mistake of treating signs and symptoms of states of nature as if they were causes, as shown in some detail in chapter 1 of Jeffrey (1983).

Newcomb's problem is an extreme and often misunderstood example of the phenomenon. Suppose a philosophy student has been exposed to the different kinds of expected value discussed above, and is a little uncertain as to which is right. A psychologist invites him to participate in an experiment. She gives the student a battery of tests and then presents him with a decision problem. He can either take a sealed envelope that he his is given to hold in his hand, or the sealed envelope together with a $10 bill. If she predicted on the basis of his tests that he take only the sealed envelope, she put $100 in it, otherwise nothing. The student believes that she is very good at predicting behavior in this situation, so that his taking only the envelope is a probabilistic symptom of the $100 being there and his also taking the $10 bill is a probabilistic symptom of the envelope being empty. If his belief is strong enough, Jeffrey's (1965) system gives taking only the envelope higher expected value than taking both the envelope and the $10 bill. This is basing expected value on a spurious correlation.

In this example, the possible acts are to take the envelope only or to take both the envelope and the bill, the consequences are monetary values received and the states are the contents of the envelope. In this example the relevant conditional chances are 0 or 1; the act and the state jointly determine the consequence. The assertability value of the conditional *"If I were to take only the envelope, I would get $100"* is here just the probability of the state "$100 in the envelope" and it is also the assertability value of *"If I were to take both, I would get $110.* The assertability value of *"If I were to take only the envelope, I would get nothing"* is just the probability of the state "Nothing in the envelope" and it is also the assertability value of *"If I were to take both, I would get $10."* In the calculation of expected value, the spurious correlation is properly ignored, and the expected value of taking the $10 as well as the envelope is indeed $10 more than that of taking the envelope alone.

Let me note again that even when the decision maker has made up her mind which act to do, we assume that the other acts retain at least infinitesimal probability capable of grounding the conditional probabilities required by our theory. There are two attitudes that one could take about such infinitesimals. The first is that they are really there as a legitimate expressions of belief that falls short of certainty. The alternative is that they are "ghosts of departed quantities" in a legitimate and useful sense. As you decide to do an act alternative to A, and $pr(A)$ approaches zero, the conditional chance of consequence C given A in state K, $pr(C| A \cap K)$ typically remains constant. The infinitesimal probability assigned to $A \cap K$ provides a record of this quantity.

In summary, the application of the Bayesian theory of subjunctive conditionals to static decision problems shows two things. The first is that nothing like the full theory of subjunctives is required here. Rather than a whole family of partitions, only one partition is required. The second is that the appropriate partition of states is already familiar from decision theory, and subjunctive considerations are already addressed implicitly by a theory like that of Savage.

V. NORMAL FORM GAMES

In a game in strategic normal form, each player faces a static decision problem. Players are again interested in subjunctive conditionals of the form:

If I were to do act A, *then consequence* C *would ensue.*

For each player, the appropriate background partition for her conditionals is that generated by the other players' acts. One player's state consists of the acts of all the other players. If we confine ourselves to pure states, act together with state determine consequence as specified in the payoff matrix.

Here, each of the n players has a subjective probability space. The points in the space must specify (at least) an n-tuple pure acts of the n players. The space is thus the same for all players, although different players may have different degree-of-belief probability measures on the space and different act, state, and consequence partitions of the space. For example, consider Prisoner's Dilemma with a clone. As in Newcomb's problem, the players may perhaps each believe her own act is diagnostic of the other player's act, without causing it. The spurious correlation disappears conditional on the states as before, and the dominant act gets higher expected value. After the players become almost certain that they will defect, each will properly assert "If I were to cooperate I would get less."

What about mixed acts? When we allow players to choose mixed acts in a game we are, strictly speaking, imbedding that game in a larger game where players' choices are of settings on random devices which execute the act. Here the conditionals of interest are:

If I were to choose this mixed act, consequence C would ensue.

The appropriate partition is that generated by other players' mixed acts. The appropriate points, or possible scenarios, in the space must specify two things. The first is a profile of mixed acts chosen by all the players, The second is a vector of the pure acts realized by the chance processes. The second component determines the vector of payoffs. Here we no longer have determinism.

For example, suppose two players are at the mixed equilibrium <1/2,1/2> in Matching Pennies:

1, -1	-1, 1
-1, 1	1,-1

and consider the four subjunctive conditionals:

If Row plays Up, he will get 1. If Row plays Down, he will get 1.
If Row plays Up, he will get -1. If Row plays Down he will get -1.

If the true member of the background partition Column plays 1/2, so there the objective value of the conditionals – the conditional chance of consequent given antecedent – is 1/2. If the players know that they are at equilibrium, the assertability values of the conditionals are also 1/2 and the expected values of Up and Down for Row are both 0.

Again we do not need a whole family of partitions (or the induced judgements of closeness of possible worlds). For the conditionals of interest, a single partition for each player is sufficient, and these partitions are generated naturally from the structure of the game.

VI. Extensive Form Games

In extensive form games for the first time we will need a (partial) family of partitions to deal with conditionals with antecedents of different strengths. Here the conditionals of interest are of the form:

IF (initial segment of a path through the game tree) THEN …

For initial segments of different lengths different partitions are appropriate. However, here the structure of the game tree guides us to a natural partial family of partitions.

This family assigns to the tautology the partition generated by the players' possible behavioral strategies. Each cell of the partition corresponds to a profile of behavioral strategies, and that profile determines the conditional chances for every point within that cell. Since the tautology supplies no information relevant to the chances, the partition supplies all relevant information.

The family assigns to a proposition specifying an initial segment of a path through the game tree the partition generated by possible behavioral strategies in the subtree continuing that initial segment. Again, the condition articulated in the antecedent of the conditional together with a cell in the corresponding partition specifies the information relevant to determining the chances of propositions about how that initial segment will be continued. The points in the relevant probability space – the possible scenarios – must specify (at least) two things: (1) a profile of players' behavioral strategies and (2) a path through the game tree. Points in the probability space may specify more, for example everything about the players' degrees of belief, degrees of belief about their degrees of belief, etc. Such rich conceptions of the possible scenarios, can be useful in formulating assumptions about the players – for instance about their knowledge, beliefs, or optimizing behavior – which restrict the possible profiles of behavioral strategies, and thus both the possible conditional chances and the possible actual plays.

Although one requires a family of partitions here (as would in a sequential one-person decision problem), it is not a total family. It is a partial family which arises naturally from the extensive form structure of the game.

VII. CONCLUSION

The Bayesian theory of subjunctive conditionals has two advantages over Stalnaker selection-function accounts in applications to games and decisions. First, it gives a more general account wherein conditional chances need not be 0 or 1. Stalnaker's theory can be recovered in the special case of determinism. The more general theory has natural application when mixed strategies come into play. In the second place, it takes as primitives families of partitions rather than selection functions or similarity orderings. These primitives already arise naturally in the theories. Furthermore, in static decisions and games in strategic normal form one needs only a single partition. In sequential decision problems and games in extensive form one does need a partial family of partitions, but it is one which arises naturally from the game theoretic structure. Thus, the most problematic aspects of the general theory are left behind and compelling core notions of a natural partial family guide the application of the theory.

I have not said anything yet about the interaction of this theory of conditionals with notions of knowledge, common knowledge and rationality. Nor have I addressed the questions of the foundations of backward and forward induction. These issues will be the central topic of another essay.

REFERENCES

R.J. Aumann, "Backward Induction and Common Knowledge of Rationality", in: *Games and Economic Behavior*, 8, 1995, pp.6-19.

C. Bicchieri, "Strategic Behavior and Counterfactuals", in: *Synthese*, 76, 1988, pp.135-169.

K. Binmore, "Modeling Rational Players I and II", in: *Economics and Philosophy*, 3, 1987, pp.179-214 and 4, 1988, pp.9-55.

B. Chellas, "Basic Conditional Logic", in: *Journal of Philosophical Logic*, 4, 1975, pp.133-228.

A. Gibbard and W. Harper, "Counterfactuals and Two Kinds of Expected Utility", in: C.A. Hooker et al. (eds.), *Foundations and Applications of Decision Theory*. Dordrecht: Reidel 1978.

J. Harsanyi and R. Selten, *A General Theory of Equilibrium Selection in Games*. Cambridge, Mass.: MIT Press 1988.

R. Jeffrey, *The Logic of Decision*. New York: McGraw Hill 1965.

R. Jeffrey, *The Logic of Decision*. 2nd rev., Chicago: University of Chicago Press 1983.

D.K. Lewis, *Counterfactuals*. Oxford: Oxford University Press 1973.

D. Nute, *Topics in Conditional Logic*. Dordrecht: Reidel 1980.

J. Pollock, *Subjunctive Reasoning*. Dordrecht: Reidel 1976.

P. Reny, "Rationality in Extensive Form Games", in: *Journal of Economic Theory*, 59, 1993, pp.627-649.

L.J. Savage, *The Foundations of Statistics*. New York: Wiley 1954.

R. Selten and U. Leopold, "Subjunctive Conditionals in Decision and Game Theory", in: W. Stegmüller et al. (eds.), *Philosophy of Economics*. Berlin: Springer 1982, pp.192-200.

B. Skyrms, *Pragmatics and Empiricism*. New Haven: Yale University Press 1984.

B. Skyrms, "Conditional Chance", in: J. Fetzer (ed.), *Probability and Causality*. Dordrecht: Reidel 1988, pp. 161-178.

R.C. Stalnaker, "A Theory of Conditionals", in: N. Rescher (ed.), *Studies in Logical Theory*. Oxford: Blackwell 1968, pp. 98-112.

R.C. Stalnaker, "Letter to Brian Skyrms", in: E. Eells and B. Skyrms (eds.), *Probability and Conditionals: Belief Revision and Rational Decision*. Cambridge: Cambridge University Press Cambridge 1994, pp.27-29.

R.C. Stalnaker, "Knowledge, Belief and Counterfactual Reasoning in Games", in: C. Bicchieri et al. (eds.), *The Logic of Norms*. New York: Oxford University Press (forthcoming).

P. Woodruff, *Notes on Conditional Logic*. Working paper University of California, Irvine, 1969.

P. Woodruff, *Partitions and Conditionals*. Working paper University of California, Irvine 1996.

Department of Philosophy
University of California at Irvine
Irvine CA 92717
USA

ISAAC LEVI

PREDICTION, BAYESIAN DELIBERATION AND CORRELATED EQUILIBRIUM

In a pair of controversy provoking papers[1], Kadane and Larkey argued that the normative or prescriptive understanding of expected utility theory recommended that participants in a game maximize expected utility given their assessments of the probabilities of the moves that other players would make. They observed that no prescription, norm or standard of Bayesian rationality recommends how they should come to make probability judgments about the choices of other players. For any given player, it is an empirical question as to whether other players are Bayes rational, judge him to be Bayes rational, etc., just as it is an empirical question as to what the other players' goals and beliefs are. Participants in the game should, of course, use all empirical information available to them about how other players have behaved and how they are likely to behave in the given context of choice. Kadane and Larkey pointed out that rational players could participate in a game meeting these requirements without the game's terminating with a Nash equilibrium or a result satisfying any other particular solution concept.

Kadane and Larkey did not deny that one could study how a participant in a game should make decisions if he made special assumptions about the other players - such as that they are expected utility maximizers. They emphasized, however, that no principle of rationality mandates that one rational player should believe that his opponents are rational. Consequently, unless it is true as a matter of empirical fact that participants in games who are expected utility maximizers have opinions that lead them to make choices satisfying the conditions of Nash equilibrium or some other such solution concept, the relevance of such solution concepts to either the making of recommendations to such participants or the descriptive study of their behavior is less obvious than it has seemed to many to be.

Consider, for example, a garden variety prisoner's dilemma as in Figure 1.

2, 2	0 ,6
6, 0	1, 1

Figure 1

If Row is an expected utility maximizer, Kadane and Larkey recommend that he should make as careful an estimate as he can, given his information of what

W. Leinfellner and E. Köhler (eds.), Game Theory, Experience, Rationality, 173–185.

Column is likely to do conditional on what he, Row, chooses; compute the expectations of his options; and choose the option maximizing expected utility. Row is supposed to be in a position to do this prior to actually reaching a decision when the determination of which option maximizes expected utility according to Row's values and beliefs can be of use to Row in reducing the set of options available to him (there are two) to a set of admissible options. Consider the following two versions of what Row and Column might think about each other prior to deciding what to do.

> *Version 1*: Row judges that what Column will do is probabilistically independent of what Row will do. In that case, it does not matter what specific probabilities Row assigns to Column's choices. The expected reward conditional on choosing Bottom is greater than the expected reward conditional on choosing Top whatever Row's probability judgments about Top's choices might be. If Column makes similar judgments about what Row will do, Column should choose right and the payoffs are 1,1. Row and Column have achieved a Nash Equilibrium.

But a coherent expected utility maximizer might make different estimates of what Column will do.

> *Version* 2: Row might judge that the probability that Column will choose left conditional on Row's choosing top and the conditional probability that Row will choose right, conditional on Column's choosing bottom, are both near 1. In that case, Row *should* choose Top. The expected utility conditional on choosing Top prior to making a choice is near 2 whereas the expected utility conditional on choosing Bottom is near 1. If Column's coherent estimates of what Row will do are probabilistically dependent on Row's choice in a parallel fashion, Column *should* choose left and the payoffs will be 2,2. Column and Row can both think about each other in this way and judge each other to be expected utility maximizers. Given that they do, they ought, according to expected utility theory, to make choices that fail to yield a Nash equilibrium.

In both versions of the prisoner's dilemma, Row and Column have rational beliefs and goals and their decisions are Bayes rational. Sometimes Bayes rational decisions lead to equilibrium and sometimes they don't.

Aumann's 1987 paper on correlated equilibrium is an explicit response to the scepticism registered by Kadane and Larkey regarding the use of solution concepts in game theory.[2] Aumann declares his aim to be to show that "the notion of equilibrium is an unavoidable consequence" of the Bayesian view of the world.[3] He claims to show that Bayes rational players choose options in a way that realizes a "correlated equilibrium". Given certain assumptions about the attitudes of the players that Aumann holds to be reasonable, Column and Row must satisfy the conditions for a correlated equilibrium (that is a certain kind of generalization of the notion of a Nash equilibrium). In particular, the following assumptions are made[4]:

> A) *Joint Action Space Prior Assumption*: Prior to deciding which of the options available to him to exercise, each player i has a subjective or credal probability distribution π_i over hypotheses fully describing "states of the world including specification of the actions he and his opponent will make. In both versions of the prisoner's dilemma, there are four states in the action space, and, prior to reaching a decision, column (row) has a probability distribution π_c (π_r) over these four states.
>
> B) *Common Knowledge of Rationality*: Prior to making their decisions, it is common knowledge among the players that each player will choose in a Bayes rational manner.
>
> C) *Common Prior Assumption*: $\pi_c = \pi_r$.
>
> D) *Ratifiability Assumption*: (to be explained later).

If assumptions (A)–(D) are endorsed, the choices of Column and Row must satisfy the conditions for a correlated equilibrium. I have no quarrel with the elegant argument Aumann offers in favor of his claim. Aumann's conclusions do, indeed, follow from his premisses. But I do not believe that his conclusions refute the Kadane-Larkey thesis, because each one of these assumptions is untenable.

I shall first explain the content of what I am labelling the Ratifiability Assumption. My task will then be to argue that either the Ratifiability Assumption offers advice incompatible with the Bayesian dictum that rational agents choose options from those available to them that maximize expected utility relative to their probability and utility judgments, or the roster of options available to a decision maker, must satisfy some very restrictive and implausible conditions. According to the first alternative, players who reach correlated equilibrium are not Bayes rational. According to the second, they may be Bayes rational under certain restricted circumstances but there will be other circumstances where Bayes rational players will not end up in a correlated equilibrium. These conclusions obtain whether or not assumptions (A)–(C) hold. Aumann's endorsement of Ratifiability alone prevents him from responding relevantly to views like those of Kadane and Larkey.

The discussion of ratifiability raises the issue of conditions under which options are available to players in a game. I shall argue that both assumptions (A) (and its corollary (C)) and (B) conflict with conditions on judgments of the availability of options. (A) and (B) both fail, because prior to reaching a decision, a rational agent should not predict what he will do even though he may predict what others will do.

According to the Bayesian injunction to maximize expected utility, player i should choose as if he had considered every option σ available to him *prior to* reaching a decision and evaluated the expected value or payoff $E(\sigma)$ to *i using all the information available to him at that time.* This expected value is computed by multiplying the value of the outcome of *i*'s choosing σ while the $n - 1$ other players make a combination of moves s for every possible combination, by the probability $\pi(\mathbf{s}/\sigma)$ of that combination of moves occurring conditional on *i*'s choosing σ. σ is admissible according to the injunction to maximize expected utility if and only if $E(\sigma) \geq E(\tau)$ for every available option τ.

This version of the injunction to maximize expected utility is an application of equation (3.5) for computing expected relative value for P. Fishburn's "model 1" or "basic decision model."[5] As Spohn[6] notes, Fishburn's model 1 provides the most general formulation of the injunction to maximize expected utility advocated by Bayesians. To be sure, the task of furnishing a method of deriving the requisite probabilities and utilities from preferences among options is not fulfilled by specifying equation (3.5).[7] But as Spohn correctly observes, the question of the adequacy of expected utility theory as either a prescriptive or a descriptive theory does not stand or fall with the availability of an all purpose theory of elicitation from preferences among acts.

Having identified an admissible option, an agent *i* in control of himself will implement the recommended σ and regard the matter as settled, having become convinced that σ is or will be implemented. In Aumann's language, he discovers in which element of his "information partition" the actual state of the world is located. Updating his prior probability distribution over the joint action space to obtain a posterior distribution over the $n - 1$ tuples of hypotheses concerning the combinations s of acts adopted by the other players given the option he has chosen, *i* can compute the expected value of his situation after having made his decision. Call this the *Bayes posterior expectation* $E_\sigma(\sigma)$ for the option σ once *i* has chosen σ and knows it. This expectation is computed using the probability distribution $\pi_\sigma(\mathbf{s})$ over hypotheses concerning what other players will choose relative to the information available to i after he has chosen σ and become certain that this move will be implemented. Conditionalization on the information that σ is implemented implies that $\pi_\sigma(\mathbf{s}) = \pi(\mathbf{s}/\sigma)$. On the assumption that *i* formed this expectation by conditionalizing on the information that σ is chosen and implemented, the Bayes posterior expectation $E_\sigma(\sigma)$ for σ once σ has been chosen should be equal to the Bayes expectation $E(\sigma)$ for σ as estimated prior to making a decision and coming to know which option as chosen.

i can also compute the expectation $E_\sigma(\tau)$ for another available option τ using for the computation $\pi_\sigma(\mathbf{s})$ and the payoffs for choosing τ instead of those for using σ. This expectation does not represent the expected value $E_\tau(\tau) = E(\tau)$ of τ when *i* has chosen τ and knows it unless $\pi_\tau(\mathbf{s}) = \pi(\mathbf{s}/\tau) = \pi(\mathbf{s}/\sigma) = \pi_\sigma(\mathbf{s})$. If this condition is satisfied for all available options, then hypotheses concerning what other agents will do are probabilistically independent of agent *i*'s choices according to *i*'s prior credal probability judgments (as in version 1 of the prisoner's dilemma). In that event, the following two conditions are equivalent:

Maximization of conditional expected utility: σ is admissible among the available options if and only if $E(\sigma) \geq E(\tau)$ for every available option τ.

Ratifiability: σ is admissible among the available options if and only if $E_\sigma(\sigma) \geq E_\sigma(\tau)$ for every available option τ.

Aumann stipulates that admissible options should satisfy the Ratifiability condition. This stipulation that options satisfying the Ratifiability condition are Bayes rational is the Ratifiability Assumption (D) announced above. The term "ratifiable" comes from Jeffrey.[8] Unlike Jeffrey, however, Aumann claims that Bayes rational choices are ratifiable choices.[9]

As just noted, this claim is noncontroversial in contexts where states are probabilistically independent of acts as is required in the proposals made by Savage and Anscombe and Aumann.[10] Such independence does not hold in the second version of the prisoner's dilemma and in many other predicaments, including cases that Aumann recognizes as realizing correlated equilibrium. Yet, Aumann seems to think that satisfying ratifiability is a necessary (and perhaps sufficient) condition for Bayes rational choice even when probabilistic independence fails. How can this requirement be squared with the demand that conditional expected value should be maximized in such cases? Maximizing conditional expected utility is a core requirement of the mainstream Bayesian tradition, as the views of Luce and Krantz, Fishburn, Spohn suggest.

A response is available to Aumann if cases of probabilistic dependence in games are treated as analogous to cases of moral hazard as understood by authors like Drèze.[11] As Drèze understands moral hazard, the decision maker faces a choice between strategy-act pairs where an act from set A is representable as a function from states to (roulette lotteries) over prizes and a strategy is an action that modifies the probability distribution over the states by selecting one distribution from a given set Π. The options available to the decision maker are all (consistent) strategy-act pairs from $\Pi \times A$. If the purchaser of fire insurance has the opportunity of insuring his property for more than its market value, he may consider whether or not also to take steps to increase the chance of a fire taking place. The option of purchasing the given policy decomposes into two components: (a) purchasing the policy (an act) and (b) adopting a policy related to fire prevention or promotion controlling the chances of fire (a strategy). Options that maximize conditional expected utility among these strategy-act pairs can be determined as follows: (1) For each a in A, let $\Pi(a)$ be the set of strategy-act pairs $\langle \pi, a \rangle$ such that π maximizes the expected utility for a among all strategies in Π. (2) Choose any $\langle \pi, a \rangle$ in a $\Pi(a)$ for which the expected utility is a maximum for all members of A.

In all such cases an optimal act relative to a strategy is going to meet the formal conditions for ratifiability. Maximizing conditional expected utility is also determinable as follows: (1′) For each π in Π, let $A(\pi)$ be the set of acts that maximize expected utility relative to π among elements of A. (2′) Choose a pair $\langle \pi, a \rangle$ belonging to an $A(\pi)$ that maximizes expected utility among strategies or distributions in Π. In order that the Ratifiability Assumption be consistent with the injunction that expected utility be maximized, each element of A should be available relative to each distribution in Π.

Whereas Drèze envisages the two components of the strategy-act pair as chosen independently of one another, Aumann thinks of the set Π of strategies as a set of probability distributions over hypotheses as to what the other agents will do, where each distribution π_i is conditioned on information e_i from a set E of possible outcomes of an experiment to be made available to the decision maker at the moment of decision. Once the agent finds out which e_i in E is true, he chooses among the available options. In so doing he is choosing a strategy-act pair. The decision maker has no choice of an element of Π, so step (2′) is inoperative. But step (1′) remains applicable.[12]

The scenario envisaged in the second version of the prisoner's dilemma is a predicament where hypotheses about what others will do (the states) are probabilistically dependent on acts. Nonetheless, the predicament cannot be understood coherently as a case of moral hazard in the sense of Drèze. Nor can maximizing conditional expected utility be made to fit the requirements of ratifiability as Aumann understands it. Suppose that ρ_i is Row's probabilistic prediction as to what Column will do, conditional on Row's choosing top, and ρ_2 is Row's probabilistic prediction, conditional on Row's choosing bottom. ρ_1 assigns 0.9 to Column's choosing left and 0.1 to Column's choosing right and ρ_2 assigns 0.1 to Column's choosing left and 0.9 to Column's choosing right. Let the set of strategies Π be the convex hull of these two probability distributions or just these two distributions. If version 2 of the prisoner's dilemma were analyzable as a case of moral hazard after the fashion of Drèze, $\langle\rho_i$, top$\rangle$ and $\langle\rho_i$, bottom$\rangle$ for $i = 1,2$ are options available to Row. Choosing bottom would be optimal against both distributions and, hence, is optimal among all available options. But maximizing conditional expected utility recommends that Row choose top in version 2 of the prisoner's dilemma. That is because Column simply does not have as options available to him $\langle\rho_1$, bottom$\rangle$ and $\langle\rho_2$, top$\rangle$ in version 2 but only $\langle\rho_1$, top$\rangle$ and $\langle\rho_2$, bottom$\rangle$. Of these two, $\langle\rho_1$, top$\rangle$ maximizes conditional expected utility. That is to say, the set of options available in version 2 does not satisfy the requirements on the availability of options suggested by Drèze for moral hazard or those suggested by Aumann's version of the Ratifiability Assumption.

Thus, understanding ratifiability as consistent with maximizing conditional expected utility is purchased at the expense of ruling out as incoherent decision problems where decision makers face options as specified in version 2. But where is the argument for this?[13]

For my part, I think version 2 is coherent. So is version 1. We need adequate analyses of both of them. For Bayesians, the Drèze and Aumann approaches fail to account for version 2. This consideration alone suffices to call into question the Ratifiability Assumption (D) and with it Aumann's claim to have shown that the play of Bayes rational agents should reveal commitment to correlated equilibrium distributions as priors.

Perhaps, however, a more compelling and Bayesian argument may be invoked to rule out version 2. Row knows that Column will choose rationally by the Common Knowledge of Rationality Assumption (B). He also is convinced that Column judges that Row will choose top conditional on Column's choosing left with high probability and likewise with bottom conditional on right; because Row believes that Column shares a common prior with him and that Column's conditional probabilities are as specified in version 2. On this basis, Row should be certain that Column will choose left. But none of the probability distributions in the set Π for Row assigns probability 1 to Column's choosing left. Here Row's convictions about Column's predictions concerning Row's choices, together with Row's assumption that Column will choose rationally rule out version 2. Ratifiability need not be assumed as a condition of Bayes rational decision making. Version 2 can be ruled out by the Common Knowledge of Rationality Assumption. Perhaps Aumann's invoking ratifiability was unnecessary in the first place.

> Not so! Consider the following version of the prisoner's dilemma.
>
> *Version 3*: Suppose Row judges that Column will play left given that Row plays top with probability 0.75, and Column will play left given that Row plays bottom with probability 0.1; and Column judges that Row will play top given that Column plays left with probability 0.75 and Row will play top given that Column plays bottom with probability 0.1; and Row and Column know this of each other. Then Row will judge that rational Column regards both choosing left and choosing right as maximizing expected utility and Column will pass a corresponding judgment regarding rational Row. If maximizing expected utility is all that is required for Bayes rationality, Row might well assign equal subjective probability to Column's choosing left and choosing right. But the resulting joint distribution is not a correlated equilibrium distribution precisely because choosing top, though optimal, is not ratifiable.

In version 3 where more than one of Column's options is Bayes Rational, Row's certainty that Column will choose rationally does not suffice to warrant certainty that Column will choose ratifiably. Common knowledge of rationality and a common prior without ratifiability does not yield a correlated equilibrium distribution prior. If we are to persuade Bayesians of the importance of equilibrium solution concepts for games, we face the daunting (in my judgment insuperable) task of convincing them of the obligatory character of ratifiability as a condition of rational choice.

Aumann proves the general claim that, if in an n-person game there is common knowledge of Bayes rationality with ratifiability built in (so that assumptions B and D obtain) and the Common Prior Assumption C is operative (so that the

Joint Action Prior Assumption A holds as well), the common prior must yield a correlated equilibrium. This is the equilibrium concept that Aumann insists is inescapable as an ingredient in Bayes views of rationality. He offers it as a response to the Kadane-Larkey skepticism regarding the importance of equilibrium concepts in game theory from a Bayesian point of view.[14]

Although Aumann's formulation of standards of Bayes rationality (i.e., ratifiability) is retrospective so that it applies to the evaluation of options by agents once the choice has been made, his very argument for correlated equilibrium requires that options be assessable prospectively as well. Prospective prescriptions concerning what the agent ought to choose are made given the information *available to the decision maker* prior to actually choosing and implementing a course of action. We assume that the agent has sufficient computational capacity or logical omniscience and self knowledge of his beliefs and goals to be able to compute expected utilities of his options in the sense of expected utility, relevant to identifying policy recommendations in the state prior to choice. The agent is supposed to then compare his evaluations with respect to expected utility for all the options the agent committed to judging has available to him.[15]

Observe, however, that if an agent is certain prior to choice that he will not implement a certain policy, then from that agent's point of view at that time the policy is not optional even if we outside observers think the agent should not have been so certain. What we think the agent ought to believe is not relevant to our assessing the rationality of his judgments, as long as his beliefs meet the standards of weak rationality that theories of expected utility require. If an agent is certain that he will not perform a certain course of action, the agent is obliged to rule out choosing that course of action as available for him as an option. It is not irrational for him, therefore, to fail to choose that course of action even if he judges that, if it were an available option for him, it would have been better for him to choose it than what, given what he believes, is best.

Recall that, retrospectively, options need to be ratifiable to be rationally chosen according to Aumann. Prospectively, there must have been a prior that is a correlated equilibrium distribution. In the case of the prisoner's dilemma, only version 1 is allowed and in that case, both Row and Column must be certain as to which option they will pick prospectively – i.e., prior to making a decision. But in that case, neither Row nor Column has a decision to make. Each has but one available option. Row does not have top available to him as an option and Column does not have left. At best we can say that deliberation aimed at identifying an admissible subset of options from among the available one's given the decision maker's goals and beliefs is vacuous. It recommends what is going to happen anyhow. In that case, there is no need for Aumann to invoke ratifiability in order to compare a given option with alternatives. There is only one option available.

There are gamelike situations where maximizing conditional expected utility ratifiably allows more than one option to be consistent with the decision maker's information prior to reaching a decision. (Version 3 of the prisoner's dilemma is an example.) However, the Assumption of Common Knowledge of Bayes Rationality (Assumption (B)) implies that each decision maker is certain that he will choose rationally so that from the decision maker's point of view every option available to him is Bayes rational. In this setting, the player cannot appeal to Bayes rationality to reduce the set of options available to him to an admissible proper subset for choice. Such reduction is always vacuous. Game theory is often advertised as taking into account how agents do or should reason to a decision taking into account the reasoning of others. Assumption (B) precludes such reasoning from leading to anything but the vacuous recommendation: Choose any one of the options available to you!

If Row and Column are to use the information available to them prior to choice to deliberate as to which of their respective options to choose, they certainly ought to have some idea of what their standards for evaluating options are and should be able to identify which of their options are the Bayes rational choices. But they need not assume that they will choose rationally and implement the Bayes rational recommendations. Indeed, if they do not do so, then they are not to that extent prevented from recognizing acts other than Bayes rational acts as available to them. But if a player does not take for granted that he will act rationally, he cannot have common knowledge that all players will act rationally. An outside observer can assume that all players will act rationally but, if deliberation is not to be vacuous, cannot attribute to the players' conviction that they personally will act rationally. Thus, in addition to doubts raised about the Ratifiability Assumption (D), the assumption of common knowledge of Bayes rationality (B) has been called into question. And it is called into question without directly objecting to the Ratifiability condition.

Assumption (A) – The Joint Action Space Prior Assumption – implies that each player assigns unconditional credal probabilities to each of his options prior to choice. However, this too leads to trouble, as W. Spohn has already observed.[16] As I have argued,[17] allowing unconditional probabilities to be assigned to options as Assumption (A) requires leads to the same embarrassing vacuous recommendation as endorsing Assumption (B) does. The rational agent must be certain prior to choice that he will choose rationally. The principles of rational choice cannot be used by the deliberating agent as standards for criticizing and evaluating his options before deciding. To avoid returning to Assumption (B), we must reject Assumption (A) as well.

Since assumption (C) – the Common Prior Assumption – presupposes Assumption (A), it too is called into question.

The objections raised against (A)–(C) apply whether one insists on the Ratifiability Assumption (D) at the outset or not. No decision maker should be certain that he will choose rationally. No decision maker should assign unconditional probabilities to hypotheses as to what he will choose. Otherwise deliberation is vacuous. It is not enough to reject correlated equilibrium distributions as priors over joint action spaces. One should abandon the idea of priors over such joint spaces altogether.

Aumann himself addresses the issue of treating "personal choice as a state variable".[18] The difficulty he addresses is that treating personal choice as a state variable deprives each participant in the game of freedom of choice. He rightly sees that a game theorist or "outside observer" could regard the participants' decisions as state variables without depriving any participant of "freedom of choice" *precisely because the outside observer is not a participant in the game and is not in the business of predicting his own choices.*

This correct observation does not, however, address the point under discussion here. If the outside observer attributes minimal rationality to the players of the game, he should not model them as endorsing unconditional probability distributions over their own acts. He can make predictions as to how each of them will choose; but no participant in the game should make predictions as to how he will choose or whether he will choose rationally and the outside observer, who assumes that each agent is rational, should not attribute such predictions to participants. But if the outside observer is going to derive a correlated equilibrium from his modelling assumptions and Aumann's correlated equilibrium theorem, the outside observer must assume that each of the participants has an unconditional probability distribution prior to choice over the joint space of acts for of the players. My contention is that if the outside observer does this, he is committed to denying that the players are capable of using the principles of rational choice to reduce the set of available options to a proper subset of admissible ones.

The issue is not whether an outside observer can regard the participants' decisions as state variables while taking these agents to be free to choose. According to those variants of so called "soft determinist" doctrine with which I am sympathetic, an outside observer can do this. But my weakened version of Spohn's thesis does not contradict such soft determinism. I question whether agent *X* can regard himself as free to choose in the sense relevant to his deliberation and regard his choice as a state variable. If *X* cannot coherently do so, the outside observer should not model *X* or any other player in the game as doing so.[19]

Aumann agrees with Kadane and Larkey that models of games should cohere with expected utility theory. He complains, however, that Kadane and Larkey "ignored the fact that a rational player must take into account how other players reason about him". In my opinion, the shoe is on the other foot. As Kadane and Seidenfeld illustrate in subtle and interesting detail,[20] player *i* can utilize whatever information *i* may have about the reasoning of other players concerning what *i* will do in making predictions as to how the other players will choose. If *i* is rational, however, not only should *i* avoid predicting his own choice; but in reasoning about how other rational agents reason about him, he should not suppose that they attribute to him predictions about his own choices. If Spohn's thesis is correct (as it certainly appears to be), it is Aumann and not Kadane and Larkey who ignores "the fact that a rational player must take into account how other players reason about him."

NOTES

1. J.B.Kadane and P.D. Larkey, "Subjective Probability and the Theory of Games," *Management Science* 28, 1982, pp.113-20 and "The Confusion of Is and Ought in Game Theoretic Contexts", Management Science 29, 1983, pp.1365-79.
2. R. Aumann, "Correlated Equilibrium as an Expression of Bayesian Rationality", in: *Econometrica*, 55, 1987, pp.1-18.
3. *op. cit.*, p.2.
4. Throughout this discussion I will consider games between two players Column and Row or *c* and *r*. However, the issues raised generalize easily enough to games between *n* players.
5. P.C. Fishburn, *Decision and Value Theory*, New York: Wiley 1964, p.43.
6. W. Spohn, "Where Luce and Krantz Do Really Generalize Savage's Decision Model", in: *Erkenntnis*, 11, 1977, pp.123-34.
7. For efforts to provide the desired theories of elicitation, see M. Balch, "On Recent Developments in Expected Utility Theory", in: M. Balch, D.McFadden and S.Wu (eds.), *Essays on Economic Behavior Under Uncertainty*, Amsterdam: North Holland 1974, pp.45-54; M.Balch/P. Fishburn, "Subjective Expected Utility for Conditional Primitives", in M.Balch et al. (eds.), *op.cit.*, pp.57-69; P.Fishburn, "On the Foundations of Decision Making Under Uncertainty", in M. Balch, et al. (eds.), *op. cit.*, pp.25-44; Luce/Krantz, "Conditional Expected Utility", in: *Econometrica*, 39, 1971, pp.253-71 and the comments by Spohn, *op.cit.*
8. R.C. Jeffrey, *The Logic of Decision*, 2nd ed., Chicago: Chicago University Press.
9. For Jeffrey, ratifiability is a supplement to the default conditional expected utility theory he favors. It applies when hands tremble - i.e., the agent is sure one option is chosen but is unsure as to which will be implemented. Aumann considers cases where the agent has chosen and is certain that his choice is (or will be) implemented. So the conditional expectations of the options not chosen are, for Aumann, "counterfactual" or, more accurately, "belief contravening". (See Aumann, *op.cit.*, p.7.) In spite of these differences, Aumann's criterion of Bayes rationality is formally similar to Jeffrey's notion of ratifiability in the respects spelled out in the text. Shin restricts the term "ratifiability" to trembling hand cases. In H.S. Shin, "Two Notions of Ratifiability", in: M. Bacharach/S. Hurley (eds.), *Foundations of Decision Theory*, Oxford: Basil Blackwell 1991, pp.242-62, Shin identifies a species of (modestly) ratifiable distributions with correlated distributions. In H.S. Shin, "Counterfactuals and a Theory of Equilibrium in Games", in: C. Bicchieri/M.L. Dalla Chiara (eds.), *Knowledge, Belief and Strategic Interaction*, Cambridge: Cambridge University Press 1992, pp.397-413, he discusses the use of closest world counterfactual reasoning in order to obtain correlated equilibria. He is noncommittal as to whether the "trembling hand" or counterfactual interpretation better captures Aumann's view even though Aumann explicitly endorses the latter. He is tempted (*op.cit.*, p.413) to equate his reconstruction of Aumann in terms of trembling hand ratifiability with his counterfactual closest world reconstruction except for the fact that the closest world counterfactual account does not sit well within the Bayesian framework - as R.C. Jeffrey rightly points out. I argue later on that a Bayesian should not assign unconditional probabilities to action statements (as Jeffrey and Shin do) in contexts where choosing to implement one of them is optional for him. Even so, Shin is right to recognize that closest world accounts of conditionals are incompatible with Bayesianism. As explained later in the text, there is an account of ratifiable decision making that appeals neither to trembles nor to nearest world counterfactual reasoning and is compatible with maximizing conditional expected utility. As explained later in the text, there is an account of ratifiable decision making that appeals neither to trembles nor to nearest world counterfactual reasoning and is compatible with maximizing conditional expected utility. So is the account of counterfactual reasoning I favor (See footnote 7). It is possible to interpret Aumann as charity requires - namely, as an orthodox conditional expected utility maximizer. This shall be explained shortly. I thank B. Skyrms for directing my attention to Shin's discussion of Aumann's version of ratifiability. Space does not permit a review of the literature on ratifiability - especially the

extensive discussion in Skyrms, "Ratifiability and the Logic of Decision", in: *Midwest Studies in Philosophy*, XV, 1990, pp.45–56.

10. L.J. Savage, *Foundations of Statistics*, New York: Wiley, 1954; 2nd revised edition, New York: Dover 1972; and F.J. Anscombe/R.J. Aumann, "A Definition of Subjective Probability", *Annals of Mathematical Statistics*, 35, 1963, pp.199–205.
11. J.Drèze, *Individual Decision Making Under Partially Controllable Uncertainty*, unpublished Ph.D. thesis, Columbia University 1958; and *Essays on Economic Decisions Under Uncertainty*, Cambridge: Cambridge University Press 1987.
12. Notice that on the assumption that the decision maker is Bayes rational, the conditional probability concerning what other players will do given the act chosen by the agent is equal to the conditional probability concerning what the other players will do given the information e_i.
13. It does not matter whether version 2 predicaments are called "prisoner's dilemmas" or not. As I understand them, they are noncooperative single shot games where the players have not made any binding commitments beforehand. The players decide "independently" of one another in the sense that what one of them does is not controlled or influenced by what the other does. The players may believe all this, may believe that their partners are Bayes rational and yet regard in a Bayes coherent fashion that the decisions of their partners are probabilistically dependent on their own decisions. K. Binmore (*Fun and Games*, London: D.C. Heath 1992, 7.5.4) and other game theorists obviously disagree, for reasons I do not understand.
14. A correlated equilibrium distribution over the joint action space of all the players is one where for every player i and every option σ available to i according to the game assigned positive prior probability and every τ available to i according to the game, $E_\sigma(\sigma) \geq E_\sigma(\tau)$. Formal definitions are found in R.J. Aumann, *op. cit.*; and "Subjectivity and Correlation in Randomized Strategies", in: *Journal of Mathematical Economics*, 1, 1974, pp.67–96 together with some discussion of intended interpretations.
15. H.S. Shin, *op.cit.*., rightly observes that if an agent has already chosen an option, any comparison of the expected value of the option chosen with other options will have to be "counterfactual" or, more accurately "belief contravening". The agent needs to suppose that instead of choosing the option he chose he had chosen one of the other options. According to Bayesian approaches to decision making, the available options are the options available *according to the agent's perspective prior to making a decision*. So the suppositional reasoning of the agent invites him first to contract his state of full belief to the state he was in prior to reaching a decision and then expand by adding an option judged to be available relative to that *prospective* belief state. This kind of suppositional reasoning differs from the "closest world" reasoning Shin favors. It recommends maximizing act conditional expected utility from the point of view prior to choice just as Bayesian decision makers are accustomed to do. Whatever the merits of Shin's account of counterfactual reasoning in the context of game theory might be, it is *not* congenial with Bayesian thinking. If Aumann understands maximizing expected utility in the manner sketched by Shin, his paper is not a response to the views of Kadane and Larkey or other Bayesian decision theorists skeptical of the centrality of the concept of equilibrium. I have sought to interpret Aumann's use of ratifiability in a manner more congenial to what appear to be his purposes. For elaboration of an account of conditionals congenial with Bayesian decision theory see I. Levi, *For the Sake of the Argument*, Cambridge: Cambridge University Press 1996.
16. W. Spohn, *op.cit.* and *Grundlagen der Entscheidungstheorie*, Scriptor Verlag 1978.
17. I. Levi, *The Fixation of Belief and Its Undoing*, Cambridge: Cambridge University Press 1991.
18. R.J. Aumann, "Correlated Equilibrium as an Expression of Bayesian Rationality", pp.8–9.

19. R.J. Aumann and A. Brandenburger ["Epistemic Conditions for Nash Equilibrium", discussion paper #94 of the Center for Rationality and Interactive Decision Theory of the Hebrew University in Jerusalem 1994 (to be published in *Econometrica*)] examine conditions under which the "conjectures" of players in a game as to what other players will do constitute a Nash equilibrium. Whatever the significance of Nash equilibria in conjectures might be, it does not seem to bear on the issues raised by Kadane and Larkey. But it is worth comparing the assumptions used for the main theorems of this paper with the assumptions used in the correlated equilibrium paper. Theorem A claims that mutual knowledge of the payoff functions, rationality of the players and conjectures are sufficient for Nash equilibrium of conjectures in the 2 person case. Mutual knowledge of rationality presupposes that each player knows that he is rational and, hence, that the difficulty raised in the text arises. In the n-person case, theorem B states that in addition to the conditions just mentioned, a common prior is available, and that the conjectures are commonly known. So the problems related to our Assumptions (A) and (C) resurface. The common prior raises another complication as well. The scheme examined by Aumann and Brandenburger stipulates that the states over which the common prior is defined include specifications of each player's conjectures concerning what the other players will do. If an agent has a prior over such states, he must be uncertain when in that prior state as to what his conjecture is. But this means that *in that prior state*, he is unsure of his probabilities. As Savage. *op.cit.*, pointed out, this leads to incoherence.
20. J.B. Kadane/T. Seidenfeld, "Equilibrium, Common Knowledge, and Optimal Sequential Decisions", in: C.Bichhieri/M.L. dalla Chiara (eds.), *Knowledge, Belief and Strategic Interaction*, Cambridge: Cambridge University Press 1992, pp.27–45.

Department of Philosophy
Columbia University
New York, NY - 10027
USA

Peter J. Hammond

Consequentialism and Bayesian Rationality in Normal Form Games*

I. Introduction

In single-person decision theory, Bayesian rationality requires the agent first to attach subjective probabilities to each uncertain event, and then to maximize the expected value of a von Neumann–Morgenstern utility function (or NMUF) that is unique up to a cardinal equivalence class. When the agent receives new information, it also requires subjective probabilities to be revised according to Bayes' rule.

In social choice theory and ethics, Harsanyi (1953, 1955, 1975a, 1975b, 1976, 1978) has consistently advocated Bayesian rationality as a normative standard, despite frequent criticism and suggestions for alternatives. In game theory, however, Bayesian rationality is almost universally accepted, not only as a normative standard, but also in models intended to describe players' actual behaviour. Here too Harsanyi (1966, 1967–8, 1977a, b, 1980, 1982a, b, 1983a, b) has been a consistent advocate. In particular, his work on games of incomplete information suggests that one should introduce extra states of nature in order to accommodate other players' types, especially their payoff functions and beliefs. Later work by Bernheim (1984), Pearce (1984), Tan and Werlang (1988), and others emphasizes how subjective probabilities may be applied fruitfully to other players' strategic behaviour as well.

In the past I have tried to meet the social choice theorists' understandable criticisms of the Bayesian rationality hypothesis. To do so, I have found it helpful to consider normative standards of behaviour in single-person decision trees. In particular, it has been useful to formulate a surprisingly powerful "consequentialist" hypothesis. This requires the set of possible consequences of behaviour in any single-person decision tree to depend only on the feasible set of consequences in that tree. In other words, behaviour must reveal a consequence choice function mapping feasible sets into choice subsets.

Previous work (Hammond, 1988a, b, 1997a, b) has applied this consequentialist hypothesis to dynamically consistent behaviour in an (almost) unrestricted domain of finite decision trees. The only restriction is that objective probabilities must all be positive at any chance node. Then, provided that behaviour is continuous as objective probabilities vary, provided that there is state independence, and provided also that Anscombe and Aumann's (1963) reversal of order axiom is satisfied, it follows that behaviour must be Bayesian rational. Moreover, null events are excluded, so strictly positive subjective probabilities must be attached to all states of the world. Of course, these arguments do not really justify Bayesian rationality; they merely

W. Leinfellner and E. Köhler (eds.), Game Theory, Experience, Rationality, 187–196.

indicate that critics and proponents of alternative theories should go beyond discussions of simple "one-shot" problems and explain how to make sequential decisions in trees.

The aim of this paper and some associated work (Hammond, 1997c) is to extend the consequentialist hypothesis from single-person decision trees to n-person games. The appropriate extension appears to be the *consequentialist normal form invariance hypothesis*, with antecedents in von Neumann and Morgenstern (1944, 1953). Yet the results concerning consequentialist behaviour in single-person decision trees rely on being able to consider, if not necessarily a completely unrestricted domain of decision trees with a fixed set of states of the world, then at least one that is rich enough. In particular, a player i's preferences over the random consequences of two different strategies are revealed by forcing i to choose between just those two strategies. Now, when such alterations in the options available to an agent occur in a single-person decision tree, there is no reason to believe that nature's exogenous "choice" will change. But as Mariotti (1996) and Battigalli (1996) have pointed out, when such changes apply to an n-person game, they typically affect that player's anticipated behaviour in a way that makes other players want to change their strategies. Then, of course, it is illegitimate to treat these other players' strategies as exogenous.

In order to surmount this difficulty, Battigalli's (1996) comment on Mariotti's paper suggests introducing, for each player i whose subjective probabilities are to be determined, one extra player i^* who is an exact copy or clone of i. Player i^* faces a variable opportunity to bet on how players other than i will play the original game, but is unable to affect the consequences available to all the other players, including i. With this useful and ingenious device, player i^* can be faced with each possible single-person decision tree in turn. This allows i^*'s subjective probabilities over strategy profiles for players other than i to be inferred. Moreover, they apply to i^*'s behaviour when facing a single-person decision problem equivalent to that which i faces in the game itself. Because i^* is an exact copy of i, it follows that i's behaviour in the original game matches i^*'s in this equivalent single-person decision problem; in particular, i will maximize subjective expected utility using i^*'s subjective probabilities.

Hence, Battigalli's device can be used to provide a consequentialist justification for behaviour in n-person games to be Bayesian rational. There is a need, however, to attach strictly positive probabilities to all other players' strategies which are not ruled out as completely impossible and so irrelevant to the game. This suggests that strictly positive probabilities should be attached to all other players' rationalizable strategies, at least — i.e., to all those that are not removed by iterative deletion of strictly dominated strategies.

In the remainder of the paper, Section II reviews in somewhat more detail the earlier results characterizing consequentialist behaviour in single-person decision trees. Then Section III defines a "consequentialist" n-person normal game form G, in which payoffs are replaced by personal consequences, together with associated families $\mathcal{G}_i$. For each $i \in N$, these consist of game forms $G(i, T)$ which differ from

G in having an extra player i^* who is an exact copy of i and faces the decision tree T. Next, Section IV introduces three different kinds of "consequentialist" player type in order to describe each player's behaviour in every family $\mathcal{G}_i$. Last, Section V explains why the consequentialist hypotheses imply that there should be subjective probabilities for each player that are attached to the profiles of all other players' strategies.

II. CONSEQUENTIALIST SINGLE-PERSON DECISION THEORY: A BRIEF REVIEW

In single-person decision theory, the basic "consequentialist" hypothesis requires that actions should be evaluated purely on the basis of their consequences. More specifically, if two decision trees face the decision maker with identical feasible sets of consequences, then behaviour in those trees should generate, or reveal as "chosen", identical sets of consequences in the two trees. This consequentialist hypothesis has strong implications when applied to an unrestricted domain of finite decision trees involving a given *consequence domain* Y, on which behaviour is required to satisfy a mild "dynamic consistency" requirement.

The first and simplest result applies to finite decision trees T in the domain $\mathcal{T}_1(Y)$ which contain decision nodes and also terminal nodes having consequences in the specified domain Y. In any tree $T \in \mathcal{T}_1(Y)$, behaviour has sure consequences in the domain Y. On the unrestricted domain $\mathcal{T}_1(Y)$, consequentialist dynamically consistent behaviour must be *ordinal* in the sense that it both reveals and maximizes a (complete and transitive) weak preference ordering R over the feasible set of consequences. This can be proved either indirectly by means of Arrow's (1959) characterization of ordinal choice, or directly — see Hammond (1977, 1988b, 1997a) for more details. Moreover, ordinality is a complete characterization of behaviour satisfying the three "consequentialist" axioms.

A second result applies to finite decision trees T in the domain $\mathcal{T}_2(Y)$ which, in addition to decision nodes and terminal nodes having consequences in the specified domain Y, also contain chance nodes at which strictly positive objective probabilities are specified.[1] In any such tree, behaviour has random consequences in the domain $\Delta(Y)$ of simple (finitely supported) lotteries on Y. Then, on the unrestricted domain $\mathcal{T}_2(Y)$, consequentialist dynamically consistent behaviour not only maximizes a weak preference ordering R over the feasible set of consequence lotteries; the independence axiom due to Marschak (1950) and Samuelson (1952) is also satisfied. Again, this is a complete characterization of behaviour satisfying the three axioms. If behaviour is also required to vary continuously as objective probabilities vary, in the sense of generating a closed graph behaviour correspondence from probabilities to decisions, then R must also satisfy a familiar continuity axiom ensuring the existence of a unique cardinal equivalence class of NMUFs on Y, all of whose expected values represent R.

A final third result is required in order to justify subjective expected utility maximization. It is necessary to consider a new class of finite decision trees T in the family of domains $\mathcal{T}_3(E,Y)$ ($\emptyset \neq E \subset S$), where Y is the consequence domain, S is a finite set of possible uncertain *states* of the world, and each non-empty $E \subset S$ respresents an *event*. Trees in each domain $\mathcal{T}_3(E,Y)$ contain "natural" nodes where nature refines a partition of the subset of E that corresponds to the set of states which remain possible. These nodes are in addition to the decision nodes, to the terminal nodes having consequences in the specified domain Y, and to the chance nodes at which strictly positive objective probabilities are specified. In any tree $T \in \mathcal{T}_3(E,Y)$, behaviour has random consequences in the domain $\Delta(Y^E)$ of simple lotteries on the Cartesian product space $Y^E = \prod_{s \in E} Y_s$ whose members $y^E = \langle Y_s \rangle_{s \in E}$ specify a function $s \mapsto y_s$ mapping states of the world $s \in E$ to consequences $y_s \in Y_s = Y$. Of course, each such y^E is what Savage (1954) calls an "act".

Now, given the family $\mathcal{T}_3(E,Y)$ ($\emptyset \neq E \subset S$) of unrestricted domains, consequentialist dynamically consistent behaviour in any tree $T \in \mathcal{T}_3(E,Y)$ must not only maximize a conditional weak preference ordering R_E over the feasible set of consequence lotteries in $\Delta(Y^E)$. In addition, R_E must satisfy the Marschak–Samuelson independence axiom. Moreover, different the orderings R_E ($\emptyset \neq E \subset S$) must together satisfy Anscombe and Aumann's (1963) extension to "random acts" of the sure thing principle originally formulated by Savage (1954). As before, this is a complete characterization of behaviour satisfying the three axioms. If behaviour is also continuous as objective probabilities vary, then for each event E there must exist a unique cardinal equivalence class of NMUFs on Y^E whose expected values all represent R_E. Moreover, under two additional axioms like those used by Anscombe and Aumann (1963) — namely reversal of order and state independence — there must exist a unique and strictly positive family of subjective conditional probabilities $P(s|E)$ ($s \in E \subset S$) satisfying Bayes' rule, together with a unique cardinal equivalence class of state-independent NMUFs on Y whose subjective conditionally expected values all represent each ordering R_E. In this sense, when supplemented by mild additional conditions, consequentialism implies the Bayesian rationality hypothesis.

III. A FAMILY OF CONSEQUENTIALIST NORMAL GAME FORMS

It would seem highly desirable to have a single integrated theory of normative decisions which applies to all n-person games, and which reduces to consequentialist single-person decision theory in the special case of one-person games "against nature". Accordingly, it appears natural to formulate the *consequentialist normal form invariance hypothesis*. This requires that, whenever two game forms have identical or "equivalent" normal forms, any player's strategic behaviour in those two games should give rise to identical sets of consequences that are revealed as chosen. The hypothesis is an obvious adaptation of a claim that figured so prominently in von Neumann and Morgenstern's book (1944, 1953). Obviously, it is also a natural extension to extensive form games of the consequentialist hypothesis that was previously advanced for decision trees. The aim of this paper is to sketch the main implications of this hypothesis, to be discussed elsewhere in more detail.

The first step in such a theory is to formulate an appropriate extension of single-person decision trees to n-person "consequentialist extensive game forms". These are like decision trees and game forms, but they differ from orthodox extensive form games in that they have consequences rather than payoffs attached to terminal nodes. This seems entirely appropriate for a theory in which the existence of a payoff function, in the form of a von Neumann–Morgenstern utility function, should be a major result rather than a questionable assumption.

However, to save space, I shall proceed directly to the definition of an associated *consequentialist normal game form*. This consists of a list

$$G = \langle N, Y^N, S^N, \phi^N \rangle$$

Here N denotes the finite set of *players*. Each player $i \in N$ is assumed to have a *personal consequence domain* Y_i which is one component of the Cartesian product $Y^N := \prod_{i \in N} Y_i$ of *consequence profiles*, and also a finite strategy space S_i making up one component of the Cartesian product $S^N := \prod_{i \in N} S_i$ of *strategy profiles*. Finally, $\phi^N : S^N \to \Delta(Y^N)$ is the *outcome function* determining the random consequence profile that results from each possible strategy profile. For each $i \in N$, let $\phi_i(s^N) \in \Delta(Y_i)$ denote the marginal distribution on Y_i that is induced by $\phi^N(s^N)$ on Y^N.

Given any such game form G, it will also be necessary to consider a family $\mathcal{G} = \{G\} \cup (\cup_{i \in N} \mathcal{G}_i)$ of game forms derived from G, where

$$\mathcal{G}_i = \{\, G(i,T) \mid T \in \mathcal{T}_3(S_{-i}, Y_i) \,\}$$

That is, for each player $i \in N$ and tree $T \in \mathcal{T}_3(S_{-i}, Y_i)$, there is a corresponding game form in $G(i,T)$ in $\mathcal{G}_i$ specified by

$$G(i,T) = \langle \{i^*\} \cup N, Y_i \times Y^N, S^T \times S^N, \bar{\phi}^{\{i^*\} \cup N} \rangle$$

As explained in the introduction, this involves one extra player i^* who is a copy of player i. So player i^*'s consequence space, like i's, is Y_i. It is assumed that player i^* effectively faces a single-person decision tree $T \in \mathcal{T}_3(S_{-i}, Y_i)$, in which the set of possible states of nature is S_{-i}. The finite set of i^*'s strategies in tree T is denoted by S^T, and the outcome function is assumed to be $\phi^T : S^T \to \Delta(Y^{S_{-i}})$. In $G(i,T)$ the outcome function is given by

$$\phi_{i^*}(s^T, s^N) := \phi^T(s^T) \quad \text{and} \quad \bar{\phi}_j(s^T, s^N) := \phi_j(s^N) \text{ (all } j \in N)$$

for all $(s^T, s^N) \in S^T \times S^N$. Note that, for each player $j \in N$, the random outcome in $G(i,T)$ of each strategy profile $s^N \in S^N$ that can be played in G is the same as the random outcome of s^N in G itself. In particular, this random outcome must independent both of T and of i^*'s choice of strategy be $s^T \in S^T$ in T.

IV. PLAYERS' TYPE SPACES

It may be useful to think of a game form as a book of rules, specifying what strategies players are allowed to choose, and what random consequence results from any allowable profile of strategic choices. So the family $\mathcal{G}$ of consequentialist game forms needs fleshing out with descriptions of players' preferences, beliefs, and behaviour. The Bayesian rationality hypothesis involves preferences represented by expected values of von Neumann–Morgenstern utility functions (NMUFs) attached to consequences. Also, beliefs take the the form of subjective probabilities attached jointly to combinations of other players' preferences, strategies, and beliefs. And behaviour should maximize subjectively expected utility. It has yet to be shown, however, that the consequentialist hypotheses imply such preferences, beliefs, and behaviour. To do so satisfactorily requires a framework for describing preferences, beliefs, and behaviour in game forms before the consequentialist hypotheses have been imposed. We shall postulate type spaces similar to those considered by Harsanyi (1967–68) in his discussion of games of incomplete information. However, each player will have three separate types, corresponding to preferences, beliefs, and behaviour respectively.

Indeed, since one cannot directly assume that preferences exist, it is necessary to consider instead, for each player $i \in N$, a *decision type* $d_i \in D_i$ which determines what is acceptable behaviour for i in any single-person finite decision tree $T \in \mathcal{T}_2(Y_i)$ having random consequences in $\Delta(Y_i)$. Of course, consequentialist normal form invariance implies the consequentialist hypotheses for single-person decision theory. So if continuity of behaviour is added to these hypotheses, we know already that each player $i \in N$ will have a unique cardinal equivalence class of NMUFs $v_i(y_i; d_i)$ on Y_i which are parametrized by their decision type d_i. Together, the list of all players' types forms a *decision type profile* $d^N \in D^N := \prod_{i \in N} D_i$.

As in orthodox game theory, each player $i \in N$ is assumed next to have beliefs or an *epistemic type* $e_i \in E_i$, with $E^N := \prod_{i \in N} E_i$ as the set of all possible *epistemic type profiles*. It will be a result rather than an assumption of the theory that all such beliefs can be represented by subjective probabilities on an appropriately defined space. For the moment, each $e_i \in E_i$ is assumed to determine parametrically player i's *strategic behaviour* in the form of a non-empty set $\sigma_i(G', d_i, e_i) \subset S_i$ defined for every game form $G' \in \mathcal{G}$ and each possible decision type d_i for player i. In orthodox game theory, $\sigma_i(G', d_i, e_i)$ is the set of i's "best responses" given the NMUF $v_i(y_i; d_i)$ and subjective probability beliefs over other players' strategies determined by e_i. The assumption that such a parameter e_i exists is without loss of generality because if necessary this parameter could be the correspondence $(G', d_i) \mapsto\!\!\!\rightarrow \sigma_i$ itself. Finally, it is also necessary to define $\sigma_{i^*}(G', d_i, e_i)$ for the copy i^* of player i in every game $G' \in \mathcal{G}_i$. Note that, because i^* is a copy of i, player i^*'s behaviour depends on i's type pair (d_i, e_i), as the above notation reflects.

From the normative point of view, each set $\sigma_i(G', d_i, e_i)$ already describes how i with decision type d_i and epistemic type e_i should play G'. However, in forming beliefs, it is not enough for player i (and also i^* if $G' \in \mathcal{G}_i$) to know the other players' sets $\sigma_j(G', d_j, e_j)$ $(j \in N \setminus \{i\})$; also relevant are the tie-breaking rules which the other players $j \in N \setminus \{i\}$ use to select one particular strategy s_j from the set $\sigma_j(G', d_j, e_j)$ whenever this set has more than one member. Accordingly, each player $i \in N$ is assumed to have in addition a *behaviour type* $b_i \in B_i$, with

$B^N := \prod_{i \in N} B_i$ as the set of all possible *behaviour type profiles*. Each $b_i \in B_i$ is assumed to determine parametrically player i's *selection rule* yielding a single member $s_i(G', d_i, e_i, b_i) \in \sigma_i(G', d_i, e_i)$ of each strategic behaviour set. The assumption that b_i exists is without loss of generality because it could be the function $(G', d_i, e_i) \mapsto s_i$ itself. Note that player i^*'s behaviour type need not be specified because i^*'s behaviour has no effect on any other player.

To simplify notation in future, define for each player $i \in N$ a combined type space $\Theta_i := D_i \times E_i \times B_i$, whose members are triples $\theta_i := (d_i, e_i, b_i)$. Note that each player's selection rule can then be expressed as $s_i(G', \theta_i)$. Let $\Theta^N := D^N \times E^N \times B^N$ be the space of combined type profiles, with typical member $\theta^N := (d^N, e^N, b^N)$, and let $\Theta_{-i} := \prod_{j \in N \setminus \{i\}} \Theta_i$ denote the set of all possible types for players other than i. A complete epistemic type $e_i \in E_i$ should then describe in full player i's beliefs about the other players' types $\theta_{-i} \in \Theta_{-i}$, including their epistemic types e_{-i}. This creates a problem of circularity or infinite regress which is an inevitable and fundamental part of modern game theory.

V. SUBJECTIVE PROBABILITIES OVER OTHER PLAYERS' STRATEGIES

First, given any game $G' = G(i, T) \in \mathcal{G}_i$, suppose that player i^* moves first, before any player $j \in N$, without these players knowing what i^* has chosen. Then after i^* has moved, G is effectively a "subgame of incomplete information". Now, given any player $j \in N$, and any combined type $\theta_j \in \Theta_j$ that player j may have, applying an obvious dynamic consistency hypothesis to the subgame G of $G' = G(i, T)$ yields the result that

$$\sigma_j(G', d_j, e_j) = \sigma_j(G, d_j, e_j) \quad \text{and} \quad s_j(G', \theta_j) = s_j(G, \theta_j)$$

In particular, for each $j \in N$, both $\sigma_j(G', d_j, e_j)$ and $s_j(G', \theta_j)$ are effectively independent of whatever player $i \in N$ is copied and of whatever tree $T \in \mathcal{T}_3(S_{-i}, Y_i)$ is given to the copy i^* of player i. So variations in i^*'s decision tree within the domain $\mathcal{T}_3(S_{-i}, Y_i)$ are possible without inducing changes in the behaviour of other players $j \in N$. This justifies applying the consequentialist hypotheses to the whole domain $\mathcal{T}_3(S_{-i}, Y_i)$ of single-person decision trees faced by i^* and so by i, while treating each $s_{-i} \in S_{-i}$ as a state of nature determined entirely outside the tree. So the usual arguments imply the existence of unique and strictly positive subjective probabilities $P_i(s_{-i})$ $(s_{-i} \in S_{-i})$ such that i^*'s decisions in trees $T \in \mathcal{T}_3(S_{-i}, Y_i)$ maximize the subjectively expected value of a von Neumann–Morgenstern utility function $v_i(y_i; d_i)$ parametrized by i's decision type $d_i \in D_i$.

It remains to consider player i's behaviour in the game form G itself. To do so, consider the special decision tree $T_i^G \in \mathcal{T}_3(S_{-i}, Y_i)$ in which the set of i^*'s strategies is S_i, equal to i's strategy set in G, and the random outcome of each strategy $s_i \in S_i$ is specified by

$$\phi_i^G(s_i)(y^{S_{-i}}) = \prod_{s_{-i} \in S_{-i}} \phi(s_i, s_{-i})(y_{s_{-i}})$$

for all $y^{S_{-i}} = \langle y_{s_{-i}} \rangle_{s_{-i} \in S_{-i}} \in Y^{S_{-i}}$. Then, at least under Anscombe and Aumann's reversal of order axiom, both the strategy set S_i and the outcome function ϕ_i^G are exactly the same as in G itself. In this case T_i^G and G are consequentially equivalent from i's (or i^*'s) point of view, so consequentialism requires i's behaviour in G to match that of i^* in T_i^G or $G(i, T_i^G)$. This implies that

$$\sigma_{i^*}(G(i, T_i^G), d_i, e_i) = \sigma_i(G, d_i, e_i).$$

It follows that player i should behave according to the hypothesis of Bayesian rationality and choose $s_i \in S_i$ to maximize subjectively expected utility based on the subjective probabilities $P_i(s_{-i})$ $(s_{-i} \in S_{-i})$ that are appropriate for all decision trees in $\mathcal{T}_3(S_{-i}, Y_i)$. Really, one should write these probabilities as $P_i(s_{-i}, e_i)$ to indicate that they represent player i's epistemic type e_i and so characterize i's acceptable behaviour sets $\sigma_i(G', d_i, e_i)$ on the domain $\mathcal{G}$ of game forms, including G itself. This is the promised consequentialist justification of the Bayesian rationality hypothesis in normal form n-person games.

NOTES

* *Acknowledgments:* November 1996 revision of a paper presented to the SITE (Stanford Institute of Theoretical Economics) summer workshop on "Game Theory: Epistemic and Other Foundational Issues". The paper is based on an earlier presentation to the International Symposium in Honor of John C. Harsanyi on "Game Theory, Experience, Rationality: Foundations of the Social Sciences, Economics and Ethics" organized by the Institute Vienna Circle in Vienna, June 12–15, 1996. I am grateful to Marco Mariotti and Pierpaolo Battigalli for most helpful discussions of this and related work.

1. Allowing zero probabilities at chance nodes yields the unacceptably strong conclusion that all lotteries in $\Delta(Y)$ should be indifferent.

REFERENCES

F.J. Anscombe and Robert J. Aumann, "A Definition of Subjective Probability", in: *Annals of Mathematical Statistics*, 34, 1963, pp.199–205.

Kenneth J. Arrow, "Rational Choice Functions and Orderings", in: *Economica*, 26, 1959, pp.121–127.

Pierpaolo Battigalli, "Comment [on Mariotti (1996)]", in: Kenneth J. Arrow, Enrico Colombatto, Mark Perlman, and Christian Schmidt (Eds.), *The Rational Foundations of Economic Behaviour*. London: Macmillan 1996, pp.149–154.

B. Douglas Bernheim, "Rationalizable Strategic Behavior", in: *Econometrica*, 52, 1984, pp.1007–1028.

B. Douglas Bernheim, "Axiomatic Characterizations of Rational Choice in Strategic Environments", in: *Scandinavian Journal of Economics*, 88, 1986, pp.473–488.

Peter J. Hammond, "Dynamic Restrictions on Metastatic Choice", in: *Economica*, 44, 1977, pp.337–350.

Peter J. Hammond, "Consequentialism and the Independence Axiom", in: Bertrand Munier (Ed.) *Risk, Decision, and Rationality: Proceedings of the 3rd International Conference on the Foundations and Applications of Utility, Risk, and Decision Theories*. Dordrecht: D. Reidel 1988a, pp.503–516.

Peter J. Hammond, "Consequentialist Foundations for Expected Utility", in: *Theory and Decision*, 25, 1988b, pp.25–78.

Peter J. Hammond, "Objective Expected Utility: A Consequentialist Perspective", in: Salvador Barberà, Peter J. Hammond and Christian Seidl (Eds.) *Handbook of Utility Theory*. Dordrecht: Kluwer Academic Publishers 1997a — in preparation.

Peter J. Hammond, "Subjective Expected Utility", in: Salvador Barberà, Peter J. Hammond and Christian Seidl (Eds.) *Handbook of Utility Theory*. Dordrecht: Kluwer Academic Publishers 1997b — in preparation.

Peter J. Hammond, "Utility Theory for Non-Cooperative Games", in: Salvador Barberà, Peter J. Hammond and Christian Seidl (Eds.) *Handbook of Utility Theory*. Dordrecht: Kluwer Academic Publishers 1997c — in preparation.

John C. Harsanyi, "Cardinal Utility in Welfare Economics and in the Theory of Risk-Taking", in: *Journal of Political Economy*, 61, 1953, pp.434–5; reprinted in Harsanyi (1976).

John C. Harsanyi, "Cardinal Welfare, Individualistic Ethics, and Interpersonal Comparisons of Utility", in: *Journal of Political Economy*, 63, 1955, pp.309–321; reprinted in Harsanyi (1976).

John C. Harsanyi, "A General Theory of Rational Behavior in Game Situations", in: *Econometrica*, 34, 1966, pp.613–634.

John C. Harsanyi, "Games with Incomplete Information Played by 'Bayesian' Players, I–III", in: *Management Science*,14, 1967–68, pp.159–182, 320–334, 486–502; reprinted in Harsanyi (1982a), chs. 6–8.

John C. Harsanyi, "Nonlinear Social Welfare Functions: Do Welfare Economists Have a Special Exemption from Bayesian Rationality?" in: *Theory and Decision*, 6, 1975a, pp.311–32; reprinted in Harsanyi (1976).

John C. Harsanyi, "Can the Maximin Principle Serve as a Basis for Morality? A Critique of John Rawls's Theory", in: *American Political Science Review*, 69, 1975b, pp.594–606; reprinted in Harsanyi (1976).

John C. Harsanyi, *Essays on Ethics, Social Behavior, and Scientific Explanation*. Dordrecht: D. Reidel 1976.

John C. Harsanyi, *Rational Behavior and Bargaining Equilibrium in Games and Social Situations*. Cambridge: Cambridge University Press 1977a.

John C. Harsanyi, "Advances in Understanding Rational Behavior", in: Robert E. Butts and Jaakko Hintikka (Eds.) *Proceedings of the Fifth International Congress of Logic, Methodology and Philosophy of Science, Vol. II*. Dordrecht: D. Reidel 1977b, pp.315–343; reprinted in Harsanyi (1976).

John C. Harsanyi, "Bayesian Decision Theory and Utilitarian Ethics", in: *American Economic Review (Papers and Proceedings)*, 68, 1978, pp.223–8.

John C. Harsanyi, "Uses of Bayesian Probability Models in Game Theory", in: D.H. Milnor (Ed.) *Science, Belief and Behaviour: Essays in Honour of R.B. Braithwaite*. Cambridge: Cambridge University Press 1980, pp.189–201; reprinted in Harsanyi (1982a), ch. 9.

John C. Harsanyi, *Papers in Game Theory*. Dordrecht: D. Reidel 1982a.

John C. Harsanyi, "Subjective Probability and the Theory of Games: Comments on Kadane and Larkey's Paper" and "Rejoinder to Professors Kadane and Larkey", in: *Management Science*, 28, 1982b, pp.120–124 and 124–125.

John C. Harsanyi, "Uses of Subjective Probability in Game Theory", in: Bernt P. Stigum and Fred Wenstop (Eds.) *Foundations of Utility and Risk with Applications*. Dordrecht:

D. Reidel 1983a, pp.297–310.

John C. Harsanyi, "Bayesian Decision Theory, Subjective and Objective Probabilities, and Acceptance of Empirical Hypotheses", in: *Synthese*, 57, 1983b, pp.341–365.

Marco Mariotti, "The Decision-Theoretic Foundations of Game Theory" in: Kenneth J. Arrow, Enrico Colombatto, Mark Perlman, and Christian Schmidt (Eds.) *The Rational Foundations of Economic Behaviour*. London: Macmillan 1996, ch. 6, pp.133–148.

Jacob A. Marschak, "Rational Behavior, Uncertain Prospects, and Measurable Utility", in: *Econometrica*, 18, 1950, pp.111–141.

John von Neumann and Oskar Morgenstern, 1944, *Theory of Games and Economic Behavior (3rd. edn.)*. Princeton: Princeton University Press 1953.

David Pearce, "Rationalizable Strategic Behavior and the Problem of Perfection", in: *Econometrica*, 52, 1984, pp.1029–1050.

Paul A. Samuelson, "Probability, Utility, and the Independence Axiom", in: *Econometrica*, 20, 1952, pp.670–678.

Leonard J. Savage, *The Foundations of Statistics*. New York: John Wiley 1954.

Tommy C.-C. Tan and Sérgio R. da C. Werlang, "The Bayesian Foundations of Solution Concepts of Games", in: *Journal of Economic Theory*, 45, 1988, pp.370–391.

Department of Economics
Stanford University, CA 94305-6072
USA

WERNER LEINFELLNER

GAME THEORY, SOCIODYNAMICS, AND CULTURAL EVOLUTION

Evolution is a game – the greatest game
in the world if you know how to play it.

I. A SHORT HISTORY OF EVOLUTIONARY GAME THEORY

Since Neumann and Morgenstern's theory of games, the debate among social scientists, economists, mathematicians, and social philosophers about what kind of theory it is has not ended. Some think that it is a new interdiscipline, some that it is a mere accumulation of gametheoretical models, such as utility theory, competitive, cooperative, collective choice models, and so on. Most of them agree that the models of game theory deal with isolated, single, and independent specific societal interactions between individuals who wish to maximize their gains and minimize their losses within prescribed constraining rules.

The situation changed abruptly when game theorists found out that most games depend on previous games, on past experiences, on their history (Friedman 1986: 80, 72) which rest on a specifically human, not always rational way of thinking. There is no doubt that we learn from our past successful and unsuccessful attempts, from iterating games, from improving and adapting rules, and exploiting random events. Here the similarity between series of causally interdependent competitive, cooperative games, serial games against nature, and evolutionary processes, is striking. The same applies when we learn to adapt losses or gains to changing rules and to selective random events in our society and environment. In the eighties, these similarities led to the new name "evolutionary game theory." More and more astonishing similarities with evolutionary processes were detected, for example: Evolutionary games follow numerous societal background rules; they are causally dependent on previously played games and on their stored rules; serial winnings resemble the measurable "fitness-increase" – but in a non-Darwinian sense. Series of temporally ordered value matrices were used to explain and successfully compute the dynamics of evolutionary serial games in animal populations (Maynard Smith 1982; Weibull 1996; Rapoport this volume; Schuster & Sigmund 1985). It is no wonder that since the eighties evolutionary game theory began to merge with Neodarwinism, with the evolution equations of modern biological evolution theory, of molecular biology, and of population genetics into a new unified theory of evolution. This new theory attempts to explain all kinds of evolutionary processes. Its methods and models in fact cover the prebiotic, molecular, and biological evolution of organ-

W. Leinfellner and E. Köhler (eds.), Game Theory, Experience, Rationality, 197–210.

isms and the evolution of animal and human behavior in their societies and environments.

When this new theory of evolutionary dynamics was applied to societal evolution, many inconsistencies within (the game-theoretical model of) the predominant Darwinian traditional evolution theory emerged. The leading role of Darwinian competition was replaced by the self-organization of societal cooperation (Sigmund 1993; Axelrod 1984: 145; Lumsden & Wilson 1981; Leinfellner 1991: 244, 1993: 55). Evolutionary game theory reached a peak when it succeeded in modelling and reformulating Darwinian selection and adaptation (Kauffman 1993: 5f.). Today, a new, twofold, mathematical-stochastic method explains the evolutionary sociodynamics of societies. One component is the dynamic evolutionary matrix method developed from evolutionary game theory (Maynard Smith 1982: 14ff; Weibull 1996: xiv; Leinfellner 1995: 149–151). The other uses differential equations, called "evolution equations" (Schuster 1985: 255; Helbing 1995: 230; Leinfellner 1984: 248–346). Both, when applied to societal and cultural evolution, explain in a statistico-causal sense why societal forms change from generation to generation; moreover, they explain the creation of new and the disappearance of old societal forms within prescribed, rulelike constraints. Kauffman's grand view (Kauffman 1993: 16, 209) that all evolutionary processes exist only "near the edge of chaos" (Kauffmann 1993: 232) entails both: that randomness subdues societal cultural evolution and that it enables the self-organization of a more and more complex democratic order. It also explains the evolution of complex molecules in prebiotic evolution (Eigen & Schuster 1979). Or, as genetic algorithm, it is, according to Holland (1995: 62ff.), a new method to explain genetic evolution.

This article tries to apply the unified theory of evolution to societal and cultural evolution in modern, democratic societies and tries to remove some hindrances and obstacles, a task already tackled by Lumbsden & Wilson, Maynard Smith, and Helbing, to mention just a few.

II. THE FIRST OBSTACLE: DO WE PLAY GAMES OR DO WE SOLVE CONFLICTS IN OUR PRESENT DEMOCRATIC SOCIETIES? A PARADIGM CHANGE.

It makes no empirical sense to speak about playing games with and in our present society. According to Selten, most games of game theory are only abstract, mathematical models of how we may compute their outcomes. Like so many abstract concepts, the application of gametheoretical models to explain societal evolution runs into almost insurmountable difficulties. The most serious is: All evolutionary processes rest on causal or statistico-causal interactions between the individuals and between the individuals and their environment. Till today, game theory offered no viable empirical 'microscopic' foundation or interpretation of these interactions (Helbing 1996: 227). This article proposes a new empirical interpretation of serial, evolutionary games, which explains how games can be interpreted as societal, empirical interactions. Solving games *in mente* or solving them theoretically does not change anything in our societal life. But when we realize our solutions of societal conflicts, we do interact empirically with our society. Only our empirical realizations cause changes in our socio-cultural evolution and in our environment. Empirical realizations of our mental or theoretical solutions of evolutionary conflicts are the real causes which push forward socio-cultural evolution.

This amounts to a paradigm change in today's evolutionary game theory from "playing games" and "gambling" to "solving and realizing serial conflicts" in our society. To solve a problem is, according to H. Simon, a theoretical effort; to solve a societal conflict includes its empirical realizations. Moreover, empirical solutions need additional, democratically accepted background rules which we impose on the few special gametheoretical rules, as in Harsanyi's social ethics (1995, 1982, 1976) or in Holland's performance system (1995: 43–90, 1986; Leinfellner 1995: 167f). Not the gametheoretical special rules, but the societal, juridical, cultural, and socio-ethical empirical rules for the realizations of theoretically possible solutions play the most important role. One began to doubt that the traditional, rational methods can be used when we realize empirically evaluations in utility theory under uncertainty and risk. Allais' paradox and most of the paradoxes which have haunted and shaken utility and game theory and related disciplines, including postmodern constructivism and many cognitive sciences, suffer from the reality syndrome. It indicates that, in reality or when we realize our solutions of societal conflicts, everything looks suddenly paradoxical and different than in the abstract theory. We may solve these paradoxes through a preventive measure: Not the abstract theory, only its empirical realization counts.

III. THE COMPLEXITY INCREASE OF THE SOCIETAL CULTURAL EVOLUTION

Since the complexity of the socio-cultural evolution increased beyond the complexity of the biological and animalic evolution, the complexity of the corresponding theories increased (Schuster 1996). Firstly, the dynamic evolutionary mechanism of the theory of socio-cultural evolution explains the societal and cultural changes by, firstly, an additional information- and rule-transfer to future generations by individual and cultural storage, learning and teaching. This can only be achieved optimally in modern democracies. Secondly, by the paradigmatic change of one-shot games to serial solutions and empirical realizations of societal conflicts; all individual preferences, evaluations, and decisions of how to solve societal conflicts can be taken into account, for example, by collective choice theory, a typical democratic theory. Thirdly, we have an ever-increasing imposition of new and stored socio-ethical background rules on gametheoretical rules in democracies. Fourthly, by virtue of the fact that our brain's intelligent faculties to model and solve societal conflicts consciously and scientifically, the brain functions as a real 'prime mover' of societal and cultural evolution. Fifthly, mental, scientific, and even creative conflict-solving is not enough: Only those mental or theoretical solutions 'survive' which turn out to be the product of practically, technically, and empirically successful processes of realizations. These realizations are the long-sought empirical interactions and at least partial causes of the socio-ethical evolution; at the same time, they serve as their empirical tests. Only realizations may become man-made, differently weighted, partial causes for successful or unsuccessful evolutionary changes in our socio-cultural settings. Sixthly, according to Weibull and Leinfellner, evolutionary game theory permits not only the explanation of the macrodynamics of societal evolution in the large, but also of the evolutionary mini-dynamics of how our intelligent brain solves mentally societal conflicts. When we – mentally or scientifically – solve societal conflicts or problems, our methods and thought processes are to be evolutionary mini-methods, too, or small-scale mental evolutionary processes. They differ from the traditional, merely rational, linear and recursive methods by their specific stochastic inferences and by the default rules (Holland 1986: 22, 153f, 268; 1995: 57f; Leinfellner 1995: 167).

For example, when we learn, when we invent and create better solutions, better empirical realizations, we have to cope with imagined randomness. When we solve societal conflicts, the mini-evolutionary methods of thinking rest in most instances not on rational deductive methods, not on classical statistical methods, but on non-linear, partly stochastic Bayesian methods under uncertainty and risk and on statistico-causal inferences and beliefs (Penrose 1997: xvii, xvi, xviii). For example, when we adapt our solutions of societal conflicts to random changes in our environment, when we empirically interpret mixed strategies, or when we cooperate, the empirical solutions drift randomly. Since evolutionary, stochastic thinking does not use exclusively rational deductive methods, it resembles – according to Simon, Leinfellner, and many brain physiologists – statistico-causal inferencing. Besides the rational, predominantly axiomatic and deductive, thinking, reserved for first-hand shaping of models, this way of evolutionary cognitive and practical thinking is our second, inborn, 'deeper' intelligent capacity for gathering and evaluating social experience under uncertainty and risk with the help of numerous causal and prescriptive default rules which are stored in our individual and cultural memory (Leinfellner 1995: 167ff.). Another reason is: Here we use statistical inferences to solve and realize our pending, future societal conflicts; and we do this only for the better of each individual's and the entire society's welfare. This describes the main attractor in the evolution of democratic societies.

The next important problem for democratic societies is the socio-ethical complexity of its stored democratic rules. For example, when we solve societal conflicts, we obey not only a vast number of democratic, juridical, social, and ethical rules, but we also follow, at the same time, very often hidden evolutionary attractors, non-teleological tendencies, such as certain overriding socio-ethical rules. Only they can increase our own individual well-being and, at the same time, society's welfare and security in a Pareto-optimal sense.

IV. SOCIETAL CONFLICTS AND CONFLICT SOLVING

Do we really gamble when we solve societal conflicts? Apparently not, because, in contrast to private conflicts, societal conflicts are always open conflicts between the individuals' egoistic and altruistic interests within their society. "Egoistic" means here regarding only the selfish interests, "altruistic" regarding also the interests of the others (Leinfellner 1993: 32–65; 1991: 229–250; 1986: 141–147; Axelrod 1984: 88f, 145f). Open societal conflicts are solved theoretically by 'optimal' solutions. Such solutions are in most cases weighted, evolutionarily stable equilibria between egoistic and altruistic, extreme solutions. This holds equally for all societal conflicts copying with random events which effect some or all members of the society, such as contagious diseases, natural catastrophes, etc. Therefore, optimal or theoretically possible Nash equilibria are only necessary conditions for optimal solutions of societal conflicts. There is no other way: Empirical realizations within democratic societies have to increase individual as well as collective well-being and welfare – not "in the long run" but in the "not too long run." In other words: They have to satisfy the Pareto principle: Optimal are solutions whose realizations increase, or at least do not decrease, all the individuals' and the society's welfare and stability. By no means can some individuals or groups in democracies of the social-welfare type profit at the expense of the other individuals or groups.

Even the best abstract framework and its mathematical solutions will not help us to solve our daily societal conflicts and problems, if their realizations violate democratic human rights or democratic rules which are based on the Pareto principle. Therefore, very often "theoretically optimal, rational, and consistent" solutions of societal conflicts, proposed by governments or specialists, are voted down by the majority of citizens who follow their democratically accepted rules.

In practice, serial realizations of societal solutions are a temporal, complex, Bayesian, evolutionary learning process, since this process unfolds always by including new empirical evidence. When we have to decide between gametheoretically sound, but multiple, equilibria (possible solutions), the usefulness of their empirical realization, not rational deductive arguments, are decisive for a society. And only democracies dispose of enough democratically accepted socio-ethical rules which follow the Pareto principle. They permit only such empirical applications of societal conflict solutions as do not violate human rights and Bayesian learning.

VI. PROBLEMS OF COEVOLUTION

What most Neodarwinians do not want to admit is that societal and cultural evolution is an empirically quite different kind of evolution than biological evolution, the reason being that we ourselves effect the empirical realizations of societal solutions, including those of today's technical, medical and, hopefully, environmental conflicts. This is one reason why we cannot reduce the evolution of societal behavior in human societies to biological genetic evolution. Another reason is that coevolution is a holistic, synergetic, complex evolutionary flow which cannot be split up into components. Its evolutionary dynamic rests not only on mutually linked interactions, but can be steered by our active realizations of societal conflict solutions, at least in part (Leinfellner 1995: 170f). Our scientific, technological, our biological, medical, and socio-cultural know-how permits us to interfere with our socio-cultural evolution to a far greater extent than, for example, with our basic genetic and biological evolution. We may change the former by the realizations of competitive and cooperative conflict solutions; we may change the causal impact of selective random events from our environments and from internal changes; and, of course, we may alter our adaptive rules for socio-cultural evolution - see example (1).

Therefore, research in societal-cultural evolution focusses today on the empirical, the psychological, the brainphysiological, and the intelligent partial causes of the socio-cultural coevolution. Recently, Damasio (1994: 173) has shown how the realizations of societal solutions under selective pressure and adaptational rules depend on additional brainphysiological emotional causes.

VII. THE SOCIO-CULTURAL EVOLUTION IS AN EVOLUTION OF SOCIETAL FORMS

As already described, a theory or the models of our societal and cultural evolution have to explain, firstly, how individuals in democratic societies solve mentally and scientifically all their societal conflicts. Secondly, theoretical solutions have to be realized; this is their ultimate empirical test. Iterations of societal solutions across generations have created an ever-increasing diversity of cultural forms. Cultural forms are *mentifacts*, such as ideas, models, theories; *artifacts* are artistic creations and creations by artisans; *technifacts* are products of our technological culture or civilization. Finally, we have patterns of behavior and of social activities, expressed by prescriptive rules and customs as cultural forms. Their respective evolutionary trajectories represent in the evolutionary models their temporal changes, their 'growth or increase' or 'extinction,' from generation to generation. The sum total of all changes in each evolutionary trajectory represents the changes in our socio-cultural evolution. Luckily, till today they increased steadily the societal welfare and stability of democratic societies.

VIII. REALIZATIONS OF SOCIETAL CONFLICTS ARE STATISTICO-CAUSAL INTERACTIONS BETWEEN THE INDIVIDUALS AND THEIR ENVIRONMENT

It is a fact that any kind of evolution rests on empirical causal interactions among the individuals, organisms and among them and their environments. However, to this day, even in gametheoretical evolutionary models, it was impossible to explain, according to Helbing (1995: 227), which 'microscopic' empirical causal interactions push forward the societal and cultural evolution. Our proposal to regard societal interactions as the empirical realizations of theoretically possible solutions of serial societal conflicts offers for the first time an empirical interpretation of all the man-made causes of socio-cultural evolution. For the author, all sciences, including the technical and the social sciences, serve only as a means to successfully realize our mentifacts, models, predictions, and technifacts empirically. Successful terminations of societal conflicts depend, on the one hand, on the conflict-solving constructive and inferencing mental activities of the individuals, and on the other hand, on their successful or unsuccessful realizations. This highlights the paradigm change from "solving games" to "solving conflicts and realizing the solutions". Therefore, societal and cultural evolution rests increasingly on individuals' readiness to solve mentally or with the help of gametheoretical models any societal conflicts, as they arise, and on the practical empirical realizations of the theoretical solutions. The successful realization of solutions serves now as the ultimate empirical criterion, even for theoretically equivalent optimal solutions, for example, when we have to choose between optimal equilibria.

Carefully executed empirical realizations will prevent most of the haunting paradoxes which plague today's utility and game theory. Sociodynamics explains not only why and how temporally ordered sequences or strings of optimal societal solutions and their successful realizations change, but also why new societal forms originate, why they multiply, replicate, improve or deteriorate, or disappear. But let's keep in mind: Each realization of a societal conflict solution is only a partial cause among others which may influence causally the channeling or start of a societal form's evolutionary trajectory. So far we have had luck, according to Gould, since no chaotic changes of preferences or disturbing random events have impeded the rise of democratic societies.

Finally: Since the successful empirical realizations of theoretical solutions of societal conflicts can change the socio-cultural evolution by creating or terminating societal forms, artifacts, mentifacts, technifacts, customs, and rules, the individuals are no longer passive onlookers of their socio-cultural evolution. They can change actively, at will, with the help of societal, political, economic, scientific and technical rules, the future course of their cultural evolution when they realize the solutions of societal conflicts. But in spite of these advantages, some individuals play a relatively passive role when their conflict solutions and realizations depend on the decisions or the free will of majority voters, on undemocratic power groups, or, in extreme cases on bad, catastrophic random events.

Thus, by cultural evolution we understand any change in any cultural forms or goods; it plays no role whether they are partially caused by the realizations of the solutions of conflicts and/or by selective random events.

The following models and their interpretations describe theoreticaly and empirically the causal connections which effect the specific changes of cultural forms or goods, and why they increase or decrease across generations.

But if we find, through intelligent and mental efforts and with the help of evolutionary models, theoretically possible optimal solutions, this is only a necessary and formal condition before we realize them. This first stage can be achieved by gametheoretical methods and/or stochastic reasoning. But without realizations, theoretically optimal solutions remain platonistic, mere mathematical constructions, as Selten has emphasized again and again. Only their confirmation by their empirical realizations – the relizations must fulfill the Pareto principle – can avert disastrous paradoxes or consequences. It is as in all realizing sciences: New technifacts have to be invented mentally or theoretically; but without their empirical realization and proof of their usefulness they remain mentifacts, products of our constructive fantasy. The cultural self-organization of new goods has multicausal origins (Leinfellner 1981: 221–260). It often arises from an unpredictable causal mix between intended, man-made realizations, provided positive random events, such as windfall profits in our economic system, support the creation of societal goods. See Bak's sandpile paradigm (Bak 1996: 33–80).

Cultural forms, by the way, come in two categories: (i) cultural, mental, and material goods; and (ii) cultural, rule-governed patterns (activities). Cultural goods and cultural patterns are the inputs and outputs of our cultural evolution. Our realized solutions may give the ongoing cultural evolution a new direction, that is, channel new evolutionary paths; but they can shape only our immediate societal future (Leinfellner 1994: 61–69). All evolutionary processes are open-ended. Thus we may interfere partially with the evolutionary macrodynamics of our culture by influencing causally certain evolutionary paths (trajectories). This may, with luck, streamline or channel for a while the sum total of all temporal changes of all mentifacts, artifacts, technifacts and societal patterns of behavior. Such a change can be expressed by default rules. Most of the time we are involved with channelling the microdynamics of single evolutionary trajectories of mentifacts, artifacts, behaviors in small units or subcultures. Evolutionary changes show up the first time as observable redistributions within subcultures before they aggregate in larger societies. Presently, all empirical tests can check only the microevolutionary trajectories in subcultures, as described in the next section, not the total macrocultural evolution. Here exists approximately the same gap as between micro- and macroeconomics.

X. DISCUSSION OF AN EMPIRICAL EXAMPLE (1)

This example deals with an everyday, 'perpetual' conflict for women: What they should wear. Interestingly, there exists already a gametheoretical, mathematical model to solve this conflict theoretically, but till today there has been no empirical societal interpretation. This empirical interpretation explains the changes of an everyday cultural form, more precisely of an artifact: the changes in the style of women's evenings robes across generations. These changes depend firstly on the above mentioned preferential conflict: 'What shall I wear today', or, 'What is fashionable'. The empirical realizations of such conflicts I took from various paintings of European women's dresses in the last three centuries. This empirical interpretation of an evolutionary dynamics shows the manifold partial causes, the psychological and mental preferences, and the statistico-causal role of the social and genetic rules which determine the daily realizations of solutions of a relatively simple cultural conflict. More specifically, the conflict is: Should women wear evening dresses with longer or shorter skirts, lower or higher neck- and waistlines? The empirical graphic representation of these changes yielded a simple geometric sinusoid curve, beginning in the year 1600, and describes the periodically changing optimal solutions and the realizations of these evolutionary conflicts. The curve confirms empirically Maynard Smith's theoretical solutions of the hawk-and-dove game, as discussed in Rapoport's contribution to the present volume. The empirical data and the model describe the cultural changes as typical periodic changes of fashions; they are restricted to the subculture of the European upper class. As data serve the temporal changes of the decolletages, of waistlines and skirt lengths in percentages of the body length during almost 300 years (until 1900). The cycles of ups and downs lasted 50 years each. The range between the lowest and highest hem-, neck-, and waistlines was determined by societal rules of decency. The cyclically recurring, alternating ups and downs characterize typical evolutionary trajectories for most cultural changes. Moreover, the European curve confirmed globally the already existing American graphical representations from American museums (for details see Leinfellner 1988).

The sinuslike curve shows the iterated, empirical changes of an artifact which was cyclic, but stable, till the thirties of our century. The same changes influenced the periods of demand-and-supply curves of the tailors' and consumers' dress market. Other partial causes of changes are the hidden social and sexual competitive preferences of women, for example, assuming women are courting men. The evolutionary competitive conflict starts when each woman tries to draw attention to herself by wearing showy dresses which differ from the other women's dresses, for instance in how much, or how little, they reveal of the body. It is clear that new changes of dress, new styles, new kinds of adornment increase a woman's social dominance and the attention from the opposite sex. Thus, skillful new innovations in the art of dressmaking and the competition among couturiers may reinforce women's and men's genetically inherited factors.

To show a little more skin gives a woman in the fashion lead an erotic advantage over the others; but cultural, cooperative customs and traditional-conservative rules prevented public nudity, at least in the last two centuries. Here you have a coevolutionary interplay of cultural and genetic rules. Societal rules determine whether decolletages go down; but in Western societies they do not plunge as far as in ancient Crete. A new deeper decolletage will draw more attention than a higher one used by conservative women. This model and its interpretation help us to explain why we have so many periodically changing styles and trends in our culture, whose evolutionary dynamics can be represented mathematically so well by gametheoretical models. At the same time they explain why women and men who use artifacts to draw attention to their appearance have to change from time to time dressing rules and the style of their outfits. In contrast, when a sexy look dominates, a more chaste fashion may draw more attention. This illustrates typical solutions of a serial, specific societal conflict which is regulated by many more or less conscious rules (customs) within a specific subculture and environment.

XI. Example (2)

This model shows, firstly, how partial causes synergetically push forward the temporal trajectories of societal forms; secondly, how the evolutionary dynamics can be represented either by a simple evolution equation of the Eigen-Schuster type (1), or by the dynamics of evolutionary gametheoretical matrices. It is shown that they yield the same solution, even though their methods are different.

(1) $dx_i/dt = (E^i - R^i)\,x_i + \Sigma_k\,\sigma_{ik}\,x_k$

Equation (1) describes a simple evolutionary dynamics caused by realizations (→) of solutions of societal conflicts and by conflicts created by random events which also may codetermine causally the temporal course of a societal form X in a subculture or in a democratic society.

x_i denotes the frequency (concentration) of a societal form X^i among other forms at a certain time point t of an evolving subculture or society. The matrix form

(2)

$$\begin{array}{llccccc}
 & Y^1 & & Y^j & & Y^n \\
 & y_1 & & y_j & & y_n \\
X^1\ x_1 & a_{11} & \dots & a_{1j} & \dots & a_{1n} \\
X^2\ x_2 & a_{21} & \dots & a_{2j} & \dots & a_{2n} \\
 & \bullet & & \bullet & & \bullet \\
X^i\ x_i & a_{i1} & \dots & a_{ij} & \dots & a_{in} \\
 & \bullet & & \bullet & & \bullet \\
X^m\ x_m & a_{m1} & \dots & a_{mj} & \dots & a_{mn}
\end{array}$$

expresses the expected values of realizations for the individuals involved, their dependencies on other individuals and on random events. Intuitively, the matrix expresses how the individuals evaluate at any time their future socio-cultural evolution. Mathematically, it can be reformulated as an evolutionary lottery which may change at any time point during the course of the socio-cultural evolution. Because of the open end of any evolution, future prospects (predictions) of societal evolution always present themselves to the single individuals as evolutionary lotteries or short-term predictions (Leinfellner 1994: 61-69). The first row (3) in the societal evaluation matrix (2), here at a time t, expresses such an expected utility of a lottery L_1 for a certain user of its societal form X^1:

(3) $W(L_1) = y_1\ x_1\ a_{11}, \ldots, y_j x_l\ a_{lj}, \ldots, y_m\ x_1\ a_{1n} = \Sigma_j y_j\ x_l\ a_{lj}.$

The first composite partial cause E^i in the equation (1) indicates any increase of the frequency of a societal form caused by realizations of solutions of societal conflicts; it is empirically, for example, a 'reproduction,' or a 'replication,' or traditing, or a creation of a societal form X^i at an initial time point t. Such successful realizations of theoretical optimal solutions increase the frequency of a societal form at least partially.

(4) R^i is the second partial causal factor in the equation (1). It indicates a diminishing of the frequency of societal forms, for example, by changing preferences and evaluations or by unsuccessful realizations.

External interferences by the government, restricting adaptive rules, laws, the influence of the media, of fashions, can be represented by both antagonistic partial causes E^i and R^i of evolutionary trajectories, for example:

E^i may depend on the partial causes A^i and Q^i in the following way:

(5) $\rightarrow E^{i,t+1} = A^{i,t} \cdot Q^i$ at the time t^{t+1}

A^{it+1} symbolizes now the actual rate of transmission or of tradition of a societal form X^i from generation to generation at the initiating period.

→ Q^i expresses then a partial change caused by a random event or a quality factor which can be increased by education, cooperative teamwork, scientific technological knowledge, and enlightenment. If $Q^i = 1$, no deviations or errors in the transmission and replication of a societal form will occur.

→ $A^i (1 - Q^i)$ is then the rate of change of a societal form caused mainly by perturbing random events and errors, Q^i. Now, R^i is composed of partial causes, too:

(6) $R^i = D^i + K^i$

This equation shows the dependency of R^i on the subfactors (partial causes) D^i and K^i. R^i may depend on D^i, the observed natural rate (→) of decomposition or vanishing of societal forms because of imminently occurring momentous random events, or simply because the life span of a societal consumer good has been exhausted.

→ K^i represents a man-made, artificially controllable, causal factor, for example, the longer life-span of usable goods, an increase of financial support for research, etc.

(7) In the formula $\Sigma_k \sigma_{ik} x_k$ expresses the transitional probability of a back-formation of a societal form X^i from another, for example reusing older societal forms, artifacts, technifacts, goods X^k instead of X^i.

Finally, we get conditions for evolutionary dynamics, depending on the selective causal factors A^i and Q^i and D^i:

(8) If $A^i Q^i > D^i$, then the average marginal growth of a societal form within a population or subculture, caused by an optimal realization of a successful solution, increases.

(9) If $A^i Q^i < D^i$, then this clearly shows a failure of any man-made interference into the evolutionary societal process.

(10) $W^i = A^i Q^i - D^i$. W^i is a typical selective causal factor and indicates an excess of growth of a certain societal form X^i, i.e., the spread of an individual societal form X^i.

From the value-theoretical interpretation in the matrix form (2) we know that the expected value of a societal form W^i for an individual i, $W^i = (L^i)$, see (3); therefore, W^i can be expressed mathematically as the expected utility of an evolutionary lottery, representing the short-term evaluation of a possible future trajectory by an individual i at the time t; it is equivalent to $\Sigma_j y_j x_l a_{lj}$, according to (3). Now W^{i*} represents the mean average value of the evolution at a certain time point and is represented by the average value of all lotteries (L) = $\Sigma_i \Sigma_j x_i a_{ij} y_j$ (Leinfellner 1984: 257).

A general default decision-rule for an optimal solution of a conflict, if aversion of risk prevails, is given by:

(11) If advantageous distributions of random events prevail and/or advantageous societal interference in the societal evolution exist, then:

(12) $W^i \geq W^{i*}$ holds. It indicates that the societal form X^i has gained the upper hand and stays relatively stable. This evolutionary stability expressed by

(12) and (14) is equal to Maynard Smith's formulation of an optimal evolutionary strategy, (13):

(13) $xa_{ij}\ e < ea_{ij}\ e$, or, if $xa_{ij}\ e = ea_{ij}\ e$, then $xa_{ij}\ x < ea_{ij}\ x$ (Maynard Smith 1981: 14ff, 198; Leinfellner 1984: 260ff). That means that a haphazard societal form arising in small numbers will be less fit then *e*, because it will lose out in any case either against *e* or against itself. In a perfect and free democratic society, consisting entirely of individuals adopting the evolutionarily stable strategy *e*, any emerging rare variants of undemocratic societal forms which adopt a different, undemocratic strategy X^e would never have a chance to increase in frequency (for more details see Maynard Smith 1982: 56, 204; Leinfellner 1984: 261).

If (12) does not happen, and if random events occur in great numbers, and/or detrimental societal realizations of solutions and/or environmental catastrophes set in, the societal form X^i may vanish from the society, and chaotic attractors may get he upper hand, as indicated by the following condition:

(13) $W^i < W^{i*}$

This indicates that the societal form X^i becomes instable and will probably cease to exist or go extinct.

This model should give some scientific insight into our abilities to interfere to our advantage or when we should not interfere into our societal evolution. Contrary to the pessimistic beliefs of some evolutionists that mankind may become extinct sooner or later, just as 90% of all species on earth before, or is completely dependent on randomness, this model and article shows something different: Once societies become democracies and technological civilizations and continue to support their scientific-technological progress, we can take our evolutionary future fate more and more into our own hands, unless global or cosmic catastrophes occur. Along with it, we must face and weigh carefully the great societal risk and responsibility as well as the great potentiality posed by our unprecedented scientific, technical, and cultural power. Thercforc, under the condition that we maintain our democratic welfare state, support our technological advances, and continue to increase our scientific knowledge and research of the evolutionary process, we have the best chance that the optimal continuation of our societal evolution is not an exception and extinction will not become the rule.

REFERENCES

R. Axelrod, *The Evolution of Cooperation*. New York 1984.

Per Bak, *How Nature works*. New York 1966.

A.R. Damasio, *Descartes' Error, Emotion, Reason and the Human Brain*. New York 1994.

M. Eigen/P.Schuster, *The Hypercycle: A Principle of Natural Self-organization*. Berlin 1979.

J.W. Friedman, *Game Theory with Applications to Economics*. New York 1986.

H. Götschl (ed.), *Revolutionary Changes in Understanding Man and Society* (Theory and Decision Library). Dordrecht 1995.
J.C. Harsanyi, "Games with Incomplete Information", in: *Zeitschrift für Wissenschaftsforschung*, 9/10, 1995, pp.5-31.
J.C. Harsanyi, *Papers in Game Theory* (Theory and Decision Library). Dordrecht 1982.
J.C. Harsanyi, *Essays on Ethics, Social Behavior, and Scientific Explanation* (Theory and Decision Library). Dordrecht 1976.
D. Helbing, *Quantitative Sociodynamics* (Theory and Decision Library). Dordrecht 1995.
J.H. Holland et al., *Induction*. Boston, MA 1986.
J.H. Holland, *Hidden Order*. Reading, MA 1995.
S.A. Kauffman, *The Origins of Order*. New York, NY 1993.
W. Leinfellner, "The New Theory of Evolution: A Theory of Democratic Societies", in: H. Götschl (ed.), *Revolutionary Changes in Understanding Man and Society* (Theory and Decision Library). Dordrecht 1995, pp.149-189.
W. Leinfellner, "Evolution und Zufall: Risiko und Lotterien", in: Forschungsgruppe Konsum und Verhalten (ed.), *Konsumentenforschung*. München.1994.
W. Leinfellner, "Ein Plädoyer für die Sozialethik", in: W. Lütterfelds (ed.), *Evolutionäre Ethik zwischen Naturalismus und Idealismus*. Darmstadt 1993, pp.32-65.
W. Leinfellner, "Traditional Ethics, Ethical Decision Theory and Evolutionary Ethics", in: *Absolute Values and the Reassessment of the Contemporary World*. New York 1991, pp.229-250.
W. Leinfellner, "Nature is the Real Designer", in: *Austria Today*, 2, 1988, pp.39-43.
W. Leinfellner, "The Prisoner's Dilemma and its Evolutionary Iteration", in: A. Diekmann/P. Mitterer (Eds.), *The Paradoxical Effects of Social Behavior*. Heidelberg 1986, pp.135-149.
W. Leinfellner, "Evolutionary Causality, Theory of Games, and Evolution of Intelligence", in: F. Wuketits (ed.), *Concepts and Approaches in Evolutionary Epistemology* (Theory and Decision Library), Boston 1984, pp.233-27.
W. Leinfellner, "Statistische Kausalität in den Sozialwissenschaften", in: J. Posch. (ed.), *Kausalität, neue Texte*. Stuttgart 1982, pp.221-260.
Ch. Lumsden/E.O. Wilson, *Genes, Mind and Culture*. Cambridge 1981.
J. Maynard Smith, *Evolution and the Theory of Games*. Oxford 1982.
A. Rapoport, *Mathematische Methoden in den Sozialwissenschaften*. Würzburg 1980.
R. Penrose, *The Language, the Small and the Human Mind*. Cambridge 1997.
R. Schuster, "How does Complexity Arise in Evolution?", in: *Santa Fe Institute working papers*, SFI 96-05-26. Santa Fe 1996.
P. Schuster/K. Sigmund, "Towards a Dynamic of Social Behavior, Strategic and Genetic Models or the Evolution of Animal Conflicts", in: *J. Soc. Biol. Struct.*, 8, 1985, pp.255-277.
R. Selten, *Models of Strategic Rationality* (Theory and Decision Library). Dordrecht 1988.
K. Sigmund, *Games of Life: Explorations in Ecology, Evolution and Behavior*. Oxford 1993.
H.A. Simon, *The Sciences of the Artificial*. Cambridge, MA, 2nd ed. 1979.
H.A. Simon, *The Theory of Scientific Discovery*. Hillsdale 1995.
J.W. Weibull, *Evolutionary Game Theory*. Cambridge, MA. 1996.

Department of Philosophy
University of Nebraska
Lincoln
USA

Ludwig Boltzmann Institute
Karl Franzens-University Graz
A–8020 Graz
Austria

DIRK HELBING

MICROSCOPIC FOUNDATION OF STOCHASTIC GAME DYNAMICAL EQUATIONS

I. INTRODUCTION

Since von Neumann and Morgenstern initiated the field of game theory,[1] it has often proved of great value for the quantitative description and understanding of competition and co-operation between individuals. Game theory focusses on two questions: 1. Which is the optimal strategy in a given situation? 2. What is the dynamics of strategy choices in cases of repeatedly interacting individuals? In this connection game dynamical equations[2] find a steadily increasing interest. Although they agree with the replicator equations of evolution theory (cf. Sec. II), they cannot be justified in the same way. Therefore, we will be looking for a foundation of the game dynamical equations which is based on individual actions and decisions (cf. Sec. IV).

In addition, we will formulate a stochastic version of evolutionary game theory (cf. Sec. III). This allows us to investigate the effects of fluctuations on the dynamics of social systems. In order to illustrate the essential ideas, a concrete model for the self-organization of behavioral conventions is presented (cf. Sec. V). We will see that the game dynamical equations describe the average evolution of social systems only for restricted time periods. Therefore, a criterium for their validity will be developed (cf. Sec. VI). Finally, we will present possible extensions to more general behavioral models and discuss the actual meaning of the game dynamical equations (cf. Sec. VII).

II. THE GAME DYNAMICAL EQUATIONS

Let $p_x(t)$, such that

$$0 \leq p_x(t) \leq 1 \qquad \text{and} \qquad \sum_x p_x(t) = 1, \tag{1}$$

denote the *proportion* of individuals pursuing the *behavioral strategy* $x \in S$ at time t. We assume the strategies considered to be mutually exclusive. The set S of strategies may be discrete or continuous, finite or infinite. The only difference will be that sums over x are to be replaced by integrals in cases of continuous sets. By A_{xy} we will denote the possibly time-dependent *payoff* for an individual using strategy x when confronted with an individual pursuing strategy y. Hence, his/her

W. Leinfellner and E. Köhler (eds.), Game Theory, Experience, Rationality, 211–224.

expected success $\langle E_x \rangle_t$ will be given by the weighted mean value

$$\langle E_x \rangle_t = \sum_y A_{xy}\, p_y(t)\,, \tag{2}$$

since p_y is the probability that the interaction partner uses strategy y. In addition, the *average expected success* will be

$$\overline{\langle E \rangle_t} = \sum_x p_x(t)\langle E_x \rangle_t = \sum_x \sum_y p_x(t) A_{xy}\, p_y(t)\,. \tag{3}$$

Assuming that the relative temporal increase $(dp_x/dt)/p_x$ of the proportion p_x of individuals pursuing strategy x is proportional to the difference between the expected success $\langle E_x \rangle_t$ and the average expected success $\overline{\langle E \rangle_t}$, we obtain the *game dynamical equations*

$$\begin{aligned} \frac{dp_x(t)}{dt} &= \nu p_x(t)\left[\langle E_x \rangle_t - \overline{\langle E \rangle_t}\right] \\ &= \nu p_x(t)\Big[\langle E_x \rangle_t - \sum_y p_y(t)\langle E_y \rangle_t\Big]\,, \end{aligned} \tag{4}$$

where the possibly time-dependent proportionality factor ν is a measure for the *interaction rate* with other individuals. According to (4), the proportions of strategies with an above-average success $\langle E_x \rangle_t > \overline{\langle E \rangle_t}$ increase, whereas the other strategies will be diminished. Note, that the proportion of a strategy does not necessarily increase or decrease monotonically. Certain payoffs are associated with an *oscillatory* or even *chaotic* dynamics[3].

Equations (4) are identical with the replicator equations from evolutionary biology. They can be extended to the *selection-mutation equations*

$$\begin{aligned} \frac{dp_x(t)}{dt} &= \nu p_x(t)\Big[\langle E_x \rangle_t - \sum_y p_y(t)\langle E_y \rangle_t\Big] \\ &+ \sum_y \big[p_y(t) w_1(y \to x) - p_x(t) w_1(x \to y)\big]\,. \end{aligned} \tag{5}$$

The terms which agree with (4) describe a selection of superior strategies. The new terms correspond to the effect of mutations, i.e. to *spontaneous* changes from strategy x to other strategies y with possibly time-dependent *transition rates* $w_1(x \to y)$ (last term) and the inverse transitions. They allow to describe *trial and error behavior* or behavioral fluctuations.

III. STOCHASTIC DYNAMICS: THE MASTER EQUATION

Let us consider a social system consisting of a constant number

$$N = \sum_x n_x(t) \tag{6}$$

of individuals. Herein, $n_x(t)$ denotes the number of individuals who pursue strategy x at time t. Hence, the time-dependent vector

$$\vec{n} = (n_1, n_2, \dots, n_x, \dots, n_y, \dots) \tag{7}$$

reflects the *strategy distribution* in the social system and is called the *socioconfiguration*. If the individual strategy changes are subject to random fluctuations (e.g. due to trial and error behavior or decisions under uncertainty), we will have a stochastic dynamics. Therefore, given a certain socioconfiguration $\vec{n}_0$ at time t_0, for the occurence of the strategy distribution $\vec{n}$ at a time $t > t_0$ we can only calculate a certain *probability* $P(\vec{n}, t)$. Its temporal change dP/dt is governed by the so-called *master equation*[4]

$$\frac{dP(\vec{n}, t)}{dt} = \sum_{\vec{n}\,'} \left[P(\vec{n}\,', t) W(\vec{n}\,' \to \vec{n}) - P(\vec{n}, t) W(\vec{n} \to \vec{n}\,') \right] . \tag{8}$$

The sum over $\vec{n}\,'$ extends over all socioconfigurations fulfilling $n_x \in \{0, 1, 2, \dots\}$ and (6).

According to equation (8), an *increase* of the probability $P(\vec{n}, t)$ of having socioconfiguration $\vec{n}$ is caused by transitions from other socioconfigurations $\vec{n}\,'$ to $\vec{n}$. While a *decrease* of $P(\vec{n}, t)$ is related to changes from $\vec{n}$ to other socioconfigurations $\vec{n}\,'$. The corresponding changing rates are proportional to the *configurational transition rates* $W(\vec{n} \to \vec{n}\,')$ of changes to socioconfigurations $\vec{n}\,'$ *given* the socioconfiguration $\vec{n}$ and to the probability $P(\vec{n}, t)$ of *having* socioconfiguration $\vec{n}$ at time t.

The configurational transition rates W have the meaning of transition probabilities per time unit and must be non-negative quantities. Frequently, the individuals can be assumed to change their strategies independently of each other. Then, the configurational transition rates have the form

$$W(\vec{n} \to \vec{n}\,') = \begin{cases} n_x w(x \to y; \vec{n}) & \text{if } \vec{n}\,' = \vec{n}_{xy} \\ 0 & \text{otherwise,} \end{cases} \tag{9}$$

i.e. they are proportional to the number n_x of individuals who may change their strategy from x to another strategy y with an *individual transition rate* $w(x \to y; \vec{n}) \geq 0$. In relation (9), the abbreviation

$$\vec{n}_{xy} = (n_1, n_2, \dots, n_x - 1, \dots, n_y + 1, \dots) \tag{10}$$

means the socioconfiguration which results after an individual has changed his/her strategy from x to y.

It can be shown that the master equation has the properties

$$P(\vec{n}, t) \geq 0 \qquad \text{and} \qquad \sum_{\vec{n}} P(\vec{n}, t) = 1 \tag{11}$$

for all times t, if they are fulfilled at some initial time t_0. Therefore, the master equation actually describes the temporal evolution of a probability distribution.

IV. APPROXIMATE MEAN VALUE EQUATIONS

In order to connect the stochastic model to the game dynamical equations, we must specify the individual transition rates w in a suitable way. Therefore, we derive the mean value equations related to the master equation (8) and compare them to the selection-mutation equations (5).

The proportion p_x is defined as the *mean value*

$$\langle f \rangle_t = \sum_{\vec{n}} f(\vec{n},t) P(\vec{n},t) \tag{12}$$

of the number $f(\vec{n},t) = n_x$ of individuals pursuing strategy x, divided by the total number N of considered individuals:

$$p_x(t) = \frac{\langle n_x \rangle_t}{N} = \frac{1}{N} \sum_{\vec{n}} n_x P(\vec{n},t) \, . \tag{13}$$

Taking the time derivative of $\langle n_x \rangle_t$ and inserting the master equation gives

$$\begin{aligned} \frac{d\langle n_x \rangle_t}{dt} &= \sum_{\vec{n}} n_x \big[P(\vec{n}',t) W(\vec{n}' \to \vec{n}) - P(\vec{n},t) W(\vec{n} \to \vec{n}') \big] \\ &= \sum_{\vec{n}} (n'_x - n_x) W(\vec{n} \to \vec{n}') P(\vec{n},t) \, , \end{aligned} \tag{14}$$

where we have interchanged $\vec{n}$ and $\vec{n}'$ in the first term on the right hand side. Taking into account relation (9), we get

$$\begin{aligned} \frac{d\langle n_x \rangle_t}{dt} &= \sum_{\vec{n}_{yx}} n_y w(y \to x; \vec{n}) P(\vec{n},t) - \sum_{\vec{n}_{xy}} n_x w(x \to y; \vec{n}) P(\vec{n},t) \\ &= \sum_{y} \big[n_y w(y \to x; \vec{n}) - n_x w(x \to y; \vec{n}) \big] P(\vec{n},t) \, . \end{aligned} \tag{15}$$

With (13) this finally leads to the *approximate mean value equations*

$$\frac{dp_x(t)}{dt} = \sum_{y} \big[p_y(t) w(y \to x; \langle \vec{n} \rangle_t) - p_x(t) w(x \to y; \langle \vec{n} \rangle_t) \big] \tag{16}$$

However, these are only exact if the individual transition rates w are independent of the socioconfiguration $\vec{n}$. In any case, they are *approximately* valid as long as the probability distribution $P(\vec{n},t)$ is narrow, so that the mean value $\langle f(\vec{n},t) \rangle_t$ of a function $f(\vec{n},t)$ can be replaced by the function $f(\langle \vec{n} \rangle_t, t)$ of the mean value. This problem will be discussed in detail later on.

Comparing the rate equations (16) with the selection-mutation equations (5), we find a complete correspondence for the case

$$w(y \to x; \vec{n}) = w_1(y \to x) + w_2(y \to x) \, n_x \tag{17}$$

with

$$w_2(y \to x) = \frac{\nu}{N} \max(E_x - E_y, 0) \tag{18}$$

and the *success*

$$E_x = \sum_y A_{xy} \frac{n_y}{N} , \tag{19}$$

since

$$\max(\langle E_x \rangle_t - \langle E_y \rangle_t, 0) - \max(\langle E_y \rangle_t - \langle E_x \rangle_t, 0) = \langle E_x \rangle_t - \langle E_y \rangle_t . \tag{20}$$

Whereas w_1 is again the mutation rate (i.e. the rate of spontaneous transitions), the additional term in (17) describes *imitation processes,* where individuals take over the strategy x of their respective interaction partner. Imitation processes correspond to pair interactions of the form

$$y + x \to x + x . \tag{21}$$

Their frequency is proportional to the number n_x of interaction partners who may convince an individual of strategy x. The proportionality factor w_2 is the *imitation rate*.

Relation (18) is called the *proportional imitation rule* and can be shown to be the best learning rule.[5]

It was discovered in 1992[6] and says that an imitation behavior only takes place if the strategy x of the interaction partner turns out to have a greater success E_x than one's own strategy y. In such cases, the imitation rate is proportional to the difference $(E_x - E_y)$ between the success of the alternative x and the previous strategy y, i.e. strategy changes occur more often the greater the advantage of the new strategy x would be.

All specifications of the type

$$w_2(y \to x) = C + \frac{\nu}{N} [\lambda E_x - (1 - \lambda) E_y] \tag{22}$$

with an arbitrary parameter λ also lead to our game dynamical equations. However, individuals would then, with a certain rate, take over the strategy x of the interaction partner, even if its success E_x is smaller than that of the previously used strategy y. Moreover, if C is not chosen sufficiently large, the individual transition rates $w \geq 0$ can become negative.

In summary, we have found a microscopic foundation of evolutionary game theory which is based on four plausible assumptions: 1. Individuals evaluate the success of a strategy as its average payoff in interactions with other individuals (cf. (19)). 2. They compare the success of their strategy with that of the respective interaction partner, basing on observations or an exchange of experiences. 3. Individuals imitate each others' behavior. 4. In doing so, they apply the proportional imitation rule (18) [or (22)].

V. SELF-ORGANIZATION OF BEHAVIORAL CONVENTIONS

For illustrative reasons, we will now discuss an example which allows us to understand how social conventions emerge. We consider the simple case of two alternative strategies $x \in \{1, 2\}$ and assume them to be equivalent so that the payoff matrix is symmetrical:

$$\left(A_{xy}\right) = \begin{pmatrix} A+B & B \\ B & A+B \end{pmatrix} . \tag{23}$$

If $A > 0$, the additional payoff A reflects the *advantage* of using the same strategy as the respective interaction partner. This situation holds, for example, in cases of network externalities like in the historical rivalry between the video systems VHS and BETA MAX.[7] Finally, the mutation rates are taken to be constant, i.e. $w_1(x \to y) = W_1$.

The resulting game dynamical equations are

$$\frac{dp_x(t)}{dt} = -2\left[p_x(t) - \frac{1}{2}\right]\left\{W_1 + \nu A p_x(t)\left[p_x(t) - 1\right]\right\} . \tag{24}$$

Obviously, they have only *one* stable stationary solution if the (control) parameter

$$\kappa = 1 - \frac{4W_1}{\nu A} \tag{25}$$

is smaller than zero. However, for $\kappa > 0$ equation (24) can be rewritten in the form

$$\frac{dp_x(t)}{dt} = -2\nu A\left[p_x(t) - \frac{1}{2}\right]\left[p_x(t) - \frac{1+\sqrt{\kappa}}{2}\right]\left[p_x(t) - \frac{1-\sqrt{\kappa}}{2}\right] . \tag{26}$$

The stationary solution $p_x = 1/2$ is unstable, then, but we have two new stable stationary solutions $p_x = (1/2 \pm \sqrt{\kappa}/2)$. That is, dependent on the detailed initial condition, one strategy will gain the majority of users although both strategies are completely equivalent. This phenomenon is called *symmetry breaking*. It will be suppressed if the mutation rate W_1 is larger than the advantage effect $\nu A/4$.

The above model allows us to understand how behavioral conventions come about. Examples are the pedestrians' preference for the right-hand side (in Europe), the revolution direction of clock hands, the direction of writing, or the already mentioned triumph of the video system VHS over BETA MAX.

It is very interesting how the above-mentioned symmetry breaking affects the probability distribution $P(\vec{n}, t) = P(n_1, n_2, t) = P(n_1, N - n_1, t)$ of the related stochastic model (cf. Fig. 1[8]). For $\kappa < 0$ the probability distribution is located around $n_1 = N/2 = n_2$ and stays small, so that the approximate mean value equations are applicable. At the so-called *critical point* $\kappa = 0$, a *phase transition* to a qualitatively different system behavior occurs and the probability distribution becomes very broad. As a consequence, the game dynamical equations do not correctly describe the temporal evolution of the mean strategy distribution anymore.

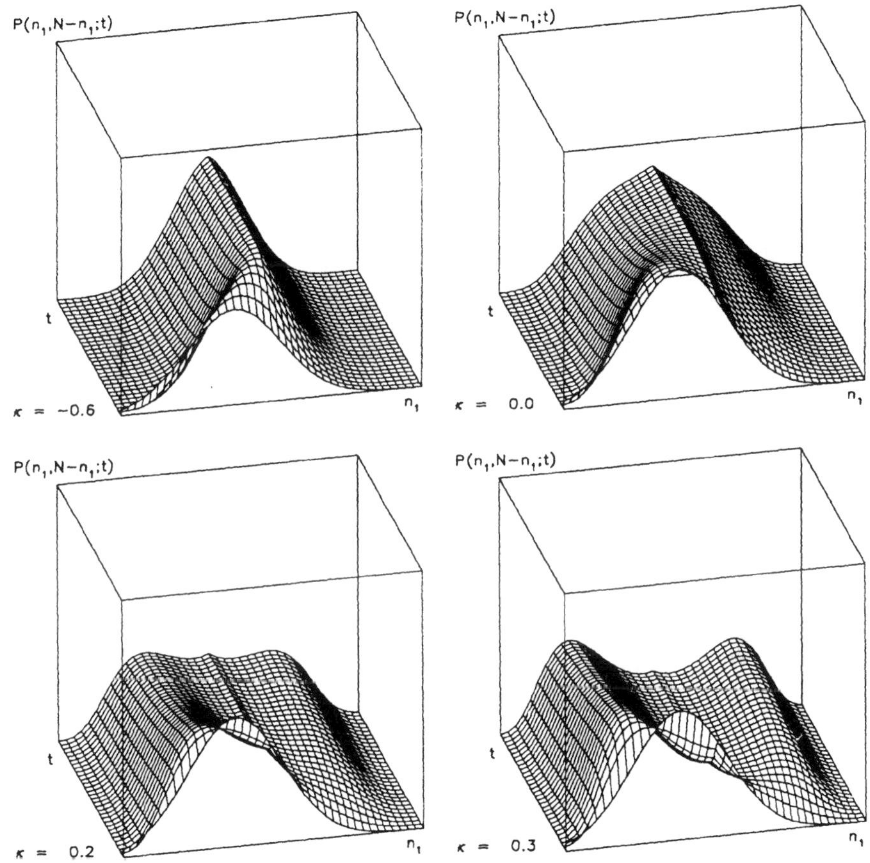

Figure 1: Probability distribution $P(\vec{n}, t) = P(n_1, N - n_1; t)$ of the socioconfiguration $\vec{n}$ for varying values of the control parameter κ according to the stochastic version of the game dynamical equations.

For $\kappa > 0$, a bimodal and symmetrical probability distribution evolves. That is, the likelihood that one of the two equivalent strategies will win throughout is much larger than the likelihood of finding approximately equal proportions of both strategies. At the beginning, the initial state or maybe some random fluctuation determines which strategy has better odds of winning. However, in the long run both strategies have exactly the same odds. It is clear that in such cases the game dynamical equations fail to describe the mean system behavior (cf. Fig. 2) which would correspond to the average temporal evolution of an ensemble of identical social systems. In cases of oscillatory or chaotic solutions of the game dynamical equations the situation is even worse.

VI. EXACT, APPROXIMATE, AND CORRECTED MEAN VALUES AND VARIANCES

In the last section we have seen that the approximate mean value equations

$$\frac{d\langle n_x\rangle_t}{dt} = M_x(\langle\vec{n}\rangle_t), \tag{27}$$

with the so-called *first jump moments*

$$M_x(\vec{n}) = \sum_{\vec{n}\,'}(n'_x - n_x)W(\vec{n} \to \vec{n}\,') \tag{28}$$

(cf. (14)), are not sufficient. This calls for corrected mean value equations and a criterion of validity for the time period. If the individual transition rates $w(x \to y; \vec{n})$ depend on the socioconfiguration, the exact mean value can only be evaluated via formula (13). This requires the calculation of the probability distribution $P(\vec{n}, t)$ and, therefore, the numerical solution of the respective master equation (8). Since the number of possible socioconfigurations is normally very large, an extreme amount of computer time would be necessary for this.

Luckily, it is possible to derive from (14) the *corrected mean value equations*

$$\frac{\partial\langle n_x\rangle_t}{\partial t} = M_x(\langle\vec{n}\rangle_t) + \frac{1}{2}\sum_y\sum_{y'}\sigma_{yy'}(t)\frac{\partial^2 M_x(\langle\vec{n}\rangle_t)}{\partial\langle n_y\rangle_t\partial\langle n_{y'}\rangle_t} \tag{29}$$

by means of a suitable Taylor approximation. This equation depends on the *covariances*

$$\sigma_{xy}(t) = \Big\langle \big(n_x - \langle n_x\rangle_t\big)\big(n_y - \langle n_y\rangle_t\big)\Big\rangle_t = \sum_{\vec{n}} \big(n_x - \langle n_x\rangle_t\big)\big(n_y - \langle n_y\rangle_t\big)P(\vec{n}, t)\,, \tag{30}$$

which can be determined by means of the *covariance equations*

$$\begin{aligned}\frac{\partial\sigma_{xx'}(t)}{\partial t} &= M_{xx'}(\langle\vec{n}\rangle_t) + \frac{1}{2}\sum_y\sum_{y'}\sigma_{yy'}(t)\frac{\partial^2 M_{xx'}(\langle\vec{n}\rangle_t)}{\partial\langle n_y\rangle_t\partial\langle n_{y'}\rangle_t}\\ &+ \sum_y\left[\sigma_{xy}(t)\frac{\partial M_{x'}(\langle\vec{n}\rangle_t)}{\partial\langle n_y\rangle_t} + \sigma_{x'y}(t)\frac{\partial M_x(\langle\vec{n}\rangle_t)}{\partial\langle n_y\rangle_t}\right]. \end{aligned} \tag{31}$$

The functions

$$M_{xy}(\vec{n}) = \sum_{\vec{n}'}(n'_x - n_x)(n'_y - n_y)W(\vec{n} \to \vec{n}') \tag{32}$$

are called the *second jump moments.*

Equations (29) and (31) build a closed system of equations, but still no exact one, since this would depend on higher moments of the form $\langle n_x n_y n_z \cdots \rangle_t$. Nevertheless, according to Figure 2 the corrected mean value equations yield significantly better results than the approximate ones. As a consequence, they are valid for a much longer time period. Suitable *validity criteria* are the *relative variances*

$$V_x(t) := \frac{\sigma_{xx}(t)}{(\langle n_x \rangle_t)^2}, \tag{33}$$

since these are a measure for the relative width of the probability distribution $P(\vec{n}, t)$. It can be shown that the covariances and all higher moments are small, if only $V_x(t)$ is much smaller than 1 for every x. Numerical investigations indicate that the approximate mean value equations begin to separate from the exact ones as soon as one of the relative variances $V_x(t)$ becomes greater than 0.04. The corrected mean value equations and covariances remain reliable as long as $V_x(t)$ is smaller than 0.12 for all x (cf. Fig. 2).

A more detailled discussion of the above matter is presented elsewhere.[9]

VII. DIVERSE GENERALIZATIONS

The behavioral model discussed above can be generalized in different respects.

Modified transition rates: The strange cusp at $n_1 = N/2$ in Figure 1, which comes from the discontinuous derivative of $w_2(x \to y)$ at $E_x = E_y$, can be avoided by using modified imitation rates:

$$w_2(y \to x) = \frac{\nu}{N} \frac{\exp(E_x - E_y)}{D_{xy}} \quad \text{with} \quad D_{xy} = D_{yx} = 2\,. \tag{34}$$

This approach agrees with relation (22) in linear approximation for $C = \nu/(2N)$ and $\lambda = 1/2$, but it always yields non-negative imitation rates. As in (18) it guarantees two essential things: 1. The imitation rate grows with an increasing gain $(E_x - E_y)$ of success. 2. If the alternative strategy x is inferior, the imitation rate is very small (but, due to uncertainty, not negligible). The results of the corresponding stochastic behavioral model are presented in Figure 3. They show the usual flatness of the probability distribution $P(n_1, N - n_1, t)$ at the critical point $\kappa = 0$, where again a phase transition occurs.

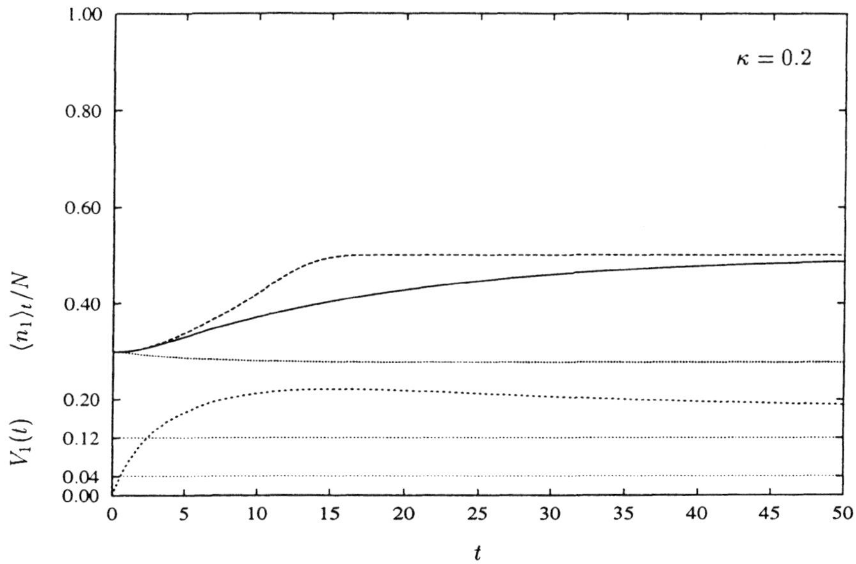

Figure 2: The numerical solutions of the approximate mean value equations (· · ·) agree with those of the exact mean value equations (—) only for a short time interval. The corrected mean value equations (– –) yield much better results, although they also deviate from the exact curves when the relative variances (- - -) become too large. Nevertheless, they describe the average long-term behavior properly.

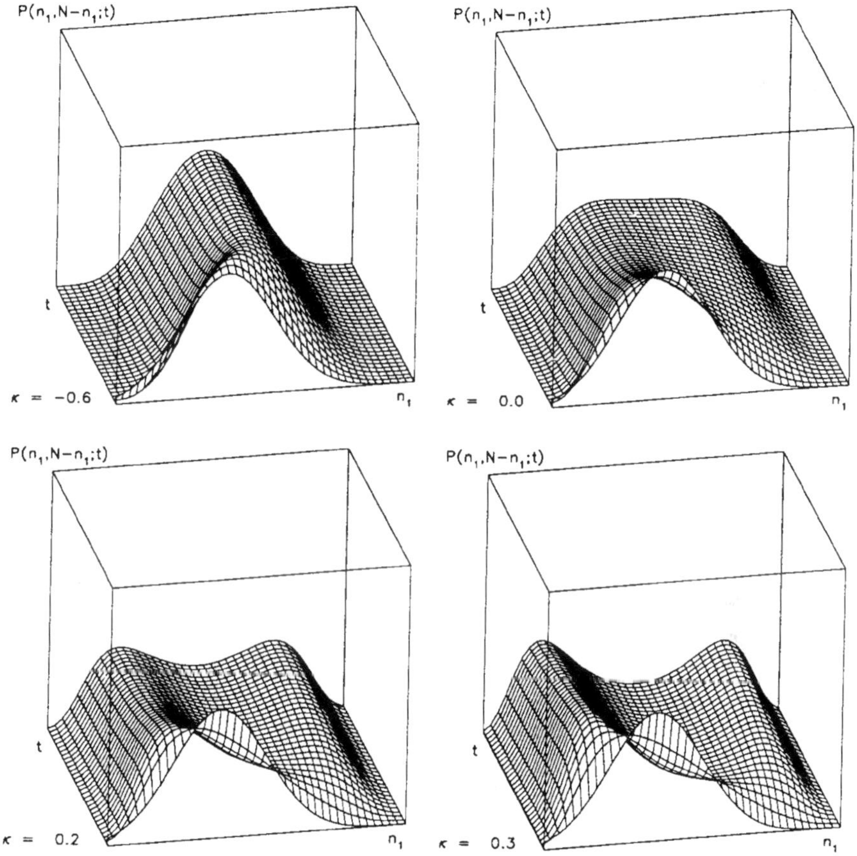

Figure 3: Probability distribution $P(\vec{n}, t) = P(n_1, N - n_1; t)$ of the socioconfiguration $\vec{n}$ according to the modified stochastic game dynamical equations.

Dynamics with expectations: The decisions of individuals are often influenced by their *expectations* $\langle E_x \rangle^*_{t'}$ about the success of a strategy x at future times $t' > t$. These will base on some kind of extrapolation of past experiences with the success of x. If expected payoffs at future times t' were to be weighted exponentially by their distance $(t' - t)$ from the present time t, one would set[10]

$$\langle E_x \rangle_t = \frac{1}{T} \int_t^\infty dt' \, \langle E_x \rangle^*_{t'} \, \exp\left(\frac{t' - t}{T} \right) . \tag{35}$$

Other kinds of pair interactions: Apart from imitative behavior, individuals also sometimes show an *avoidance behavior*

$$x + x \rightarrow y + x \, , \tag{36}$$

especially if they dislike their interaction partner (so-called 'snob effect'). This can be taken into account by an additonal contribution to the individual interaction rates:

$$w(y \rightarrow x; \vec{n}) = w_1(y \rightarrow x) + w_2(y \rightarrow x) \, n_x + w_3(y \rightarrow x) \, n_y \, , \tag{37}$$

where w_3 denotes the *avoidance rate.*

Several subpopulations: Sometimes one has to distinguish different *subpopulations* a, i.e. different kinds of individuals. This is necessary if not all individuals have the same set S of strategies.[11] A similar thing holds if the social system considered consists of competing groups, where only individuals of the same group behave cooperatively. The generalized behavioral equations are[12]

$$\frac{dp^a_x(t)}{dt} = \sum_y \left[p^a_y(t) w^a(y \rightarrow x; \langle \vec{n} \rangle_t) - p^a_x(t) w^a(x \rightarrow y; \langle \vec{n} \rangle_t) \right] \tag{38}$$

with individual interaction rates of the form

$$w^a(y \rightarrow x; \vec{n}) = w^a_1(y \rightarrow x) + \sum_b \left[w^{ab}_2(y \rightarrow x) \, n^b_x + w^{ab}_3(y \rightarrow x) \, n^b_y \right] . \tag{39}$$

Inclusion of memory effects: If the strategy distribution at past times $t' < t$ influences present decisions in a non-Markovian way, the approximate mean value equations have the form

$$\frac{dp^a_x(t)}{dt} = \sum_y \int_{-\infty}^t dt' \left[p^a_y(t') w^a_{t-t'}(y \rightarrow x; \langle \vec{n} \rangle_{t'}) - p^a_x(t') w^a_{t-t'}(x \rightarrow y; \langle \vec{n} \rangle_{t'}) \right] . \tag{40}$$

For example, in cases of an exponentially decaying memory, one would have

$$w^a_{t-t'}(x \rightarrow y; \langle \vec{n} \rangle_{t'}) = w^a(x \rightarrow y; \langle \vec{n} \rangle_{t'}) \, \frac{1}{\tau} \exp\left(\frac{t - t'}{\tau} \right) . \tag{41}$$

VIII. SUMMARY AND CONCLUSIONS

We have found a microscopic foundation for the game dynamical equations which are based on a certain kind of imitative behavior. Moreover, a stochastic version of evolutionary game theory has been formulated. It allowed us to understand the self-organization of social conventions as a phase transition which is related with symmetry breaking. Moreover, we have seen that the game dynamical equations correspond to approximate mean value equations. Normally, they agree with the mean-value equations of stochastic game theory for a certain time period only, which can be determined by calculating the relative variances. For an improved description of the average system behavior we have derived corrected mean value equations which require the solution of additional covariance equations.

The interpretation of the game dynamical equations proceeds by reformulating these in terms of a *social force model,*[13] assuming a continuous strategy set:

$$\frac{dx_\alpha(t)}{dt} = f_1(x_\alpha) + \sum_{\beta(\neq\alpha)} f_2(x_\alpha, x_\beta) + \text{fluctuations}. \tag{42}$$

The force term

$$f_1(x_\alpha) = \int dx\,(x - x_\alpha)\, w_1(x_\alpha \to x) \tag{43}$$

delineates spontaneous strategy changes by individual α, whereas

$$\begin{aligned} f_2(x_\alpha, x_\beta) &= (x_\beta - x_\alpha)\, w_2(x_\alpha \to x_\beta) \\ &+ \int dx\,(x - x_\alpha)\, w_3(x_\alpha \to x)\, \delta(x_\alpha - x_\beta) \end{aligned} \tag{44}$$

is the *interaction force* which originates from individual β and influences individual α. Here, $\delta(x - y)$ denotes *Dirac's delta function* (which yields a contribution for $x = y$ only). According to (42), the game dynamical equations describe the *most probable strategy changes* rather than the average (representative) evolution of a social system. Therefore, they neglect the effects of fluctuations on the system behavior.

A more detailled discussion of the results presented in this paper is available elsewhere[12,13].

REFERENCES

1. J. von Neumann/O. Morgenstern, *Theory of Games and Economic Behavior*. Princeton: Princeton University Press 1944.
2. P. Taylor/L. Jonker, "Evolutionarily Stable Strategies and Game Dynamics", in: *Math. Biosciences*, 40, 1978, pp.145–156; J. Hofbauer and K. Sigmund, *Evolutionstheorie und dynamische Systeme*. Berlin: Parey 1984.
3. W. Schnabl/P.F. Stadler/C. Forst/P. Schuster, "Full Characterization of a Strange Attractor. Chaotic Dynamics in Low-Dimensional Replicator Systems", in: *Physica D*, 48, 1991, pp.65–90.

4. W. Weidlich/G. Haag, *Concepts and Models of a Quantitative Sociology. The Dynamics of Interacting Populations.* Berlin: Springer, 1983.
5. K. H. Schlag, "Why Imitate, and if so, How? A Bounded Rational Approach to Multi-Armed Bandits", Discussion Paper No. B-361, Department of Economics, University of Bonn.
6. D. Helbing, "Interrelations between Stochastic Equations for Systems with Pair Interactions", in: *Physica A*, 181, 1992, pp.29–52.
7. W. B. Arthur, "Competing Technologies, Increasing Returns, and Lock-In by Historical Events", in: *The Economic Journal* 99, 1989, pp. 116–131.
8. For illustrative reasons, a small number of individuals ($N = 40$) and a broad initial probability distribution have been chosen. In each picture, the box is twice as high as the maximal occuring value of the probability.
9. D. Helbing, "A Stochastic Behavioral Model and a 'Microscopic' Foundation of Evolutionary Game Theory", in: *Theory and Decision* 40, 1996, pp. 149–179.
10. N. S. Glance and B. A. Huberman, "Dynamics with Expectations", in: *Physics Letters A* 165, 1992, pp. 432–440.
11. J. Hofbauer and K. Sigmund, *The Theory of Evolution and Dynamical Systems.* Cambridge: Cambridge University Press 1988.
12. D. Helbing, *Quantitative Sociodynamics. Stochastic Methods and Models of Social Interaction Processes.* Dordrecht: Kluwer Academic 1995.
13. D. Helbing, *Stochastische Methoden, nichtlineare Dynamik und quantitative Modelle sozialer Prozesse,* 2nd edition. Aachen: Shaker 1996.

Institute of Theoretical Physics
Pfaffenwaldring 57/III
D–70550 Stuttgart
Germany

IMMANUEL M. BOMZE

UNIFORM BARRIERS AND EVOLUTIONARILY STABLE SETS*

I. INTRODUCTION

A key concept in evolutionary game theory is that of *evolutionary stability* [21, 20]. Originally defined for pairwise contests, i.e. symmetric two-person games, a mixed strategy x is said to be evolutionarily stable if it is a best reply to itself, and, moreover, is a better reply to any alternative best reply y than this is to itself. It has been shown that evolutionary stability has important implications for population dynamics based on evolutionary selection. In particular, an evolutionarily stable strategy, viewed as a population state in a finite game in which individuals are "programmed" to pure strategies, is *asymptotically stable* in the corresponding replicator dynamics [26].

However, many games of interest in economics do not possess any strategy which is evolutionarily stable, because the strict inequality in the criterion of evolutionary stability is never met. Therefore researchers have turned to weaker stability concepts such as neutral stability or evolutionarily stable sets (see e.g. [15, 17, 3, 30, 4]).

A mixed strategy x in a symmetric two-person game is said to be *neutrally stable* [20] if it is a best reply to itself and is not a worse reply to any alternative best reply y than this is to itself. Clearly, this criterion is not sufficient for asymptotic stability in the replicator dynamics, since, if, for instance, all payoffs are identical, then all strategies are neutrally stable but also stationary in the replicator dynamics. Hence, no perturbation of a state then induces a movement back towards the state. However, any neutrally stable strategy, viewed as a population state, is *Lyapunov stable* in the replicator dynamics, as applied to pairwise interactions. This result was established already by [28], where also the notion of an *evolutionarily stable set* (ES set) was introduced. G is said to be an ES set if all $x \in G$ are evolutionarily stable against any outsider $y \notin G$, but only neutrally stable against insiders $z \in G$. If one drops the latter requirement, one arrives at the notion of an ES* set recently introduced in [31].

In this study an alternative approach to these concepts is provided, emphasising the notion of barriers. The *invasion (neutrality) barrier* of a mixed strategy x against another mixed strategy y is the largest population share ε_y of y-strategists such that, for all smaller positive population shares ε, x earns a higher (not lower) payoff than y in the post-entry population mix $z = (1 - \varepsilon)x + \varepsilon y$.

The paper is organised as follows. Section II briefly reviews different known solution concepts considered in evolutionary game theory, with an emphasis on scores (or score functions) and barriers. In a novel approach, the notion of an ES set is then

W. Leinfellner and E. Köhler (eds.), Game Theory, Experience, Rationality, 225–243.

introduced and the role of barriers protecting sets rather than single strategies is discussed; in particular, how this approach is related to the recent work by Balkenborg and Schlag [1] on the structure of ES sets.

Section III focuses on uniformity of the protecting barriers, introducing the new notion of uniformly immune sets. The hierarchy of the concepts considered is specified, and related to external stability concepts like that of an ES* set. External stability has the appealing feature of ignoring the interaction between insiders. But it will be demonstrated that at least for thin externally stable sets (which are desired to predict behaviour as precisely as possible), these concepts coincide with those introduced previously.

Section IV deals with the standard case of pairwise contests. For this class of models studied most frequently in the literature, it is well known (see, e.g. [11]) that invasion and neutrality barriers for single strategies are always uniform. Here this result is established also for set-valued concepts. Moreover, an explicit description of the structure of ES sets is provided along the lines of [28] and [14].

Section V establishes the relationship of score and barriers properties to the behaviour under the replicator dynamics, and concludes with an example of an asymptotically stable set consisting entirely of neutrally stable strategies, although is not an ES set.

II. Barriers in evolutionary stability notions

Consider a large population of individuals each of whom is genetically or socially "programmed" to some pure strategy in a symmetric and finite n-player game $G = (N, \Theta, u)$. Here $N = \{1, 2, ..., n\}$ is the set of player positions, $M = \{1, \dots, m\}$ the set of pure strategies to each player position, and $\Theta = \Delta^n$ is the polyhedron of mixed-strategy profiles, where $\Delta = \{x \in \mathbb{R}^m_+ : \sum_{i\in M} x_i = 1\}$. For any pair of mixed strategies $x, y \in \Delta$, let $u(x, y)$ denote the average payoff to mixed strategy x when all other players play the same mixed strategy y. A strategy x is called a **(symmetric) Nash (equilibrium) strategy**, if $u(x, x) \geq u(y, x)$ for all $y \in \Delta$. Denoting transpose by $'$, u can be written as

$$u(x, y) = \sum_i x_i \varphi_i(y) = x'\varphi(y),$$

for some Lipschitz continuous function $\varphi : \Delta \to \mathbb{R}^m$, the **return function**. In this notation, x is a Nash strategy if and only if $(x - y)'\varphi(x) \geq 0$ for all $y \in \Delta$.

In the standard case of pairwise interactions, i.e., when $n = 2$, we have $u(x, y) = x'Ay$ where $A = (a_{ij})_{i,j\in M}$ is the payoff matrix to player position 1 (hence a_{ij} is the payoff to pure strategy i when played against pure strategy j). In this special case the return function is linear, $\varphi(y) = Ay$. For any i.i.d. randomisations $y \in \Delta$ in a symmetric n-player game ($n \geq 2$) with pure-strategy payoff function π, we have

$$\varphi_i(y) = \sum_{j_2\in M} \cdots \sum_{j_k\in M} \pi(i, j_2, \dots, j_k) y_{j_1} \cdots y_{j_k},$$

where $j_2, \ldots, j_k$ are the pure strategies for the other $n-1$ players. The function φ_i is thus polynomial, and it is non-linear if $n > 2$. The latter case is called *playing the field* in [20]. In general, the return $\varphi_i(x)$ to an i-individual within an x-population may depend on x in a complicated way, because the individual's payoff depends on the overall state of the population, such as in a model for traffic congestion or in the sex-ratio game [23], where $\varphi(x) = [1/x_1, 1/x_2]'$ if $\varepsilon \leq x_1 \leq 1-\varepsilon$, with continuous (e.g. piecewise constant) continuation outside this region. Note that Lipschitz continuity of the return function is usually required just for technical reasons, e.g. to ensure existence and uniqueness of trajectories under the replicator dynamics. From a game theory perspective, non-polynomial return functions may be used to model correlations over individual randomisations.

For the readers' convenience, first the key notions for pointwise evolutionary stability are briefly reviewed, cf. [11]. Let $x \in \Delta$ represent a pure-strategist population state, and let $y \in \Delta$ denote any other population state. Suppose a population share ε of mutants enters the population which initially is in state x, and suppose the mutant subpopulation is in state y. Then the post-entry population state is $z = (1-\varepsilon)x + \varepsilon y \in \Delta$. The average payoff to the sub-population of mutants is $u(y, z) = y'\varphi(z)$ and the average payoff to the residual, "incumbent", subpopulation is $u(x, z) = x'\varphi(z)$. We now introduce the notion of a **score function** which compares both these average payoffs. This argumentation is typical for the evolutionary approach of transferring a collective learning mechanism to individual behaviour (for related discussion see, e.g. [19, 32, 27, 29, 16, 2, 24, 25, 9, 8]) and holds irrespective of whether or not there is correlation in the sense used above.

Definition 1
Let $x, y \in \Delta$ be two strategies.
Then the **score function** *of y versus x is given by*

$$h_x(y, \varepsilon) = (y - x)'\varphi(z) = u[y, (1-\varepsilon)x + \varepsilon y] - u[x, (1-\varepsilon)x + \varepsilon y], \quad \varepsilon \in [0, 1].$$

By linearity of $u(y, z)$ in y we can express the score of any convex combination representing post-entry mixtures, $y^\lambda = (1-\lambda)x + \lambda y$ of x and y, versus any other such convex combination y^μ through the score $h_x(y, \varepsilon)$ as follows:

$$h_{y^\mu}(y^\lambda, \varepsilon) = (\lambda - \mu) h_x(y, \varepsilon\lambda + (1-\varepsilon)\mu). \tag{1}$$

Next we define two important variants of evolutionary **barriers** as in [11].

Definition 2
Define the **neutrality barrier** *of x against y as:*

$$b_x(y) = \inf(\{\varepsilon \in (0,1) : h_x(y, \varepsilon) > 0\} \cup \{1\}).$$

Similarly, the **invasion barrier** *of $x \in \Delta$ against any strategy $y \in \Delta$ is defined as:*

$$c_x(y) = \inf(\{\varepsilon \in (0,1) : h_x(y, \varepsilon) \geq 0\} \cup \{1\}).$$

This defines b_x and c_x as functions from Δ to $[0, 1]$. Clearly $b_x(y) \geq c_x(y)$ for all y by definition. If the graph of the score $h_x(y, .)$ intersects the ε-axis transversally at its smallest positive zero, then both $c_x(y)$ and $b_x(y)$ are located either there or at 0, and thus coincide. At first thought, this might be the generic situation, but in many games of interest some scores $h_x(y, \varepsilon) = 0$ for all ε also, so that $b_x(y) = 1 > 0 = c_x(y)$ and hence the barriers do not always coincide; see Example 1 below.

Let us say that $h_x(y, .)$ has the **zero-transversality property (ZTP)** if $h_x(y, .)$ either is constantly zero; or satisfies the transversality condition addressed above; or does not change sign on the whole interval $(0, 1)$ in the strict sense (then $b_x(y) = c_x(y) = 0$ or 1 by definition). For strictly positive invasion barriers $c_x(y) > 0$ we have the following observation which particularly applies to the standard case of pairwise contests, where scores are affine in ε (note that both assumptions in (a) and (b) below are met if scores are either strictly monotone or constant, cf. Example 2 below):

Proposition 1
Assume that $c_x(y) > 0$.
(a) If $h_x(y, .)$ *is monotone, then* $u(y, x) - u(x, x) = h_x(y, 0) < 0$*;*
(b) if $h_x(y, .)$ *has the ZTP, then* $b_x(y) = c_x(y)$.

Proof: obvious. □

For the significance of the notions introduced above under a dynamical aspect see Section 5. Now we are in a position to formalize the notions of evolutionary and neutral stability in the spirit of Taylor and Jonker [26]:

Definition 3
(a) x *is neutrally stable (TJ) if and only if* $b_x(y) > 0$ *for all* $y \in \Delta$*;*
(b) x *is evolutionarily stable (TJ) if and only if* $c_x(y) > 0$ *for all* $y \in \Delta$ *with* $y \neq x$.

Note that in important classes of models (see, e.g. [15, 17, 3, 30, 4]) the set of evolutionarily stable strategies is empty while there still exist neutrally stable strategies. One such class is that of derived games as treated in [23, 13, 9, 14]. Before specifying a concrete example let us note that every neutrally stable strategy necessarily is a Nash strategy.

Example 1: Consider the by now classical hawk-dove game introduced in [21]. If we rescale payoffs to reduce calculational effort, the payoff matrix for this 2×2-game reads $\begin{bmatrix} 0 & 1 \\ 1 & 0 \end{bmatrix}$. The unique evolutionarily stable state then is the polymorphism $p = [1/2, 1/2]'$. Now suppose that in addition to the Hawk and Dove strategies, individuals are capable of playing a mixed strategy corresponding to p, i.e. to imitate the (statistical) relative frequencies by personal (in this case equal) probabilities to display either Hawk or Dove. The strategies for this derived game are represented by the matrix of stacked state vectors

$$C = \begin{bmatrix} 1 & 0 & 1/2 \\ 0 & 1 & 1/2 \end{bmatrix}.$$

The resulting evolutionary mixed strategy game consists again of pairwise contests and has now $m = 3$ strategies while the payoff structure in reduced form is given by (cf. [23, 13, 9, 14])

$$A = C' \begin{bmatrix} 0 & 1 \\ 1 & 0 \end{bmatrix} C = \begin{bmatrix} 0 & 1 & 1/2 \\ 1 & 0 & 1/2 \\ 1/2 & 1/2 & 1/2 \end{bmatrix} .$$

Now consider the set $G = \{y \in \Delta : y_1 = y_2\}$. Every $x \in G$ is neutrally stable, but satisfies, according to formula (3) below, $h_x(y, \varepsilon) = 0$ for all $\varepsilon > 0$ and all $y \in G = \{y \in \Delta : y_1 = y_2\}$. Consequently, $c_x(y) = 0 < 1 = b_x(y)$. The set G forms (the prototype of) an evolutionarily stable set (see below). Indeed, it consists of Nash strategies and $u(x, y) - u(y, y) = (x - y)'Ay = (y_1 - y_2)^2/2$ if $x \in G$ and $y \in \Delta$. The easiest way to establish all assertions is to use Theorem 4(e) together with Theorem 1(b). Seen from a dynamic perspective, every $x \in G$ is Lyapunov-stable, but not asymptotically stable under the replicator dynamics (see Section 5): trajectories starting in other $y \in G$ near x would not be attracted by x, but necessarily stay in a neighbourhood of x [11]. Since there are no (other) asymptotically stable points, this game has no evolutionarily stable strategies. On the other hand, the set G as a whole is attractive (i.e. asymptotically stable), according to Theorem 5. The phase portrait under the replicator dynamics is depicted in Figure 1. For more (counter)examples in connection with neutrally stable strategies and evolutionarily stable sets under the perspective of replicator dynamics, see the "zoo" in [7]. Note that the error concerning phase portrait 26 there is now corrected in Example 3 below.

While in some sense the notions introduced above may be viewed as most closely related to the original purpose of evolutionary game theory, namely to predict stability of polymorphism of behaviour, several authors proposed different evolutionary stability criteria. For the hierarchy between these see [11]. Here let us only mention that x is called **uninvadable** [10] if x is protected by a global positive invasion barrier:

$$\bar{c}_x = \inf\{c_x(y) : y \in \Delta, y \neq x\} > 0 ;$$

whereas x is called **unbeatable** [11], if x is protected by a global positive neutrality barrier:

$$\bar{b}_x = \inf\{b_x(y) : y \in \Delta\} > 0 .$$

As is stressed in [11], it is these uniform immunity concepts which guarantee dynamical properties (asymptotic stability for uninvadable, and Lyapunov stability for unbeatable strategies, under the replicator dynamics). Note that Hamilton's [18], or Maynard Smith's [20] notion of uninvadability, and Mesterton-Gibbons's [22] notion of unbeatability, do not coincide with the terminology used here.

Next let us treat analogous concepts for stable sets instead of single strategies. Given such a set $G \subseteq \Delta$ it seems reasonable to require protection against invasion only against outsiders whereas neutrality against insiders would be sufficient. Hence we define the set barrier through the point barriers by means of case distinction between insiders and outsiders:

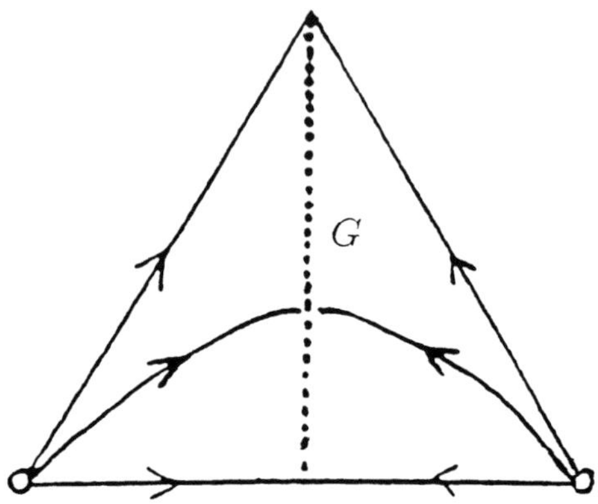

Figure 1: Replicator dynamics of the game in Example 1.

Definition 4
Let G be a non-empty subset of strategies. For an inside strategy $x \in G$ and an arbitrary strategy $y \in \Delta$ define the G-barrier as follows:

$$\varepsilon_G(x, y) = \begin{cases} c_x(y) & \text{if} \quad y \notin G\,; \\ b_x(y) & \text{if} \quad y \in G\,. \end{cases}$$

Then G is said to be an **evolutionarily stable set** *if $\varepsilon_G(x, y) > 0$ for all $x \in G$ and all $y \in \Delta$.*

One can show that this concept coincides with the notion of an ES set recently introduced by Balkenborg and Schlag [1], although their definition is different (cf. Section V). The approach used here avoids any topological assumptions and stresses the concept of barriers. For instance, it is immediately clear that the set barriers coincide with the neutrality, or invasion barrier, respectively, if one considers the whole strategy space $G = \Delta$, or the singleton $G = \{x\}$, respectively. Hence x is evolutionarily stable (TJ) if and only if the singleton $G = \{x\}$ is an ES set.[1] Furthermore, any x belonging to an arbitrary ES set G is automatically neutrally stable (TJ). This observation generalizes a result in [6].

To distinguish this from Thomas' [28] definition which requires closedness of G, we henceforth abbreviate the notion introduced above with **ES set (BS)**, while we use the abbreviation **ES set (Th)** for the following paraphrase of Thomas' original definition:

Definition 5
Let G be a non-empty subset of strategies. G is said to be an **ES set (Th)** *if and only if*
(i) G is closed; and
(ii) for all $x \in G$, there is a neighbourhood U of x in Δ such that $c_x(y) = 1$ if $y \in U \setminus G$ while $b_x(y) = 1$ if $y \in U \cap G$.

Next it is shown that, in some sense, the neutrality barrier $b_x(y)$ is not essential in Definition 4. To make this precise, we introduce a quantity which always is positive and is independent from the set G considered: for all x and y in Δ, put

$$\varepsilon^*(x,y) = \begin{cases} c_x(y) & \text{if} \quad c_x(y) > 0\,; \\ 1 & \text{if} \quad c_x(y) = 0\,. \end{cases}$$

By definition, this quantity coincides with $\varepsilon_G(x,y)$ except for the critical case where x and y both belong to G and $0 < c_x(y) < b_x(y)$. Note that the other two cases $c_x(y) = 0$ for $y \notin G$, and $b_x(y) = c_x(y) = 0$ for $y \in G$, cannot pertain if G is an ES set (BS). Indeed, we have the following characterisation for such sets:

Proposition 2
Let G be a nonempty set of strategies. Then G is an ES set (BS) if and only if

$$\varepsilon_G(x,y) \geq \varepsilon^*(x,y) \quad \text{for all} \;\; x \in G \quad \text{and all} \;\; y \in \Delta$$

(in particular this means $b_x(y) = 1$ if $c_x(y) = 0$). Furthermore if scores have the ZTP, then G is an ES set (BS) if and only if $\varepsilon_G(x,y) = \varepsilon^*(x,y)$ for all $x \in G$, $y \in \Delta$. This particularly applies to pairwise contests.

Proof: Sufficiency is obvious from definitions. To show necessity, observe that for any x belonging to an ES set G, $\varepsilon_G(x,y) = c_x(y) = \varepsilon^*(x,y)$ if $y \notin G$; and $\varepsilon_G(x,y) = b_x(y) \geq c_x(y) = \varepsilon^*(x,y)$ if $y \in G$ with $c_x(y) > 0$. Furthermore, we prove that $c_x(y) = 0$ implies $b_x(y) = 1$ (in other words, the score $h_x(y,\varepsilon) = 0$ for all ε). Indeed, by definition of $c_x(y)$ there are $\varepsilon_\nu \searrow 0$ with $h_x(y,\varepsilon_\nu) \geq 0$ for all $\nu = 1, 2, \ldots$. Thus (1) with $\mu = 0$ yields for all $\lambda > 0$ and all ν so large that $\varepsilon_\nu/\lambda \in (0,1)$

$$h_x(y^\lambda, \frac{\varepsilon_\nu}{\lambda}) = \lambda\, h_x(y,\varepsilon_\nu) \geq 0$$

and still $\varepsilon_\nu/\lambda \searrow 0$, so that $c_x(y^\lambda) = 0$. Hence necessarily $y^\lambda \in G$ for all $\lambda > 0$ with $y^\lambda \in \Delta$, even if $\lambda > 1$. Now suppose that for some such λ we had $h_x(y,\lambda) > 0$. By continuity, $h_x(y,\mu) > 0$ for all μ in an interval containing λ. Fix such a $\mu > \lambda$ and consider an arbitrary $\varepsilon \in [0,1)$. From (1) we get, interchanging y^λ and y^μ,

$$h_{y^\lambda}(y^\mu,\varepsilon) = (\mu - \lambda)\, h_x(y, \varepsilon\mu + (1-\varepsilon)\lambda) > 0\,,$$

showing $b_{y^\lambda}(y^\mu) = 0$, a contradiction to $y^\lambda \in G$. Repeating the argument for the hypothetical case $h_x(y,\lambda) < 0$ (but taking $\mu < \lambda$ this time), we have established that the score function of y versus x identically vanishes, and hence $\varepsilon_G(x, y) = b_x(y) = 1 = \varepsilon^*(x, y)$ if $y \in G$ with $c_x(y) = 0$. Finally, if scores have the ZTP, then by Proposition 1(b) $b_x(y) = c_x(y)$ if $c_x(y) > 0$, so that $\varepsilon_G(x, y) = \varepsilon^*(x, y)$ in this case for any ES set G and all $x \in G$, $y \in \Delta$. □

Inspection of the above proof reveals that whenever x and y belong to an ES set and $c_x(y) = 0$, then also $c_y(x) = 0$, and all affine combinations y^λ ($\lambda \in \mathbb{R}$ arbitrary) with $y^\lambda \in \Delta$ share this property too. Indeed, for $\lambda < 0$ with $y^\lambda \in \Delta$ put $\delta = \frac{1}{1-\lambda} \in [0, 1]$ so that $x = (1 - \delta)y + \delta y^\lambda$ is now between y and y^λ. Thus (1) entails

$$0 = h_x(y, \varepsilon) = \delta h_y(y^\lambda, \varepsilon\delta) \quad \text{for all} \quad \varepsilon \in [0, 1]\,,$$

which means $c_y(y^\lambda) = 0$ and therefore $y^\lambda \in G$. Repeating the arguments above for y and y^λ instead of x and y yields the assertion.

We close this section by a helpful observation which in some sense complements the case considered above:

Lemma 1
Suppose that scores are monotone. Let G be an ES set (BS). Then for any x and y both belonging to G, the relation $c_x(y) > 0$ implies $y_i = 0$ for some i with $x_i > 0$.

Proof: By the preceding proposition $c_x(y) > 0$ implies also $c_y(x) > 0$. Monotonicity of $h_y(x, \varepsilon)$ yields therefore $(x - y)'\varphi(y) = h_y(x, 0) < 0$. Now suppose that $y_i > 0$ whenever $x_i > 0$. Then there is a $\lambda > 1$ such that still $y^\lambda \in \Delta$. But then

$$h_y(y^\lambda, 0) = (y^\lambda - y)'\varphi(y) = (1 - \lambda)(x - y)'\varphi(y) > 0\,,$$

so that by continuity of h_y we have $b_y(y^\lambda) = 0$, in contradiction to $y \in G$. □

III. Global Immunity of Evolutionarily Stable Sets

As argued in [11], uniformity of pointwise barriers $c_x(y)$ or $b_x(y)$, i.e. positivity of $\overline{c}_x$ or $\overline{b}_x$, is important for establishing stability properties under the replicator dynamics. Similarly, global immunity of an ES set G using uniformity of set barriers is now introduced. In Section V it will turn out that this uniformity is, again, the decisive property which guarantees asymptotic stability of the set under consideration.

While the uniform stability notions introduced in [1] can be rephrased in present terminology to impose uniform positivity of $\varepsilon_G(x, y)$, there is an appealing alternative using the quantity $\varepsilon^*(x, y)$: indeed, the existence of a positive lower bound for $\varepsilon^*(x, y)$ on $G \times \Delta$ means nothing else than that there is a global immunity against invasion by y if x is immune against y at all.

Definition 6
Let G be an ES set (BS). Then G is said to be a **uniform ES set** *(UES set) if and only if*

$$\overline{\varepsilon_G} = \inf\{\varepsilon_G(x, y) : x \in G, y \in \Delta\} > 0\,;$$

whereas G is said to be **uniformly immune** *if and only if*

$$\overline{\varepsilon_G^*} = \inf\{\varepsilon^*(x, y) : x \in G, y \in \Delta\} > 0\,.$$

Of course, $\overline{\varepsilon_G^*}$ depends on G although $\varepsilon^*(x, y)$ does not. Both quantities, $\overline{\varepsilon_G^*}$ and $\varepsilon^*(x, y)$, do not involve the interior neutrality barriers any more. So the next step would naturally be to ignore the interaction inside the set totally, and just require external stability. This is the motive for the following notions, all in the spirit of an ES* set introduced by Weibull [31]. While the latter incorporates the same topological notions as Definition 5, the other notions are exclusively based upon the various barriers.

Definition 7
Let G be a non-empty subset of Δ.
(a) G is said to be an **externally ES set** *if and only if*
$c_x(y) > 0$ for all $x \in G$ and all $y \in \Delta \setminus G$.
*(b) A strategy $x \in G$ is said to be G-***externally uninvadable** *if and only if*
*x is protected by a positive global G-*external *invasion barrier $\overline{e}_{G,x}$, i.e. $c_x(y) \geq \overline{e}_{G,x} > 0$ for all $y \in \Delta \setminus G$.*
(c) G is said to be **uniformly externally immune** *if and only if*
(i) G is an externally ES set; and (ii) $\overline{\varepsilon_G^*} > 0$. Then in particular there is a uniform global external invasion barrier $\overline{e}_G = \inf_{x \in G} \overline{e}_{G,x}$.
(d) G is said to be an **ES* set** *if and only if*
(i) G is closed; and (ii) for all $x \in G$ there is a neighbourhood U of x in Δ such that $c_x(y) = 1$ for all $y \in U \setminus G$.

Most of the relations between these and the previously introduced concepts are immediate. For a full account on these see the hierarchy described in Theorem 2 below. Note that $G = \Delta$ satisfies the conditions trivially for all notions from Definition 7, which require nothing about internal interactions. So in order to make more precise predictions on behaviour, one could require minimality. But if G is thin in the topological sense, then it will turn out that the external stability concepts coincide with their respective counterparts among evolutionary stability, so that one does not have to worry about internal interactions in this case. However, if G has a non-empty interior, then the concepts are indeed different. Unfortunately, at least in pairwise contests with up to $m = 3$ strategies I have encountered up to now no example where G satisfies one of the above conditions and has a non-empty interior but does not cover the whole strategy space.

For the sake of transparency, let us break the investigation of relations between all notions introduced so far into several parts.

Theorem 1
(a) If G is uniformly immune, then G is an UES set. The converse holds if scores have the ZTP, in particular in contests with n players using i.i.d. randomisations.
(b) Every x belonging to an UES set G is unbeatable as well as G-externally uninvadable.

Proof: obvious by Proposition 2, and by $\min\{\bar{b}_x, \bar{e}_{G,x}\} \geq \overline{\varepsilon_G}$. □

As noted already in the previous section, the distinction between UES sets and uniform immunity becomes important only if $0 < c_x(y) < b_x(y)$ for some $x \in G$ and $y \in G$. This critical case plays an unfortunate role in a lacuna in the arguments of [1], who claim that every UES set is an ES set (Th). Thus the following result has to incorporate the stricter notion of uniform immunity.

We now come to the hierarchy of the stability concepts introduced up to now. In the following proof we need some more notation designating subfaces of the simplex which represent the set of strategies using only a (fixed) part of the pure strategies. Let $I \subseteq M$ and put

$$\Delta_I = \{x \in \Delta : x_i = 0 \quad \text{if} \quad i \notin I\};$$

as well as the relative interior of Δ_I,

$$\Delta_I^\circ = \{x \in \Delta_I : x_i > 0 \quad \text{if} \quad i \in I\}.$$

Theorem 2
Let G be a non-empty subset of Δ and consider the following properties:
(a1) G is uniformly immune;
(b1) G is an ES set (Th);
(c1) every $x \in G$ is unbeatable and G-externally uninvadable;
(d1) G is an ES set (BS);
(a2) G is uniformly externally immune;
(b2) G is an ES* *set;*
(c2) every $x \in G$ is G-externally uninvadable;
(d2) G is an externally ES set.
Then (a1) ⇒ (b1) and (c1) ⇒ (d1). If all scores have the ZTP, then (b1) ⇒ (c1). Furthermore, we have (x1) ⇒ (x2) for all x=a,b,c,d.
If G has a dense complement in Δ, then also the following converse implications hold: (x2) ⇒ (x1) if x=a,b,c; finally, (d2) implies that G consists entirely of Nash strategies.

Proof: Let us start showing (a1) ⇒ (b1). We first prove that G is closed. Consider a sequence $x^\nu \in G$ approximating a strategy $x \in \Delta$. Without loss of generality assume that $x^\nu \in G \cap \Delta_I^\circ$ for all $\nu = 1, 2, \ldots$. With the help of the Thinning Lemma in the appendix of [10], we are able to extract a subsequence z^ν of x^ν such that the affine hull H of $\{z^\nu : \nu = 1, 2, \ldots\}$ satisfies $H \cap \Delta \subseteq G \cap \Delta_I$. But of course also $x \in H \cap \Delta$ since H is closed due to finiteness of the dimension of Δ, and since x can be approximated by the sequence z^ν. Hence $x \in G$. Next we establish property (ii) in Definition 5. To this end, fix an $x \in G$ and define

$$U = \{(1-\varepsilon)x + \varepsilon y : y \in \Delta, \, 0 \leq \varepsilon < \overline{\varepsilon_G^*}\}.$$

Now if $c_x(y) = 0$ for some $y \in \Delta$, then by Proposition 2, $b_x(y^\lambda) = 1$ and $y^\lambda \in G$ for all $\lambda \in [0, 1]$. A fortiori we have $b_x(z) = 1$ if $z = y^\varepsilon \in U$. On the other hand, if

$c_x(y) > 0$, then $c_x(y) \geq \overline{\varepsilon_G^*}$ which means also $c_x(z) = 1$ whenever $z = y^\varepsilon \in U$, because of

$$h_x(z, \delta) = \varepsilon h_x(y, \delta\varepsilon) < 0$$

for all $\delta \in [0, 1]$ if $0 < \varepsilon < \overline{\varepsilon_G^*} \leq c_x(y)$. Hence G is an ES set (Th).

To show (b1) $\Rightarrow$ (c1), fix an arbitrary $x \in G$ and choose $\delta > 0$ so small that the neighbourhood U of x satisfies $U \supseteq \{z \in \Delta : \|z - x\| < \delta\}$. Consider an arbitrary $y \in \Delta$. Then for every $z = y^\varepsilon$ with $0 < \varepsilon < \delta$ we have $z \in U$ and thus $b_x(z) = 1$, which means $h_x(y, \varepsilon) = \varepsilon h_x(z, 1) \leq 0$, by continuity. Therefore we obtain $\overline{b}_x \geq \delta$. We now claim that $y \notin G$ yields $h_x(y, \varepsilon) < 0$ if $0 < \varepsilon < \delta$. Indeed, put $\lambda = \inf\{\mu \in [0, 1] : y^\mu \notin G\}$. Since G is closed, we have $y^\lambda \in G$, but arbitrarily close to λ there are $\mu > \lambda$ with $y^\mu \notin G$. Hence by property (ii) we also have $c_{y^\lambda}(y^\mu) = 1$ entailing via (1)

$$h_x(y, \mu') = (\mu - \lambda)^{-1} h_{y^\lambda}(y^\mu, \varepsilon) < 0 \qquad (2)$$

for all $\varepsilon < 1$, i.e. for all $\mu' = \varepsilon\mu + (1 - \varepsilon)\lambda < \mu$. On the other hand, we have by definition of λ $z = y^\varepsilon \in G$ for all $\varepsilon \in [0, \lambda]$ as well. Thus $\overline{b}_z > 0$. as already established, hence z is unbeatable and a fortiori is a Nash strategy. But then for all positive $\varepsilon \leq \min\{\lambda, \delta\}$ we arrive at

$$0 \geq \frac{1}{\varepsilon}[u(x, z) - u(z, z)] = -h_x(y, \varepsilon) \geq 0\,.$$

Now if λ were strictly positive, the score would vanish on the whole interval $(0, \lambda)$, contradicting ZTP and $h_x(y, \mu') < 0$. Hence $\lambda = 0$. Now suppose that $y^\mu \in G$ for some μ with $0 < \mu < \delta$. As above we would obtain that arbitrarily close to μ there are $\varepsilon < \mu$ with $y^\varepsilon \notin G$ and therefore via property (ii) $c_{y^\mu}(y^\varepsilon) = 1$, entailing via (1) and continuity

$$h_x(y, \varepsilon) = (\varepsilon - \mu)^{-1} h_{y^\mu}(y^\varepsilon, 1) \geq 0\,,$$

which is absurd since $\varepsilon < \mu < \delta$ yields, again using (1),

$$h_x(y, \varepsilon) = \frac{1}{\varepsilon'} h_x(y^{\varepsilon'}, \frac{\varepsilon}{\varepsilon'}) < 0$$

if we choose ε' such that $\varepsilon < \varepsilon' < \delta$ and $y^{\varepsilon'} \notin G$. Therefore we arrive at $y^\mu \notin G$ if $0 < \mu < \delta$, which shows by the same arguments as above the desired relation $h_x(y, \varepsilon) < 0$ if $0 < \varepsilon < \delta$ and therefore $\overline{e}_{G,x} \geq \delta$.

The implications (c1) $\Rightarrow$ (d1) and (x1) $\Rightarrow$ (x2) are immediate from the definitions. Therefore it only remains to prove the implications under the assumption that G has dense complement. But in this case any $y \in G$ can be approximated by a sequence $y_\nu \in \Delta \setminus G$. Then

$$h_x(y, \varepsilon) = \lim_{\nu \to \infty} h_x(y_\nu, \varepsilon) \leq 0$$

for all $\varepsilon \leq \overline{e}_{G,x}$ (or $\varepsilon < 1$, if $y \in U$, respectively), which is enough to establish the claimed (x2) $\Rightarrow$ (x1) for x=a,b,c. To show the last implication, we exploit the similar limiting argument for $\varepsilon = 0$, in order to obtain $u(y,x) - u(x,x) = h_x(y,0) \leq 0$ for all $y \in \Delta$ and all $x \in G$. □

The following counterexample shows that without any assumption on the scores, one cannot infer that every ES set (Th) is also an ES set (BS) (cf. Proposition 2.28 in [1]).

Example 2: Consider an – admittedly complicated – correlated interaction between individuals having $m = 2$ pure strategies at their disposal. Clearly, the payoff structure of the game is completely characterised by the score function $h_x(y,\varepsilon)$ if x is one and y is the other pure strategy. Now for

$$h_x(y,\varepsilon) = \begin{cases} 0 & \text{if } 0 \leq \varepsilon \leq \frac{1}{2}\,, \\ -\exp(-[\varepsilon - \frac{1}{2}]^{-1}) & \text{if } \frac{1}{2} < \varepsilon \leq 1\,, \end{cases}$$

the set $G = \{y^\lambda : 0 \leq \lambda \leq \frac{1}{2}\}$ is obviously an ES set (Th) but not an ES set (BS) since $c_x(y) = 0$. In other words, $\varepsilon_G(x,y) = 0 < 1 = \varepsilon^*(x,y)$, cf. Proposition 2.

IV. (UNIFORMLY) EVOLUTIONARILY STABLE SETS IN PAIRWISE CONTESTS

In this section we focus on pairwise contests, where we have $\varphi(y) = Ay$ for some $n \times n$ payoff matrix A. First observe that the scores can be written as

$$\begin{aligned} h_x(y,\varepsilon) &= (1-\varepsilon)[u(y,x) - u(x,x)] + \varepsilon[u(y,y) - u(x,y)] \\ &= (y-x)'Ax + \varepsilon(y-x)'A(y-x)\,. \end{aligned} \tag{3}$$

In order to show that all evolutionary stability notions considered so far coincide, first the structure of an ES set for pairwise contest is established explicitly (cf. [28], [14]).

Theorem 3
If G is an ES set (BS) in a pairwise conflict and $G \cap \Delta_I^\circ$ is not empty, then $G \cap \Delta_I^\circ$ is the relative interior of a polytope G_I generated by the intersection of an affine subspace with Δ_I. Moreover this polytope G_I is contained entirely in G.

Proof: Assume that $G \cap \Delta_I^\circ$ is not empty, and pick an arbitrary but fixed point $x_I \in G \cap \Delta_I^\circ$. Lemma 1 at the end of Section 2 shows that $c_x(y) = c_y(x) = 0$ whenever x and y both belong to $G \cap \Delta_I^\circ$. This entails via Proposition 2, $h_x(y,\varepsilon) = 0$ for all ε, which means $(y-x)'Ax = 0$ and also $(y-x)'A(y-x) = 0$. In particular, this holds for $x = x_I$. Now consider the tangent space T_I of Δ_I at x_I, which is a linear subspace described by

$$T_I = \{v \in \mathbb{R}^n : \sum v_i = 0 \text{ and } v_j = 0 \text{ if } j \notin I\}\,.$$

Denote by P_I the orthoprojection matrix onto T_I. For every $y \in \Delta_I$ and any $\lambda \geq 0$, we have $v = \lambda(y-x) \in T_I$, and vice versa. Since $x_I \in \Delta_I^\circ$ is a Nash strategy, we have $v'Ax_I = 0$. Furthermore, $b_{x_I}(y) > 0$ for any $y \in \Delta_I$ entails that $h_{x_I}(y, \varepsilon)$ does not increase in ε, so that necessarily $v'Av = \lambda^2(y - x_I)'A(y - x_I) \leq 0$ for all $v \in T_I$. But then the matrix $P_I'(A + A')P_I$ is symmetric and negative semidefinite, so that for all $y \in G \cap \Delta_I^\circ$ we have $P_I'(A + A')P_I(y - x_I) = o$, the zero vector, due to the fact that the corresponding quadratic form $(y - x_I)'A(y - x_I)$ vanishes. On the other hand, any point $y \in \Delta_I$ with $P_I'(A + A')P_I(y - x_I) = o$ satisfies $h_{x_I}(y, \varepsilon) = 0$ for all ε, and hence $y \in G$. Therefore $G \cap \Delta_I^\circ \subset G_I \subseteq G \cap \Delta_I$, where G_I is the polytope

$$G_I = \{y \in \Delta : y - x_I \in T_I \text{ and } P_I'(A + A')P_I(y - x_I) = o\}.$$

This concludes the proof. □

Since the strategy space Δ consists of the union of its vertices and all Δ_I° with $I \subseteq M$ incorporating at least two pure strategies, we obtain as a consequence of the above theorem the following (not necessarily disjoint) decomposition into polytopes:

$$G = \bigcup\{G_I : \emptyset \neq I \subseteq M\},$$

where we additionally define $G_{\{i\}} = \{e_i\}$, if a vertex e_i of Δ belongs to G, and putting $G_I = \emptyset$ if either $I = \{i\}$ and $e_i \notin G$ or if $G \cap \Delta_I^\circ = \emptyset$ and I has more than one element.

We now show that in pairwise contests, most of the evolutionary stability notions for sets coincide:

Theorem 4

Let G be a non-empty subset of Δ and assume pairwise contests. Then the following assertions are all equivalent:
(a) G is uniformly immune;
(b) G is an UES set;
(c) G is an ES set (Th);
(d) G is an ES set (BS);
(e) G consists entirely of Nash strategies and for all $x \in G$ and $y \in \Delta$ with $u(y, x) = u(x, x)$, we have $u(y, y) \leq u(x, y)$ where equality implies $y \in G$.

Proof: (d) and (e) are equivalent from formula (3).
Furthermore Theorem 1 yields equivalence of (a) and (b). In view of Theorem 2 it remains to show that (e) implies (b). Now consider the previously established decomposition of G into the polytopes G_I with interior points x_I. It is straightforward to show that $b_x(y) = \min\{1, r_y(x)\}$, where the function

$$r_y(x) = \frac{(x - y)'Ax}{(x - y)'A(y - x)} \tag{4}$$

is defined for all fixed y and for all $x \in G_I$ with $(x - y)'Ay \leq 0$. On $\Delta_I^\circ \cap G$ we have $x'Ax = x_I'Ax$, so that both expressions in numerator and denominator in (4) are affine in x. Hence $r_y(x)$ is for fixed y quasiconcave (see, e.g. [12]) on its domain which is a polytope contained in $G_I \subseteq G$, and therefore attains its minimum at one of the finitely many vertices $\{v_1, \ldots, v_k\}$ of this polytope. Now we know that every $v_i \in G$ is neutrally stable and hence unbeatable, since for pairwise contests these notions coincide [11] . Hence for arbitrary $y \in \Delta$, we have $b_{v_i}(y) \geq \bar{b}_{v_i}$ and therefore $b_x(y) \geq \bar{b}_I = \min\{\bar{b}_{v_i} : 1 \leq i \leq k\} > 0$. For the same reason we obtain for any $x \in G_I$ and any $y \notin G$ by definition $c_x(y) > 0$, hence also $c_x(y) = b_x(y) \geq \bar{b}_I$. Since there are finitely many faces I, we finally obtain $\varepsilon_G(x,y) \geq \overline{\varepsilon_G} = \min\{\bar{b}_I : I \subseteq M\} > 0$. □

V. Dynamic significance of scores and barriers

Replicator dynamics were introduced in [26] to model evolution of behaviour in intraspecific conflicts, and formalise the idea that the growth rate $\dot{y}_i/y_i$ is equal to the (dis)advantage $\varphi_i(y) - y'\varphi(y)$ measured by incremental fitness relative to the average performance within the population in state $y = [y_1, \ldots, y_m]'$:

$$\dot{y}_i = y_i[\varphi_i(y) - y'\varphi(y)]\,, \qquad i \in M\,. \tag{5}$$

Now suppose for the moment that there is competition only between the several post-entry mixes $y^\varepsilon = (1-\varepsilon)x + \varepsilon y$ of a fixed incumbent state x, and a fixed mutant state, y, so that evolution over time is completely described by a dynamics for the mutant frequency, $\varepsilon = \varepsilon(t)$. But then the dynamics for this reads, as has been established and justified, e.g. in [5],

$$\dot{\varepsilon} = \varepsilon(1-\varepsilon)h_x(y,\varepsilon)\,.$$

Hence the sign of the scores completely govern the dynamics of the mutant fraction. For instance, $\varepsilon = 0$ (which corresponds to the incumbent state x) is asymptotically stable (which means $\varepsilon(t) \searrow 0$ as $t \nearrow \infty$, provided $\varepsilon(0)$ is small enough) if and only if $c_x(y) > 0$, and x is Lyapunov stable (which means that $\varepsilon(t)$ does not leave an arbitrary neighbourhood of 0, provided $\varepsilon(0)$ is small enough) if $b_x(y) > 0$. However, if scores do not have the ZTP, then the reverse implication does not hold, and this is the (only) reason why the definition of neutral stability given by [1] does not coincide with Definition 3.

But let us return now to the full-support replicator dynamics. Rather than reviewing the well-known results on dynamic stability of single strategies (see, e.g. [11], Figures 2 and 3), we immediately specify the main result on stable sets here:

Theorem 5
Every ES* *set is asymptotically stable.*

Proof: see [31], Proposition 3.13. □

As argued in [11] for single strategies, here the uniformity of set barriers is also the decisive property which ensures asymptotic stability under the replicator dynamics. This becomes immediately clear when looking at Theorems 1 and 2. The model most prominent in the literature, the classical pairwise conflict, obscures this relation, since in the latter subclass uniformity is automatically guaranteed (Theorem 4).

We also conclude this section by a counterexample. It is well known [33] that there are asymptotically stable strategies which need not be even unbeatable, so that the converse of the above result cannot hold. But since every thin ES* set is an ES set, Theorems 2 and 5 tell us that every such set is an asymptotically stable set consisting entirely of unbeatable strategies. So one might wonder whether perhaps every asymptotically stable set consisting entirely of unbeatable strategies is in turn an ES (or ES*) set. In fact, for $m = 3$ pure strategies, the classification of the replicator dynamics' flow for pairwise conflicts given by [6] does not provide a counterexample for this conjecture. Note that the opposite assertion on phase portrait number 26 in [7] is unfortunately wrong. However, consider the following pairwise conflict with three pure strategies (which is obviously the minimal strategy number for this purpose, since for $m = 2$ evolutionary and asymptotic stability coincide, as the discussion at the beginning of this section shows):

Example 3: Let the payoff matrix be

$$A = \begin{bmatrix} 0 & 1 & 1 \\ 5-\sqrt{32} & 0 & 3 \\ 1 & 1 & 0 \end{bmatrix}.$$

Then $\overline{x} = \frac{1}{9-\sqrt{32}}\left[1, 7-\sqrt{32}, 1\right]'$ is neutrally stable (TJ) and therefore unbeatable, since the matrix A yields a negative semidefinite quadratic form on the tangent space T_M of Δ at $\overline{x}$ (cf. Section 4). Indeed, for all $v \in T_M$ we have

$$v'Av = -\left[v_1, v_2\right] \begin{bmatrix} 2 & 2\sqrt{2} \\ 2\sqrt{2} & 4 \end{bmatrix} \begin{bmatrix} v_1 \\ v_2 \end{bmatrix} \leq 0 .$$

Since the latter matrix is singular, we can easily construct a strategy $\overline{y} = \overline{x} + v$ with $\overline{y}'A\overline{y} = \overline{x}'A\overline{y}$ although $\overline{y} \neq \overline{x}$ and (necessarily) $\overline{y}'A\overline{x} = \overline{x}'A\overline{x}$, namely with $v = \lambda[\sqrt{2}, -1, 1-\sqrt{2}]'$ for some suitably small $\lambda > 0$. Therefore $\overline{x}$ is not evolutionarily stable (or uninvadable), and hence the singleton $\{\overline{x}\}$ is no ES* set. On the other hand, the flow is as depicted in Figure 2, cf. phase portrait number 9 in [6]: every trajectory starting in the relative interior converges to $\overline{x}$. Observe that with the specified matrix A, every fixed point of the replicator dynamics is hyperbolic (i.e. all real parts of the eigenvalues of the linearization around a fixed point do not vanish), so that from a dynamical point of view this is a robust case.

We now enrich this to a game with $m = 4$ pure strategies in such a way that the replicator flow yields a foliation of the (three-dimensional) four-simplex with the previous phase portrait along the new coordinate direction, rendering an infinite asymptotically stable set G consisting entirely of unbeatable states, although G is not an ES* set. See Figure 3.

To this end, for any $y = [y_1, y_2, y_3, y_4]' \in \mathbb{R}^4$ denote by $\hat{y} = [y_1, y_2, y_3]'$ and put $\varphi_i(y) = (1 - y_4)(A\hat{y})_i$ for $i = 1, 2, 3$, whereas $\varphi_4(y) = \hat{y}'A\hat{y}$. Since φ is a

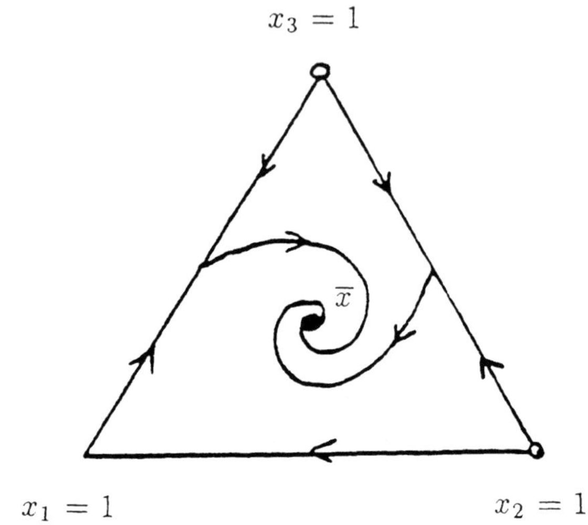

Figure 2: Replicator dynamics of the 3×3-game in Example 3.

polynomial of degree two, this return function can emerge from frequent encounters of $n = 3$ individuals with an appropriate pure-strategy payoff function π, see Section 2. With $\overline{x}$ as above, consider the set

$$G = \{x \in \mathbb{R}^4_+ : \sum_{i=1}^{4} x_i = 1 \text{ and } \hat{x} = (1 - x_4)\,\overline{x}\} .$$

This set is asymptotically stable under (5), because by definition $\dot{y}_4 = 0$ and hence

$$\frac{d}{dt}\left[\log\left(\frac{\hat{y}_i}{1 - y_4}\right)\right] = [\log y_i]^{\cdot} =$$
$$= (1 - y_4)^2 \left[\left(A\frac{\hat{y}}{1 - y_4}\right)_i - \left(\frac{\hat{y}}{1 - y_4}\right)' A\frac{\hat{y}}{1 - y_4}\right], \; i = 1, 2, 3 ,$$

which establishes Figure 3. On the other hand, G cannot be an ES set because with $\overline{y}$ as above, we have $c_x(y) = 0$ although $y = [\overline{y}, 0]' \notin G$ while $x = [\overline{x}, 0]' \in G$. Finally, straightforward calculation shows that for every $x \in G$ and every y, there is some $v \in T_M$ as above such that

$$h_x(y, \varepsilon) = \varepsilon(1 - x_4)^2(1 - y_4)\, v'Av \leq 0 \quad \text{for all } \varepsilon > 0 ,$$

showing $b_x(y) = 1$ and hence unbeatability of x.

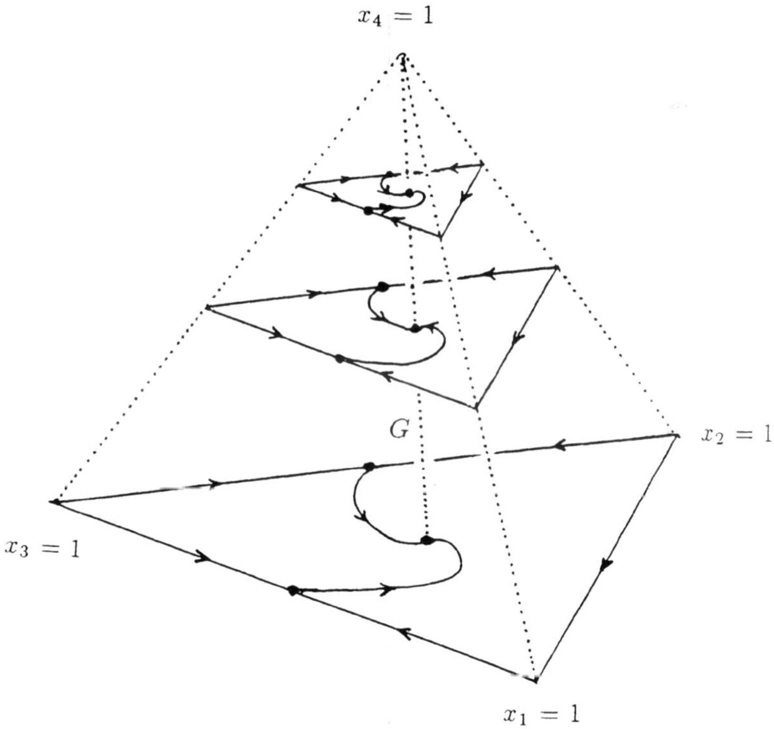

Figure 3: Replicator dynamics of the $4 \times 4 \times 4$-game in Example 3.

NOTES

* I am indebted to the participants of the 1995 conference on Dynamic Evolutionary Game Theory in Biology and Economics, Wilfried Laurier University and University of Guelph, Ontario, for stimulating discussions, and in particular to J.W. Weibull for valuable hints. I also benefited from important suggestions by anonymous referees and editors.

1. This property is shared by all stability notions for sets presented in this paper.

REFERENCES

[1] D. Balkenborg/K. Schlag, "On the interpretation of evolutionarily stable sets," SFB Discussion Paper B-313, University of Bonn 1995.

[2] A. Banerjee/J.W. Weibull, "Evolutionary selection and rational behaviour", in: A. Kirman/M. Salmon (eds.), *Rationality and learning in economics*. Blackwell 1992, pp.343-363.

[3] K. Binmore/L. Samuelson, "Evolutionary stability in repeated games played by finite automata", in: *J. Econ. Theo.*, 57, 1991, pp.278–305.

[4] A. Blume/Y.-G. Kim/J. Sobel, "Evolutionary stability in games of communication", in: *Games and Econ. Behav.*, 5, 1993, pp.547–575.

[5] I.M. Bomze, "Dynamical aspects of evolutionary stability", in: *Monatsh. Math.*, 110, 1990, pp.189–206.

[6] I.M. Bomze, "Lotka-Volterra equation and replicator dynamics: a two-dimensional classification/new issues in classification", in: *Biol. Cybern.*, 48, 1983, pp.201–211/72, 1995, pp.447–453.

[7] I.M. Bomze, "Dynamical aspects of neutral stability", extended abstract in: R. Cressman/G. Hines (eds.), *Dynamic evolutionary game theory in biology and economics, Conference Report August 15-19, 1995*, W.Laurier Univ. and Univ. of Guelph, 1995.

[8] I.M. Bomze, "Evolutionary stability is not a foolish game", in: *Central Europ. J. Oper. Res. Econ.*, 4, 1996, pp.25–56.

[9] I.M. Bomze/E. van Damme, "A dynamical characterization of evolutionarily stable states", in: *Ann. Operations Res.*, 37, 1992, pp.229–244.

[10] I.M. Bomze, "Uniform barriers and evolutionary stable sets", ISOC Technical Report TR 95–11R (June 1996), University of Vienna, 1996.

[11] I.M. Bomze/J.W. Weibull, "Does neutral stability imply Lyapunov stability?", in: *Games and Econ. Behav.*, 11, 1995, pp.173–192.

[12] I.A. Bykadorov, "On quasiconvexity in fractional programming", in: S.Komlósi/T.Rapcsák/S.Schaible (eds.), *Generalized convexity*. Berlin, Heidelberg, New York: Springer Verlag 1992, pp.281–293.

[13] R. Cressman, "Strong stability and density-dependent evolutionarily stable strategies", in: *J.theor.Biol.*, 145, 1990, pp.319–330.

[14] R. Cressman, *The Stability Concept of Evolutionary Game Theory: A Dynamic Approach*. Berlin, Heidelberg, New York: Springer Verlag 1992.

[15] J. Farrell/R. Ware, "Evolutionary stability in the repeated prisoners' dilemma", in: *Theor. Pop. Biol.*, 36, 1988, pp.161–166.

[16] D. Friedman, "Evolutionary games in economics", in: *Econometrica*, 59, 1991, pp.637–666.

[17] D. Fudenberg/E. Maskin, "Evolution and cooperation in noisy repeated games", in: *Amer. Econ. Rev.*, 80, 1990, pp.274–279.

[18] W.D. Hamilton, "The genetical evolution of social behaviour", in: *J. Theor. Biol.*, 7, 1964, pp.1–52.

[19] W.G.S. Hines, "Strategy stability in complex populations", in: *J. Appl. Prob.*, 17, 1980, pp.600–610.

[20] J. Maynard Smith, *Evolution and the Theory of Games*. Cambridge, UK: Cambridge Univ. Press 1982.

[21] J. Maynard Smith/G. Price, "The logic of animal conflict", in: *Nature*, 246, 1973, pp.15–18.

[22] M. Mesterton-Gibbons, *An Introduction to Game-Theoretic Modelling*. New York: Addison-Wesley 1992.

[23] K. Sigmund, "Game dynamics, mixed strategies, and gradient systems", in: *Theor. Pop. Biol.*, 32, 1987, pp.114–126.

[24] J.M. Swinkels, "Evolutionary stability with equilibrium entrants", in: *J. Econ. Theo.*, 57, 1992, pp.306–332.

[25] J.M. Swinkels, "Evolution and strategic stability: from Maynard Smith to Kohlberg and Mertens", in: *J. Econ. Theo.*, 57, 1992, pp.333–342.

[26] P. Taylor/L. Jonker, "Evolutionarily stable strategies and game dynamics", in: *Math. Biosci.* 40, 1978, pp.145–156.

[27] B. Thomas, "Evolutionary stability: states and strategies", in: *Theor. Pop. Biol.*, 26, 1985, pp.49–67.

[28] B. Thomas, "On evolutionarily stable sets", in: *J. Math. Biol.*, 22, 1985, pp.105–115.

[29] B. Thomas, "Evolutionarily stable sets in mixed strategist models", in: *Theor. Pop. Biol.*, 28, 1985, pp.332–341.

[30] K. Wärneryd, "Cheap talk, coordination, and evolutionary stability", in: *Game Theo. and Econ. Behav.*, 5, 1993, pp.532–546.

[31] J.W. Weibull, *Evolutionary Game Theory.* Cambridge, Mass.: MIT Press 1995.

[32] F.J. Weissing, *Populationsgenetische Grundlagen der evolutionären Spieltheorie.* Thesis, Univ. Bielefeld 1983.

[33] E.C. Zeeman, "Dynamics of the evolution of animal conflicts", in: *J. Theor. Biol.*, 89, 1981, pp.249–270.

Department of Statistics, Operations Research and Computer Science
Vienna University
Universitätsstrasse 5
A-1010 Vienna
Austria

JOSEF HOFBAUER

EQUILIBRIUM SELECTION VIA TRAVELLING WAVES

I. INTRODUCTION

Consider a game with multiple equilibria, say two strict Nash equilibria for simplicity. Suppose that in one part of the world one of these equilibria has become the standard, while the other part has somehow agreed to use the other equilibrium. This different evolution was made possible by an 'iron curtain' separating the two parts. Suppose that now the iron curtain falls, and exchange of ideas and migration of players becomes possible. What will happen to the equilibria chosen? Which equilibrium will survive this process? Intuition suggests that the 'dominant' equilibrium will drive out the other equilibrium in form of a *travelling wave front*.

The problem of equilibrium selection is as old as the concept of equilibrium: Already Nash (1950) faced it in his analysis of the bargaining problem and proposed a solution for this class of games. The most prominent and complete theory of equilibrium selection is due to Harsanyi and Selten (1988), who define the concept of risk-dominance based on the tracing procedure.

In this note, I will present a new, dynamic approach to equilibrium selection, modelled after the above story. The dynamic model is *deterministic*, in terms of *reaction diffusion (RD) equations*. The reaction term in this partial differential equation (PDE) is one of the standard game dynamics, like the replicator equation or best response dynamics; the diffusion part models the migration process. Outside game theory, such RD models with travelling waves appeared first in the seminal works of Fisher (1937) and Kolmogorov et al. (1937). A good introduction to this topic is Fife (1979). For games, such RD models were first considered by Hutson and Vickers (1992), for symmetric 2×2 games, as spatial models of evolutionary games. They allow diffusion rates to depend on the strategy, which gives an additional twist to the game. In the present paper, I assume that each strategy and each player are equally able to diffuse (migrate), in order to keep the model as simple as possible, and to introduce no new parameters. Only the normal form of the game should determine the outcome.

The present approach to equilibrium selection seems to be unique, being based on *deterministic* dynamics. All existing dynamic equilibrium selection theories are based on *stochastic* models, like Kandori et al. (1993), Kandori and Rob (1995), Young (1993), Feinberg (1994). The models closest to the present one are those of Ellison (1993) and Blume (1995), who use a stochastic model of a spatial game (on a discrete lattice).

In Section II, I consider the case of two players (and two actions per player) and derive the risk-dominant solution. Section III contains some further classes of

W. Leinfellner and E. Köhler (eds.), Game Theory, Experience, Rationality, 245–259.

games, now with $n \geq 3$ players. For unanimity games our method turns out to be consistent with Nash's bargaining theory, while for two person games with incomplete information (which are formally equivalent to n person potential games with linear incentives), we do not quite obtain the 'generalized Nash property' proposed by Harsanyi and Selten (1972) – see also Selten (1995) –, but something similar.

II. TWO PLAYERS

The model

We consider a 2×2 bimatrix game, which we can write in a normalized form

$$\begin{pmatrix} a_1, a_2 & 0, 0 \\ 0, 0 & b_1, b_2 \end{pmatrix} \tag{2.1}$$

For diagonal entries $a_i, b_i > 0$ this is a coordination game or unanimity game, which has two strict equilibria $\mathbf{A}, \mathbf{B}$ and a mixed equilibrium E. Let p_i denote the frequency of the B-strategy for the i-th player. The payoffs for the i-th player playing A_i, or B_i, resp., are given by $\pi_i(A_i, p_{-i}) = a_i(1 - p_{-i})$ and $\pi_i(B_i, p_{-i}) = b_i p_{-i}$, where we denote $\{i, -i\} = \{1, 2\}$ as usual. Equating these expressions yields the mixed equilibrium $E = (k_1, k_2)$ where $k_i := \frac{a_{-i}}{a_{-i}+b_{-i}}$. In these coordinates, we can identify the equilibria $\mathbf{A} = \mathbf{0}$, and $\mathbf{B} = \mathbf{1}$.

Recall the notion of *risk-dominance* introduced by Harsanyi and Selten (1988, see esp. p. 86). For the game (2.1), $\mathbf{A}$ is risk dominant over $\mathbf{B}$, if

$$a_1 a_2 > b_1 b_2, \tag{2.2}$$

i.e., $\mathbf{A}$ has the higher Nash product, or equivalently, if the mixed equilibrium E is closer to $\mathbf{B}$, i.e.,

$$k_1 + k_2 > 1. \tag{2.3}$$

Let us consider now two player populations distributed in space (which is assumed one-dimensional for simplicity). Hence $p_i = p_i(x, t)$ are now functions of space $x \in \mathbb{R}$ and time t. Then our model equations read

$$\frac{\partial p_i}{\partial t} = f_i(p) + d\frac{\partial^2 p_i}{\partial x^2}. \tag{2.4}$$

The second term on the right hand side of (2.4) models random migration of players at a uniform, strategy- and player-independent rate $d > 0$, see Fife (1979, ch. 1).[1]

The first term, the reaction term, $f(p)$, models the local interaction of players and the resulting adaptations of their strategy according to their local experience. We will use the best response dynamics, $f(p) = BR(p) - p$, to model this interaction. This dynamics was introduced by Gilboa and Matsui (1991) and further studied by Hofbauer (1994). It models the situation that a certain proportion of players at each spot x switches to the locally best reply at any time t. For the 2×2 game (2.1) this means (with $\{i, -i\} = \{1, 2\}$ again)

$$\begin{aligned} f_i(p) &= -p_i \quad \text{if} \quad p_{-i} < k_{-i} \\ &= 1 - p_i \quad \text{if} \quad p_{-i} > k_{-i}. \end{aligned} \tag{2.5}$$

Other choices for the reaction term would be to take $f(p)$ as the replicator dynamics, i.e.,

$$f_i(p) = p_i(1 - p_i)(-a_i + (a_i + b_i)p_{-i}), \tag{2.6}$$

or more general selection, imitation, or myopic adjustment dynamics, satisfying

$$f_i(p) < 0 \Leftrightarrow \pi_i(A, p_{-i}) > \pi_i(B, p_{-i}). \tag{2.7}$$

In the best response dynamics, we have

$$BR_i(p) = 1_+(\pi_i(B, p) - \pi_i(A, p)), \tag{2.8}$$

where 1_+ denotes the Heaviside function: $1_+(u) = 1$ for $u > 0$, and $1_+(u) = 0$ for $u < 0$.

Our choice of the best response dynamics here is based on two reasons. First, it is probably the most natural dynamics in game theory: It is the simplest dynamics for modelling *rational* players, and equilibrium selection theories are traditionally based on assumptions of extreme rationality. In contrast to many other dynamics it depends only on the best reply structure of the game. Second, this choice has a big mathematical advantage: The RD equation (2.4) is piecewise linear, and hence explicitly solvable, in contrast to other choices of the reaction term. [2]

The travelling wave

Now we look for a travelling wave solution of (2.4), i.e., a solution of the form $p(x,t) = P(x - ct)$. The constant c is called the wave speed. Then the function $P(z)$, i.e., the shape of the wave, has to satisfy the following system of second order ordinary differential equations (ODE):

$$d\frac{d^2P}{dz^2} + c\frac{dP}{dz} + f(P) = 0 \tag{2.9}$$

This solution should connect the two equilibria $\mathbf{A}, \mathbf{B}$, which leads to the 'boundary conditions'

$$P(-\infty) = \mathbf{0}, \qquad P(\infty) = \mathbf{1}. \tag{2.10}$$

Furthermore, we assume that each component $P_i(z)$ is a monotonically increasing function of z. At points of discontinuity of f, pieces of P should be glued together in a C^1 way.

Theorem 1. *There exists a unique[3] monotone travelling wave for (2.4) that connects the two strict equilibria, i.e., satisfies (2.9) and (2.10). The wave speed $c > 0$, if and only if $k_1 + k_2 > 1$, i.e., $\mathbf{A}$ risk-dominates $\mathbf{B}$.*

Hence, along this special solution of the RDE, the dominating equilibrium (i.e., $\mathbf{A}$, if $c > 0$, and $\mathbf{B}$, if $c < 0$) *drives out* the other equilibrium.

The stability properties of such a 'bistable wave' will be discussed in section 4.

If $c = 0$, then $p(x,t) = P(x)$ is a stationary solution, or a 'standing wave'. This occurs iff the equilibria are equally risky.

The proof

The linearized form of (2.9) is

$$d\frac{d^2P}{dz^2} + c\frac{dP}{dz} - P = 0, \tag{2.11}$$

which has the general solution $P(z) = Ae^{\lambda z} + Be^{-\mu z}$, where $\lambda > 0$ and $-\mu < 0$ are the solutions of the quadratic equation $dx^2 + cx - 1 = 0$, i.e.,

$$\lambda = \frac{-c + \sqrt{c^2 + 4d}}{2d} \quad \text{and} \quad -\mu = \frac{-c - \sqrt{c^2 + 4d}}{2d}. \tag{2.12}$$

It will be convenient to work with λ, μ instead of c, d in the following. Since $c = (\mu - \lambda)d$ we have to show $\mu > \lambda$ in order to get a positive wave speed $c > 0$.

Near $\mathbf{A}$, i.e., for $P_i \le k_i$, (2.5), (2.9) and (2.10) implies

$$P_i(z) = A_i e^{\lambda z} \tag{2.13}$$

Near $\mathbf{B}$, i.e. for $P_i \ge k_i$, (2.5), (2.9) and (2.10) implies

$$P_i(z) = 1 - B_i e^{-\mu z}\ . \tag{2.14}$$

In the intermediate region, say $P_1 > k_1, P_2 < k_2$ (w.l.o.g., we assume that the curve $P(t)$ runs through the lower right quarter; otherwise interchange indices 1 and 2, i.e., interchange the players, or reflect the unit square along the main diagonal), we know from (2.5): $f_1(P) = -P_1$, and $f_2(P) = 1 - P_2$ and hence the solutions continue uniquely from the neighboring region as

$$P_1(z) = A_1 e^{\lambda z}, \quad P_2(z) = 1 - B_2 e^{-\mu z}\ . \tag{2.15}$$

Now one has to fit P_i in a C^1 way at the merging point z_i:

$$\begin{aligned} P_i(z_i) &= A_i e^{\lambda z_i} = 1 - B_i e^{-\mu z_i} \\ P_i'(z_i) &= \lambda A_i e^{\lambda z_i} = \mu B_i e^{-\mu z_i} \end{aligned} \tag{2.16}$$

which implies

$$A_i = \frac{\mu}{\lambda+\mu} e^{-\lambda z_i}, \quad B_i = \frac{\lambda}{\lambda+\mu} e^{\mu z_i}. \tag{2.17}$$

The merging values z_i are determined from (2.5) as the values defined by $P_i(z_{-i}) = k_i$. They satisfy $z_1 \geq z_2$ since the curve $P(z)$ runs through the lower right quarter by assumption. This results in the following two equations

$$P_1(z_2) = A_1 e^{\lambda z_2} = k_1$$
$$P_2(z_1) = 1 - B_2 e^{-\mu z_1} = k_2 \ .$$

Together with (2.17) we obtain

$$\begin{aligned} \frac{\mu}{\lambda+\mu} e^{\lambda(z_2-z_1)} &= k_1 \\ \frac{\lambda}{\lambda+\mu} e^{\mu(z_2-z_1)} &= 1-k_2 \end{aligned} \tag{2.18}$$

Hence

$$\frac{P_1(z_2)}{1-P_2(z_1)} = \frac{k_1}{1-k_2} = \frac{\mu e^{\mu(z_1-z_2)}}{\lambda e^{\lambda(z_1-z_2)}}. \tag{2.19}$$

Since $z_1 \geq z_2$ and the function $f(r) = re^{rz}$ is monotonically increasing in r, if $z \geq 0$, the expression in (2.19) is larger than 1 if and only if $\mu > \lambda$, which means $c > 0$. This shows the desired result for the direction of the travelling wave.[4]

It remains to prove *existence and uniqueness* of the wave $P(x - ct)$. (2.18) together with $z_2 < z_1$ implies the estimate

$$k_1 < \frac{\mu}{\lambda+\mu} < k_2.$$

With the abbreviation $q = \frac{\mu}{\lambda+\mu}$, (2.18) can be simplified to

$$e^{\frac{\lambda\mu}{\lambda+\mu}(z_1-z_2)} = (\frac{q}{k_1})^q = (\frac{1-q}{1-k_2})^{1-q}$$

This equation has a unique solution $q \in (k_1, k_2)$ since the left expression is increasing in $q \in [k_1, k_2]$ from 1 upwards, and the right one is decreasing down to 1. Now q determines uniquely μ and λ (since $\lambda\mu = \frac{1}{d}$ holds additionally), and $z_1 - z_2$ through (2.18). This determines the wave uniquely (up to translation), and shows also its existence. □

Remark. One could give a simpler proof of Theorem 1 (apart from the existence and uniqueness of the wave), by showing directly that the dominance relation derived from the direction of the travelling wave satisfies the three axioms of Harsanyi and Selten (1988, p. 86) that characterize risk-dominance for 2×2 games: The first two axioms of invariance with respect to isomorphisms and best reply invariance are trivially satisfied. The third axiom of payoff monotonicity can be shown rather easily with the help of comparison arguments from PDEs.

However, this approach does not help with more complicated games, as there is no general axiomatic approach to equilibrium selection.

III. $n \geq 3$ PLAYERS

Here I consider some classes of binary, bipolar n-person games, to use the terminology of Selten (1995): Every player has two pure strategies A and B, $0 \leq p_i \leq 1$ are the corresponding frequencies. The pure strategy profiles $\mathbf{A}$ and $\mathbf{B}$, corresponding to $p = \mathbf{0}$ and $p = \mathbf{1}$, are strict Nash equilibria. The RDE (2.4) is now a system of n equations ($i = 1, \ldots n$), $f_i(p) = BR_i(p) - p_i$, with $BR_i(p) \in \{0, 1\}$ (or $= [0, 1]$) denoting the pure (or the set of all mixed) best reply (replies) for player i to a population in state $p \in [0, 1]^n$. Since the second order ODEs in (2.9) decouple, the formulas (2.13)-(2.17) hold again for a monotone travelling wave, i.e.,

$$P_i(z) = \phi(z - z_i) \quad (i = 1, \ldots, n), \tag{3.1}$$

with

$$\begin{aligned} \phi(z) &= \frac{\mu}{\lambda + \mu} e^{\lambda z} \quad \text{for} \quad z \leq 0 \\ &= 1 - \frac{\lambda}{\lambda + \mu} e^{-\mu z} \quad \text{for} \quad z \geq 0 \end{aligned} \tag{3.2}$$

As in (2.19) we obtain the crucial property

$$P_i(z_j) + P_j(z_i) > 1 \Leftrightarrow \mu > \lambda \Leftrightarrow c > 0 \tag{3.3}$$

for any two indices i, j.

Note that this holds for all binary bipolar games, whose best reply structure is monotone:

$$p \leq q \Longrightarrow BR(p) \leq BR(q). \tag{3.4}$$

The details of the game, i.e., its best reply structure, is only needed to determine the merging values z_i: They are determined as the values for which the best response to $P(z)$ switches from A to B, or where $BR_i(P(z_i))$ is not unique.

Now I turn to some important special classes of games with a 'simple' best reply structure. In this section I do not bother to prove existence and uniqueness of the travelling wave, but see section 4. Here I present only some (rather simple) arguments that determine the sign of the wave, in order to demonstrate the power of the method in finding concrete criteria for dominance of equilibria.

1. Unanimity games: The Nash property

n-person unanimity games (or pure coordination games) are the direct generalization of the 2-person game (2.1): Each of the n players has two strategies, A, and B. If all players unanimously choose A (resp., B) then player i gets payoff $a_i > 0$ (resp., $b_i > 0$), otherwise, if there is no coordination, payoff is 0 to everybody. Nash's bargaining theory selects the strategy with the higher 'Nash product': *If* $a_1 \cdots a_n > b_1 \cdots b_n$ *then strategy* $\mathbf{A}$ *is the preferred outcome over* $\mathbf{B}$.

Practically all equilibrium selection theories aim at this so-called 'Nash property', which is the direct generalization of (2.2). In particular, the tracing procedure of Harsanyi and Selten (1988, see pp. 213 ff) leads to it. The one proposed here enjoyes it as well.

Theorem 2. *In an n-person unanimity game,* $\mathbf{A}$ *drives out* $\mathbf{B}$ *along a monotone travelling wave solution (i.e., $c > 0$) if and only if* $\mathbf{A}$ *has a larger Nash product than* $\mathbf{B}$.

Proof. For unanimity games,

$$BR_i(p) = 0 \Leftrightarrow a_i \prod_{j \neq i} (1 - p_j) > b_i \prod_{j \neq i} p_j. \tag{3.5}$$

Hence the merging values z_i satisfy the equations

$$a_i \prod_{j \neq i} (1 - P_j(z_i)) = b_i \prod_{j \neq i} P_j(z_i). \tag{3.6}$$

Multiplying these n equations, and applying the inequalities $1 - P_j(z_i) < P_i(z_j)$ from (3.3), we obtain

$$\prod_i a_i > \prod_i b_i. \tag{3.7}$$

This completes the proof. □

2. Potential games with linear incentives

A binary game has linear incentives (see Selten, 1995), if the difference between the payoffs for the two strategies of a player is a linear function of the probabilities p_j of the other players. (Every two person game has linear incentives, but not every n-person game.) Hence the set of profiles $p \in [0, 1]^n$ for which player i is indifferent between his two strategies can be expressed in the form

$$s_i = \sum_{j=1}^{n} \alpha_{ij} p_j, \tag{3.8}$$

with $\alpha_{ii} = 0$. I assume further that

$$\alpha_{ij} = \alpha_{ji} > 0, \tag{3.9}$$

and, w.l.o.g.,

$$\sum_{i,j} \alpha_{ij} = 2. \tag{3.10}$$

The symmetry of the 'interaction matrix' α_{ij} implies that the game is a 'potential game' in the sense of Monderer and Shapley (1996). The positivity of its entries implies the monotonicity of the best reply structure (3.4). The following result is another generalization of Theorem 1 (take $n = 2$ and $\alpha_{12} = \alpha_{21} = 1$). The dominance relation now generalizes (2.3) in a natural way.

Theorem 3. *In a bipolar binary n-person game satisfying (3.8-3.10),* **A** *drives out* **B** *along a monotone travelling wave, if and only if* $\sum_{i=1}^{n} s_i > 1$.

Proof. The merging values z_i satisfy the equations

$$s_i = \sum_{j=1}^{n} \alpha_{ij} P_j(z_i). \tag{3.11}$$

Hence (3.3) implies in the case $c > 0$,

$$S = \sum_i s_i = \sum_{i,j} \alpha_{ij} P_j(z_i) > \sum_{i,j} \alpha_{ij}(1 - P_j(z_i)) = 2 - \sum_{i,j} \alpha_{ji} P_j(z_i) = 2 - S$$

or $S > 1$. □

Remark. Potential games allow a simple way for equilibrium selection, see Monderer and Shapley (1996, p.134): Choose the equilibrium that maximizes the potential. For the above class of games, $V(p) = \frac{1}{2}\sum_{i,j} \alpha_{ij} p_i p_j - \sum_i s_i p_i$ is a potential. $V(\mathbf{A}) = V(\mathbf{0}) = 0$ and $V(\mathbf{B}) = V(\mathbf{1}) = \frac{1}{2}\sum_{i,j} \alpha_{ij} - \sum_i s_i = 1 - \sum_i s_i$, hence $V(\mathbf{A}) > V(\mathbf{B})$ iff $\sum_i s_i > 1$. Therefore this method of equilibrium selection is consistent with Theorem 3.

The normalization (3.10) allows an interpretation of α_{ij} as the probability for choosing the (unordered) pair $\{i, j\}$ from all n players. This interpretation leads to

3. Two-person unanimity games with incomplete information

As shown in detail in Selten (1995), such games can be reduced to potential n-person games with linear incentives, considered in section III.2. Such games run as follows: First, out of n given players, two are chosen with above probabilities α_{ij}. (This is the incomplete information part of the game.) The two selected players

i, j play a 2×2 unanimity game as in Section II, with payoffs a_i, b_i, and a_j, b_j, respectively. (The restriction to only two strategies **A**, **B** does not reduce generality.) Selten (1995) shows that such games reduce formally to a binary, bipolar potential game with linear incentives, the functions in (3.8) given as

$$\alpha_i \frac{a_i}{a_i + b_i} = \sum \alpha_{ij} p_j, \tag{3.12}$$

where $\alpha_i = \sum_j \alpha_{ij}$. Hence Theorem 3 yields

Theorem 4. *Along a monotone travelling wave solution (2.9), (2.10) of (2.4) for a two-person unanimity game with incomplete information,* **A** *drives out* **B** *if and only if*

$$\sum_{i=1}^{n} \alpha_i \frac{a_i}{a_i + b_i} > 1. \tag{3.13}$$

A more symmetric way to express (3.13) is

$$\sum_{i=1}^{n} \alpha_i \frac{a_i}{a_i + b_i} > \sum_{i=1}^{n} \alpha_i \frac{b_i}{a_i + b_i}. \tag{3.14}$$

In contrast, Selten's (1995) axiomatic approach[5] selects strategy **A** according to a 'generalized Nash property', namely

$$\prod_i a_i^{\alpha_i} > \prod_i b_i^{\alpha_i}. \tag{3.15}$$

Güth (1992) selects **A** according to a different generalization of the Nash property, namely iff

$$\prod_i a_i > \prod_i b_i. \tag{3.16}$$

The generalized Nash product criterion (3.15) which goes back to Harsanyi and Selten (1972) was already criticized by Myerson (1984). He argues that the probabilities α_{ij} should occur as factors of the payoffs instead as exponenents. Our criterion (3.13) seems to heed this point.[6] I suggest interpreting the condition (3.13) in the following way. It can be rewritten as

$$\sum_{i,j=1}^{n} \alpha_{ij} \left(\frac{a_i}{a_i + b_i} + \frac{a_j}{a_j + b_j} - 1 \right) > 0 . \tag{3.17}$$

This sum can be interpreted as the *expected advantage of* **A** *over* **B** in an incomplete information game, since the term in the parenthesis,

$$\frac{a_i}{a_i + b_i} + \frac{a_j}{a_j + b_j} - 1 = k_j + k_i - 1 = \frac{a_i a_j - b_i b_j}{(a_i + b_i)(a_j + b_j)},$$

can be considered as a measure for the advantage of **A** over **B** in the 2×2 coordination game (2.1) between the two players i and j (recall (2.2) and (2.3)).

It may be worth to compare the two different dominance criteria (3.13) and (3.15) in more detail. Consider for simplicity the case of equal probabilities $\alpha_{ij} = \frac{2}{n(n-1)}$. Then (3.15) and (3.16) depend only on the (unordered) set of the payoffs a_i and b_i, respectively, as in an n-person unanimity game from (3.1). In contrast, (3.13) in its equivalent form (3.17) takes care of the pairwise matchings occurring after the sampling of the two players.

To be more specific, consider the special case of $n = 3$ players, and let the payoffs be just a cyclic permutation:

$$(a_1, a_2, a_3) = (u, v, w) = (b_2, b_3, b_1) \tag{3.18}$$

Then generalized Nash products (3.15), (3.16) would consider **A** and **B** to be equally strong, whereas according to (3.13), **A** drives out **B** iff

$$\frac{u}{u+w} + \frac{v}{v+u} + \frac{w}{w+v} > \frac{3}{2}, \tag{3.19}$$

or equivalently

$$(u - v)(v - w)(w - u) > 0. \tag{3.20}$$

E.g., if $u = 0$ (or close), then **A** drives out **B** if $w > v$. This makes sense, since if player 3 is discarded from the game then **A** and **B** are equally strong, if player 2 is dropped, then **B** is stronger and if player 1 is dropped then **A** is stronger. However, the advantage of **A** over **B** in the third case exceeds the advantage of **B** over **A** in the second case (at least if we measure the advantage as suggested above, and not simply with Nash products).

This example shows another new feature of our dominance relation defined by (3.13): It is – in stark contrast to (3.15) or (3.16) – not transitive. As an example, consider a third strategy **C** in the above example, whose payoffs complete the cyclic symmetry in (3.18), i.e., $(c_1, c_2, c_3) = (v, w, u)$. Then (3.19) or (3.20) implies that **A** drives out **B**, **B** drives out **C**, and **C** drives out **A**.

Finally, the discrepancy between the different dominance criteria (3.13) and (3.15) disappears when we compute the optimal solution. Consider the bargaining problem with incomplete information, where the payoffs a_i, b_i are supposed to satisfy $\sum_{i=1}^n a_i = \sum_{i=1}^n b_i = 1$, i.e., the n players have to agree to a splitting of a certain amount of money, whose value is normalized to 1.

Theorem 5. *In a bargaining game with incomplete information, described by the matrix (α_{ij}) satisfying (3.9-3.10), consider the following way of splitting the amount of 1:*

$$a_i = \frac{\alpha_i}{2}.$$

This solution is optimal in the following two ways:
1. It drives out (in the travelling wave sense (3.13)) every other division b with $\sum_i b_i = 1$.
2. It maximizes the generalized Nash product (3.15) of Harsanyi and Selten.

Proof. Both statements are easy applications of Jensen's inequality. The second is well-known. For the first, use the concavity of the function $f(x) = \frac{1}{1+x}$:

$$\sum_{i=1}^{n} a_i \frac{a_i}{a_i + b_i} = \sum_{i=1}^{n} a_i \frac{1}{1 + \frac{b_i}{a_i}} = \sum_{i=1}^{n} a_i f(\frac{b_i}{a_i}) \geq f(\sum_{i=1}^{n} a_i \frac{b_i}{a_i}) = f(1) = \frac{1}{2}. \square$$

4. Symmetric games

Another, particularly simple class are symmetric binary n-person games, which are invariant under interchanging any two players i and j. Here, the diagonal $p_1 = p_2 = \ldots = p_n$ is invariant, and also the corresponding wave will enjoy this symmetry. Here $BR(p) = 0$ if $s(p) < 0$, where s is a certain polynomial of degree (at most) $n - 1$. If the game is bipolar, then $s(0) < 0 < s(1)$. We further assume that s has a unique zero $\hat{p}$ in the interval $[0, 1]$. This corresponds to a symmetric mixed equilibrium of the game.

Theorem 6. *For a bipolar binary symmetric games with a unique mixed symmetric equilibrium $\hat{p}$, **A** drives out **B** along the unique symmetric monotone travelling wave if and only if $\hat{p} > \frac{1}{2}$.* [7]

Proof. The unique symmetric monotone travelling wave solution takes the form (3.1) or (3.2) (the value of z_1 is arbitrary, due to translation invariance, see footnote 2). From the glueing value $P_1(z_1) = \phi(0) = \frac{\mu}{\mu+\lambda} = \hat{p}$ the claim follows immediately. [8] $\square$

One can compute the wave speed c explicitely in this case: $c = \frac{2\hat{p}-1}{\sqrt{\hat{p}(1-\hat{p})}}\sqrt{d}$.

We remark that for the present class of games our method selects precisely the long-run equilibrium of Kandori et al. (1993) and the resistant-dominant equilibrium of Güth and Kalkofen (1989), while other equilibrium selection methods, like that of Harsanyi and Selten (1988), Güth (1992), or Carlsson and van Damme (1993), lead to different results for $n \geq 3$. For this see the illuminating comparison for a prominent example, the stag hunt game, in the latter paper.

Since for such symmetric binary games the reaction dynamics acts on a one-dimensional state space, namely the unit interval, we can also easily discuss how a different choice of the reaction term, like (2.6), would change the equilibrium selected. For a scalar equation (2.9), the direction of the bistable travelling wave can be determined from the following criterion, see Fife (1979).

$$c > 0 \quad \Leftrightarrow \quad \int_0^1 f(P)dP < 0 \tag{3.22}$$

The choice $f(p) = BR(p) - p$ leads again to Theorem 6, with a different proof. For the replicator dynamics, one has $f(p) = p(1-p)s(p)$. For two person games, s has degree one, so $s(p) = \gamma(p - \hat{p})$ (with $\gamma > 0$), and (3.22) leads to $\hat{p} > \frac{1}{2}$, i.e., the same criterion as with the best response dynamics. However for $n \geq 3$ players, simple examples show that replicator and best response dynamics lead to different selection criteria: the travelling waves connecting **A** and **B** may run in opposite directions!

IV. Remarks on Existence, Uniqueness and Stability

For the $n \geq 3$ player games I have not considered the question of existence and uniqueness of the travelling wave. It could be done as for 2 players, but it is more tedious. On the other hand, *existence of a monotone travelling wave* follows essentially from a general result of Mischaikow and Hutson (1993) and Volpert et al. (1994) on bistable waves in reaction diffusion systems whose reaction term generates a monotone flow. This monotonicity assumption holds for all games considered here, see (3.4). The only technical problem is that these authors assume that the reaction term $f(p)$ is sufficiently smooth, while in (2.4), with the best response dynamics, it is discontinuous. A way to overcome this is to approximate the best reply dynamics by smooth dynamics, i.e., approximate 1_+ in (2.8) by a smooth monotone increasing function. Any of these approximating dynamics has a monotone travelling wave. Together with suitable apriori estimates for the shape and the speed of these waves (which can be found in the above papers) an Ascoli-Arcela type compactness argument will yield the existence of a monotone travelling wave solution also for the discontinuous limiting equation (2.4).

Volpert (1994) show also *uniqueness* of monotone bistable waves, but again this does not apply directly to our system (2.2). In any case, Theorems 2,3,4,6 above show that even if there were several monotone travelling waves, they would all have the same direction, i.e., the sign of the wave speed would be the same.

I conclude with some remarks on the *stability* of the travelling waves. This can be achieved by constructing comparison solutions.[9] It turns out that besides the travelling waves satisfying the boundary conditions (2.10), there are other travelling wave solutions that have no direct physical relevance, but are good for comparison results. Instead of (3.1) we consider

$$P_i^+(z) = \phi(z - z_i) + \varepsilon_i e^{\lambda z} \tag{4.1}$$

with ϕ given as in (3.2). These are solutions of (2.9), subject to the right choice of the merging values z_i. If $\varepsilon_i > 0$, (4.1) is a monotone, unbounded wave, which approaches the bounded wave satisfying (2.10) from above in the limit $\varepsilon_i \to 0$.

Also the speed c_+ of P^+ (which is smaller than c) approaches the speed c of the bounded travelling wave. Analogously, we obtain travelling waves

$$P_i^-(z) = \phi(z - z_i) - \delta_i e^{-\mu z} \ . \tag{4.2}$$

Now any initial function $p(x, 0)$ that can be squeezed between these two, i.e., $P^-(x - x_1) \leq p(x, 0) \leq P^+(x + x_2)$ for some values x_1, x_2, in particular any function satisfying $p(x, 0) = 0$ for $x \leq x_3$, and $p(x, 0) = 1$ for $x \geq x_4$, obeys

$$P^-(x - x_1 - c_- t) \leq p(x, t) \leq P^+(x + x_2 - c_+ t)$$

for all $t > 0$ by the monotonicity properties of our RD equations, and hence moves asymptotically with the same speed c as the bounded monotone travelling wave. Hence **A** drives out **B** along any such solution, if $c > 0$. This means that if initially the dominating equilibrium strategy occupies (or predominates on) one half of the spaceline, then it will take over and invade the other half line with speed c.

This result could be strengthened further, as in Fife (1979, Prop. 4.11): Prevalence of the dominating strategy on a sufficiently large compact interval (instead of a half line as above) is already sufficient for its spreading and taking over.

V. CONCLUSION

I have proposed a new method for equilibrium selection in strategic games, based on a spatio-temporal model, combining a diffusion (migration) process with a local best reply dynamic. One Nash equilibrium **A** dominates another, **B**, if there is a travelling wave solution, along which **B** is driven out by **A**, as $t \to \infty$. As demonstrated for a number of examples, one can with rather little effort explicitly compute the dominance relationship. For the basic class of n-person unanimity games, the method is consistent with Nash products and hence with most previous equilibrium selection methods. For other classes of games different criteria emerge.

Admittedly, the method has its limits: First, it seems to work well only for games whose best reply structure enjoys some monotonicity properties. Second, it enables only a pairwise or bilateral comparison of dominance relationship, preferably between strict equilibria.[10] Hence it does not provide a complete dynamic equilibrium selection theory: In the presence of several candidates of equilibria, with cycles in the dominance relation, I cannot say anything.

NOTES

1. To be more precise, consider the densities $\rho_{A_i}(x, t)$ and $\rho_{B_i}(x, t)$: $\int_a^b \rho_{A_i}(x, t)dx$ denotes the number of A_i players in the interval $[a, b]$ at time t. It is these densities that satisfy a reaction-diffusion equation of the form (2.4). Then the density of the total i population, $\rho_i(x, t) = \rho_{A_i}(x, t) + \rho_{B_i}(x, t)$, satisfies a diffusion equation $\frac{\partial \rho_i}{\partial t} = d\frac{\partial^2 \rho_i}{\partial x^2}$, and hence converges under mild assumptions to a spatially constant: $\rho_i(x, t) \to \rho_i$, as $t \to \infty$, i.e., each player population is equally spread over the line $\mathbb{R}$. Then the frequencies $p_i(x, t) = \frac{\rho_{B_i}(x,t)}{\rho_i}$ satisfy (2.4).

2. An extension of the present model and results on the replicator equation will be considered in Hofbauer et al. (1996).
3. Since (2.9) is autonomous, $P(z+z_0)$ is a solution for any fixed z_0, if $P(z)$ is a solution of (2.9). By uniqueness, we mean up to this trivial kind of nonuniqueness.
4. One can get more information about the location of the curve $P(z)$ in the unit square. If this curve runs through the lower right quarter, i.e., stays below E, then the slopes near **0** and **1** satisfy $\frac{P_2(z)}{P_1(z)} = e^{\lambda(z_1-z_2)} < \frac{k_2}{k_1}$ for $z < z_2$, and $\frac{1-P_2(z)}{1-P_1(z)} = e^{\mu(z_2-z_1)} > \frac{1-k_2}{1-k_1}$ for $z > z_1$. Since $z_2 < z_1$ in this case, this implies $k_1 < k_2$. Hence, the curve $P(z)$ lies between the main diagonal and the equilibrium E.
5. Selten (1995, section 3.3) presents an example for which (3.15) leads to an opposite dominance relationship than the tracing procedure. For this example also (3.13) and (3.15) lead to opposite results.
6. I do not know whether (3.13) is compatible with Myerson's resolution.
7. This condition reminds of the concept of $\frac{1}{2}$-dominance: $\mathbf{A}$ is $\frac{1}{2}$-dominant if it is a best reply against all mixtures that place weight at least $\frac{1}{2}$ on A. This is a rather robust sufficient (but in general not a necessary) condition for dominance, see Ellison (1995). Also in the cases considered above, it implies dominance along the travelling wave: For unanimity games, $\frac{1}{2}$-dominance means $a_i > b_i$ for all i. For potential games (3.8), it means $s_i > \frac{\alpha_i}{2}$.
8. In other words, Theorem 6 follows immediately from (3.3) again, with $i = j = 1$.
9. Mischaikow and Hutson (1993, Theorem 7.2) construct sub- and super-solutions from the travelling wave and derive from them the Lyapunov stability of the wave in the uniform topology on $\mathbb{R}$. However, their construction depends on the continuity of the reaction term $f(.)$, and unfortunately does not work here.
10. An extension to mixed equilibria that are stable in the best response dynamics is possible. However, such equilibria are rare for asymmetric games.

Acknowledgment. This research was sponsored by the Austrian FWF, Project P 10717-MAT.

REFERENCES

L. E. Blume, "The Statistical Mechanics of Best-Response Strategy Revision", in: *Games and Economic Behav.*, 11, 1995, pp.111-145.

H. Carlsson/E. van Damme, "Equilibrium Selection in Stag Hunt Games", in: K. Binmore/A. Kirman/P. Tani (eds.), *Frontiers of Game Theory*. Cambridge, MA: MIT Press 1993, pp.237-253.

G. Ellison, "Learning, Local Interaction, and Coordination", in: *Econometrica*, 61, 1993, pp.1047-1071.

G. Ellison, "Basins of Attraction, Long Run Equilibria, and the Speed of Step-by-Step Evolution", *Preprint*, 1995.

Y. Feinberg, "Evolutionary Selection of an Equilibrium", Preprint, *Hebrew University Jerusalem*, 1994.

P. Fife, *Mathematical Aspects of Reacting and Diffusing Systems* (Springer Lecture Notes in Biomathematics, 28), New York: Springer 1979.

R. A. Fisher, "The Wave of Advance of Advantageous Genes", in: *Ann. Eugenics*, 7, 1937, pp.353-369.

I. Gilboa/A. Matsui, "Social Stability and Equilibrium", in: *Econometrica*, 59, 1991, pp.859-867.

W. Güth, "Equilibrium Selection by Unilateral Deviation Stability", in: R. Selten (ed.), *Rational Interaction, Essays in Honor of John C. Harsanyi*. Berlin: Springer, 1991, pp.161-189.

W. Güth/B. Kalkofen, *Unique Solutions for Strategic Games (Springer Lecture Notes in Economics and Mathematical Systems, 328)*, 1989.

J. C. Harsanyi/R. Selten, "A Generalized Nash Solution for 2-Person Bargaining Games With Incomplete Information", in: *Management Science*, 18 (5), Part II, 19972, pp.80-106 .

J. C. Harsanyi/R. Selten, *A General Theory of Equilibrium Selection in Games*. Cambridge, MA: MIT Press 1988..

J. Hofbauer, "Stability for the Best Response Dynamics", *Preprint*, Vienna 1994.

J. Hofbauer/V. Hutson/G.T. Vickers, "Travelling Waves for Games in Economics and Biology", in: *Proc. Conf. WCNA-96* Athens, Greece. To appear.

V. C. L. Hutson/G. T. Vickers, "Travelling Waves and Dominance of ESS's", in: *J. Math. Biology*, 30, 1992, pp.457-471.

M. Kandori/G. J. Mailath/R. Rob, "Learning, Mutation and Long Run Equilibria in Games", in: *Econometrica*, 61, 1993, pp.29-56.

M. Kandori/R. Rob, "Evolution of Equilibria in the Long Run: A General Theory and Applications", in: *J. Economic Theory*, 65, 1995, pp.383-414.

A. N. Kolmogorov/I. G. Petrovskij/N.S. Piskunov, "Studies of the Diffusion Equation, Combined with Increase in the Amount of Matter and its Applications to a Problem in Biology", in: *Bul. Moskovskovo Gos. Univ., Mat. Mekh (= Bull. Moscow Univ, Math. Mech.)* 1:6, 1937 pp.1-26. - English translation in: V.M. Tikhomirov (Ed.): *Selected works of A.N.Kolmogorov*, Vol. I, Kluwer: Dordrecht 1991, pp.242-270.

K. Mischaikow/V. Hutson, "Travelling Waves for Mutualist Species", in: *SIAM J. Math. Anal.*, 24, 1993, pp.987-1008.

D. Monderer/L. Shapley , "Potential Games", in: *Games Econ. Behav.*, 14, 1996, pp.124-143.

R. B. Myerson, "Two-Person Bargaining Problems with Incomplete Information", in: *Econometrica*, 52, 1984, pp.461-487.

J. F. Nash, "The Bargaining Problem", in: *Econometrica*, 18, 1950, pp.361-382.

R. Selten, "An Axiomatic Theory of a Risk Dominance Measure for Bipolar Games with Linear Incentives", in: *Games Economic Behavior*, 8, 1995, pp.213-263.

A. I. Volpert/V. A. Volpert/V. A. Volpert, *Traveling wave solutions of parabolic systems*. Providence, RI: American Mathematical Society 1994.

H. P. Young, "The Evolution of Conventions", in: *Econometrica*, 61, 1993, pp.57-84.

Institut für Mathematik
Universität Wien
Strudlhofgasse 4
A-1090 Vienna
Austria

ANDREAS FLACHE
RAINER HEGSELMANN

RATIONAL VS. ADAPTIVE EGOISM IN SUPPORT NETWORKS: HOW DIFFERENT MICRO FOUNDATIONS SHAPE DIFFERENT MACRO HYPOTHESES*

I. INTRODUCTION

In the following we study the evolution of support networks among egoists who differ widely in their degree of neediness, are free to choose their partners, and do so in opportunistic ways. No central authority is involved. The question we address is to what degree and under what aspect it shapes the structure of emerging solidarity networks whether we model egoistic actors as *rational* actors in a game theoretical sense or as *adaptive* actors, i.e. learning beings following a simple learning mechanism which is driven by success and failure.[1] The central modelling ideas to address that question are:

- Support relationships are modelled by a *2-person support game*.
- The social space in which a network can emerge is a *2-dimensional cellular grid*. Actors can play the support game simultaneously with all their neighbours. Not all cells are occupied. From time to time players have the *opportunities to migrate*.
- In *one* model we make our actors *rational* beings, in a *second* one we make them *adaptive* agents. Obviously we use a *cellular automata* framework of modelling (cf. Hegselmann 1996, 1996b). Especially because of its simplicity that framework has often proved fruitful for analysing micro/macro relations.

Plainly we use a *cellular automata* framework of modelling (cf. Hegselmann 1996, 1996b). Especially because of its simplicity, that framework has often proved fruitful for analysing micro/macro relations. In the following we explain the models in more detail. Section (II) deals with the support game, the basic facts of playing and migrating in our 2-dimensional world are explicated in (III), and the structure of rational and adaptive actors in (IV) and (V), respectively. Results are then presented in (VI) and discussed in (VII). Section (VIII) puts forward some conclusions and suggests directions for further research.

II. THE SUPPORT GAME

In the support game both players are characterised by the degree to which they need support, their degree of *neediness* n $(0 < n < 1)$. For simplicity, we assume that

W. Leinfellner and E. Köhler (eds.), Game Theory, Experience, Rationality, 261–275.

	C		D	
C	$R_{ij} = G_{ij} - L_{ij}$	$R_{ji} = G_{ji} - L_{ji}$	$S_{ij} = -L_{ij}$	$P_{ji} = G_{ji}$
D	$T_{ij} = G_{ij}$	$S_{ji} = -L_{ji}$	$P_{ij} = 0$	$P_{ji} = 0$

Table 1. Payoff matrix of the constituent support game

every actor living in the cellular world belongs to one of some *neediness classes.* The neediness of a member of class i is n_i. The support game played between a member of class i and a member of class j is fully characterised by the corresponding combination of neediness classes, ij. The more needy an actor Ego is the more she profits from receiving help from her neighbour Alter, and the less she is capable of providing help herself. High neediness of Alter limits the amount of help Ego can receive, while it increases the amount of help Ego will provide in case she follows the cooperative strategy with respect to Alter. Formally, we model i's gain from receiving help from j, G_{ij}, and i's loss from giving help to j, L_{ij}, as follows:

$$\begin{aligned} G_{ij} &= n_i(1 - n_j)B \\ L_{ij} &= (1 - n_i)n_jE \end{aligned} \tag{1}$$

where the parameters B and E are positive constants. B indicates the benefit i derives from receiving help when she is maximally needy ($n_i = 1$), while j is fully capable of helping i ($n_j = 0$). E quantifies the effort i invests helping j when i is not needy herself ($n_i = 0$), while j is maximally needy ($n_j = 1$). The players' payoffs derive from a particular combination of their help decisions. What matters is the net benefit of gains minus losses where gains and losses, both ensue from the decisions taken. Actors who cooperate (C) incur the corresponding loss while their partners earn the gain from receiving help. Correspondingly, a player who defects (D) incurs no loss but her partner receives a help of zero. Table 1 shows the payoff matrix of the constituent support game. Following the standard in the game theoretical literature, we denote the payoffs corresponding with the outcomes CC, DC, CD and DD as **R**eward, **T**emptation, **S**ucker's payoff and **P**unishment, respectively.

Double indices are used because the payoff of a member of class i in a pairing with a member of class j depends on *both* actors' classes. One can easily verify that $T_{ij} > R_{ij}$ and $P_{ij} > S_{ij}$. Therefore, *not* to help is a *dominant* strategy in the constituent support game. For mutual help (C,C) to be *mutually profitable* we need to have for both players

$$G_{ij} > L_{ij} \Leftrightarrow n_i(1 - n_j)B > n_j(1 - n_i)E. \tag{2}$$

Iff (2) is fulfilled, does the constituent support game with its dominant strategy 'not to help' turn into a prisoners' dilemma with an *inefficient* solution, because exactly then does it hold that $R_{ij} > P_{ij}$. Therefore, we denote this condition the *PD-condition*. This condition provides a minimum criterion for evaluating the adequacy of a model of egoistic decision making in the support game: Neither a pair of rational egoists nor a pair of adaptive egoists should form a stable solidarity relationship unless it is *mutually profitable*, i.e. unless the PD-condition is met. The other way round: The PD-condition specifies the range of pairings *ij* for which mutual solidarity may arise between selfish players, while cooperation is at the same time susceptible to the free rider problem.[2] It is well known that this problem can be solved when selfish actors face a *repeated* interaction, where they can exchange help on the basis of *reciprocity*. As to *rational* actors, they may refrain from exploiting others at a given time, because they anticipate the consequence of losing others' help in the future. *Adaptive* players may learn in the course of repeated interactions that helping others is rewarded by reciprocated help, which in the long run stabilises their cooperative tendencies. Both rational anticipation and adaptive learning require *a sufficient duration of the interaction*, however.

III. BASIC FACTS OF PLAYING AND MIGRATING IN THE TWO MODELS

In this section we put together the basic architecture of playing and migrating in the two models. Firstly, it is assumed that the individuals belong to *nine different neediness classes n*. We have $n_1 = 0.1, n_2 = 0.2, \ldots, n_9 = 0.9$. They live on the surface of a *torus*, so that the world can be represented as a checkerboard without borders. The *interaction neighbourhood* is supposed to be a von Neumann neighbourhood, i.e. neighbours are the neighbouring cells to the north, south, east and west. Not all cells are occupied by individuals. An initial distribution of individuals and empty cells is *created by chance*. All classes have the same number of members. In each period a random generator allocates migration options with an *exogenously set probability*. Such options can be used, but it is not a must. Migration is possible within a certain *migration window*, but *only to vacant destination cells*. The options are evaluated and/or used in a *sequential* order according to the results of a lottery. After that a new period starts. The migration window, an odd-sized square for an individual, which has an option in the center, has the same size for all individuals, for instance 5×5. Each actor, rational or adaptive, instantly *knows the neediness class of all his (real or potential) neighbours* and is informed about the payoffs of the support game.

IV. A MODEL OF (BOUNDEDLY) RATIONAL EGOISM IN THE SUPPORT GAME

As to the decisions of rational egoists in the solidarity game we assume that rational actors support each other whenever *Trigger* or *Tit for tat* strategies achieve an equilibrium when played against each other. For that to be the case the probability *alpha* for continuing the relation in the next period must not be lower than a certain threshold value α^*. The equilibrium condition (cf. Friedman 1986, Taylor 1987) is given by

$$\alpha \geq \frac{T_{ij} - R_{ij}}{T_{ij} - P_{ij}} = \frac{L_{ij}}{G_{ij}} = \alpha^*_{ij} \quad \text{for both players.}^3 \qquad (3)$$

Equation (3) has a straightforward interpretation. The threshold continuation probability equals the ratio of losses to gains of mutual support. Mutual support with *j*

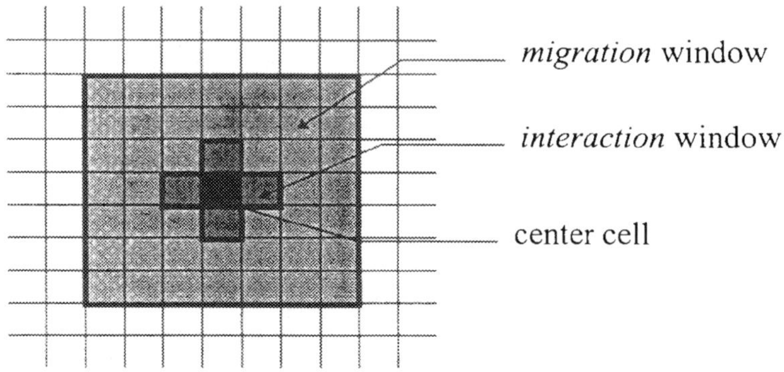

Figure 1: Basic windows.

is always unattractive from i's point of view when losses exceed gains. When gains exceed losses, some continuation probability $\alpha \leq 1$ exists for which i is willing to cooperate with j. Notice that according to (3), α^*_{ij} depends on both actors' neediness and the constants B and E. The probability of getting migration options is used to calculate a pessimistic proxy for having a further round of play in the next period. "Pessimistic" means the individuals assume that every option will be used to leave. With q as the probability for getting a migration option we therefore have a probability of stability $\alpha = (1 - q)^2$. Everyone decides whether conditions (2) and (3) are fulfilled. Two neighbours will support each other[4] iff both conditions are met.

All individuals know that – relative to their own neediness class – they are in a *best social position* when they are surrounded by four individuals which are members of the best class willing to engage in support relationships with them. On the other hand, there are worst *social positions* being surrounded by empty cells only or by only those individuals with whom support relations are not possible. An individual will be satisfied with a position which offers *a certain fraction of the difference between the payoff at the best and the worst social position*. This fraction is called *minimum level*. *Satisfied individuals* will migrate only if they can get a position which is at least as good as the position they have. *Dissatisfied individuals* will seize every migration option and accept even worse new locations in order to ultimately find a better position by further moves which are accessible from that new location.

These assumptions make it obvious that our rational actors do not decide all problems they are facing along the lines of an anticipating rationality: One can argue that the decisions whether to help or not to help are made rationally. These decisions rely on an estimation of the continuation probability, however, that is only a very rough and pessimistic guess. Moreover, the decisions about where to go do not even try to anticipate what will happen with the neighbourhood actors enter.

V. A MODEL OF ADAPTIVE EGOISM IN THE SOLIDARITY GAME

The model of adaptive egoism in the solidarity game comprises two main components. The first component modelling decisions in the iterated support game will be discussed in (V.1); the second component modelling migration decisions is explicated in (V.2).

Solidarity decisions in the support game

We model solidarity decisions on basis of a conventional Bush-Mosteller stochastic learning model (Bush and Mosteller 1955, see also Macy 1990, 1993). Our model consists of three components, the decision rule, the evaluation rule and the learning rule. We discuss these components in turn.

Decision rule
Actor *i* cooperates in some constituent support game with a member of neediness class *j* with probability P_{ijt}, her cooperation propensity. The time index *t* refers to the *t*'th iteration of the support game that *i* played with some member of neediness class *j*, indicating that in the learning process the cooperation propensity may change. Every inhabitant *i* of the cellular world memorises nine cooperation propensities corresponding to the nine neediness classes in the population. For simplicity, we assume that actors initially have no particular disposition for either cooperation or defection ($\forall i, j : p_{ij0} = 0.5$).

Evaluation rule
Player *i* evaluates the payoff attained in iteration *t* relative to an expectation level, in order to adapt her cooperation propensity in the light of the experience gathered in *t*. A player is satisfied, when the payoff is equal to or exceeds her expectation level, while she is dissatisfied otherwise. We assume that actors are capable of distinguishing between partners of different neediness classes. Accordingly, actors evaluate the payoffs they attain in the pairing *ij* relative to an expectation level that reflects the range of possible payoffs in this particular pairing. More precisely, we model *i*'s expectation level in an interaction with *j* as $\frac{1}{2}(T_{ij} + S_{ij})$, the midpoint of the interval of possible payoffs $[S_{ij}, T_{ij}]$. Formally, *i*'s satisfaction with the payoff u_{ijt} is then computed as follows:

$$s(u_{ijt}) = u_{ijt} - \frac{1}{2}(T_{ij} + S_{ij}) \tag{4}$$

where u_{ijt} represents the payoff obtained in the *t*'th iteration ($u_{ijt} \in T_{ij}, R_{ij}, P_{ij}, S_{ij}$). The satisfaction $s(u_{ijt})$ is rescaled to the interval $[-l, +l]$, yielding the reinforcement $r(u_{ijt})$ for cooperation:

$$r(u_{ijt}) = \frac{2lr(u_{ijt})}{T_{ij} - S_{ij}} \tag{5}$$

reinforcement:		$r(u_{ijt}) \geq 0$	$r(u_{ijt}) < 0$
strategy	C	$r(u_{ijt})(1 - p_{ijt})$	$r(u_{ijt})p_{ijt}$
player i in t	D	$-r(u_{ijt})p_{ijt}$	$-r(u_{ijt})(1 - p_{ijt})$

Table 2: Change of cooperation propensity Δp_{ijt} as a function of i's action and the related inforcement

where l represents the learning rate ($0 \leq l \leq 1$). The larger l, the stronger is the reinforcement corresponding with a positive or negative satisfaction.

Learning rule:
After every iteration of the game, players adapt their cooperation propensities. The learning rule implies that the propensity to repeat cooperative behaviour increases, when i cooperated and experienced a satisfactory payoff, or when she defected and obtained a payoff falling below her expectation level. Conversely, i reduces her propensity to cooperate when she experienced a negative reinforcement for cooperation, or she defected and this defection was rewarded. Formally, the adapted propensity $p_{ij,t+1}$ is computed according to:

$$p_{ij,t+1} = p_{ijt} + l\Delta p_{ijt} \tag{6}$$

where Table 2 shows how the propensity increment Δp_{ijt} results from the outcome of iteration t.

The adaptation of propensities described by Table 2 follows two principles. First, the larger the reinforcement $r(u_{ijt})$ a player experienced, the more this will increase or decrease her propensity to cooperate, all other things being equal. Second, to guarantee that the adaptation rule only generates valid propensities, the change function flattens when propensities approach the extremes of the propensity interval, 0 and 1, respectively.[5]

PD-condition
Above we argued that it is a minimum requirement for the adequacy of a decision model of egoistic behaviour that full cooperation can only arise when the PD-condition is satisfied. It follows from the evaluation rule and the learning rule that the stochastic learning model meets this requirement. To explain, it is easily seen from Equation (2) that the outcome of mutual cooperation (CC) yields a negative reinforcement $r(u_{ijt})$ for at least one of the two players when the PD-condition is not met. As a consequence, this player will reduce her cooperation propensity whenever in some iteration both players cooperate. Hence, whenever the PD-condition is not met, it is impossible that full cooperation stabilises for both players.

Coop-condition

Both for the rational egoist model and for the adaptive egoist model the PD-condition is a *necessary* condition for the possibility of mutual cooperation in the support game. The COOP-condition instead specifies a minimum continuation probability α^*_{ij} that is *sufficiently* large so that mutual cooperation is the predicted result of egoistic players' behaviour. The rational egoist model, as we showed in (4), predicts that whenever the PD-condition is met for a pairing *ij* there is some continuation probability $\alpha(\alpha^*_{ij} \leq \alpha < 1)$ such that it is rational to follow a conditionally cooperative strategy in the iterated support game. This result is reflected by the stochastic learning model. The stochastic learning model implies that a pairing *ij* of adaptive egoists will attain full cooperation within a certain number of iterations whenever the PD-condition applies. The PD-condition implies that both players receive a positive reinforcement when the outcome of iteration *t* is bilateral cooperation (CC) or bilateral defection (DD). In this case, both players' cooperation propensities increase by some positive amount Δp_{ijt} (or Δp_{jit} from *j*'s point of view). After a sufficiently large, but finite, number of consecutive occurrences of CC or DD, both players' propensities are close enough to 1 to 'lock in' to full commitment to cooperation (see Note 5). Once this occurs, defection probabilities are zero and both players' full commitment to cooperation never changes again. In terms of Markov chain theory, bilateral full cooperation ($p_{ijt} = p_{jit} = 1$) is the unique absorbing state of the stochastic learning process and this state is reachable from every other combination of players' propensities by a sequence of CC or DD outcomes. In (7) below, we use this property of the learning model to assess for every pairing *ij* the number of iterations that is required on average to attain the absorbing state. This allows us to derive numerically a minimum continuation probability α^{**}_{ij} that must be given so that it is likely that mutual cooperation will be attained.

Adaptive migration decisions in the solidarity game

Like rational egoists, the adaptive players we model take their migration decisions by evaluating free social positions in their migration window relative to their present social position. In contrast to rational egoists, however, adaptive players estimate the attractiveness of a potential solidarity partner *k* on the basis of the payoffs they experienced in previous interactions with actors of *k*'s neediness class. More precisely, whenever *i* interacted with a member of neediness class *k*, she computes a weighted mean of the payoff she attained in this particular interaction and of her previous estimate of the opponent's attractiveness. The second difference between the migration decisions pertains to the evaluation of neighbourhoods. In contrast with the rational egoist model, adaptive players are not capable of anticipating which neediness classes are at all willing to cooperate with them. As a consequence, the minimum level for adaptive players is computed using the assumption that the best social position attainable is where the focal actor is surrounded by 4 members of the class with the lowest neediness, whom she succeeds to exploit continuously. Correspondingly, the worst social position from an adaptive player's point of view is the position where she is continuously exploited by 4 neighbours who are exclusively members of the neediness class with the highest neediness. With exception of this difference in the evaluation of social positions, adaptive egoists apply the same decision mechanism that rational egoists do.

VI. RESULTS

In this section, we apply both simulation models to a scenario where the solidarity network is characterised by a relatively high degree of stability[6]. We assume that actors receive on average a migration option only in every 20th period, i.e. with a probability of 0.05 per period. This corresponds with a continuation probability of $\alpha = 0.903$. All individuals initially are randomly distributed in the cellular grid. Furthermore, we initialise the simulation as follows. We assume a von Neumann neighbourhood as interaction window. The migration window has a size of 5×5, with the given position in the center. The world is a torus with 21×21 cells, populated by 9 neediness classes with 35 individuals per class. This results in a ratio of occupied cells to empty cells of 315/136. With respect to the payoffs we assume that $B = 4$ and $E = 1$, yielding a *benefit/effort* ratio of 4/1. The minimum level, finally, is 50% of the payoff difference between best and worst position. Figures 2a and 2b below show the patterns of the solidarity networks between rational egoists and adaptive egoists, respectively, as they typically emerge from the initial random configuration after 1000 simulation periods.

For interpretation of these figures notice the following:

- White cells are empty cells.
- Round cells are dissatisfied with their social position (i.e. get less than the expected 50% of the difference between worst and best social position). Squared cells are satisfied actors.
- Short white lines connecting two actors indicate bilateral cooperation.
- Different patterns and grey levels represent different neediness classes as described to the legend at the bottom of the figures. The neediness of Class 1 is $n_1 = 0.1$, neediness of Class 2 is $n_2 = 0.2$, etc.

The simulations illustrated by Figure 2a demonstrate that 1) an *extensive solidarity network* can arise even between egoistic actors, 2) this network is characterised by a *segregation pattern* that we call the core-periphery structure of solidarity. The extensiveness of the solidarity network is obvious from the fact that almost every member of the population has one or more support relationships with some other actor.

In this scenario, the average individual has 3.52 solidarity partners (mean value of 10 replications with a standard deviation of 0.06). In particular, even most of the 'unattractive' actors with high neediness are embedded in solidarity relations. On closer inspection, however, the network reveals a distinct segregation pattern. Clusters of members of the least needy Class 1 tend to be in the core of the network, around which a belt has emerged that is primarily formed by members of

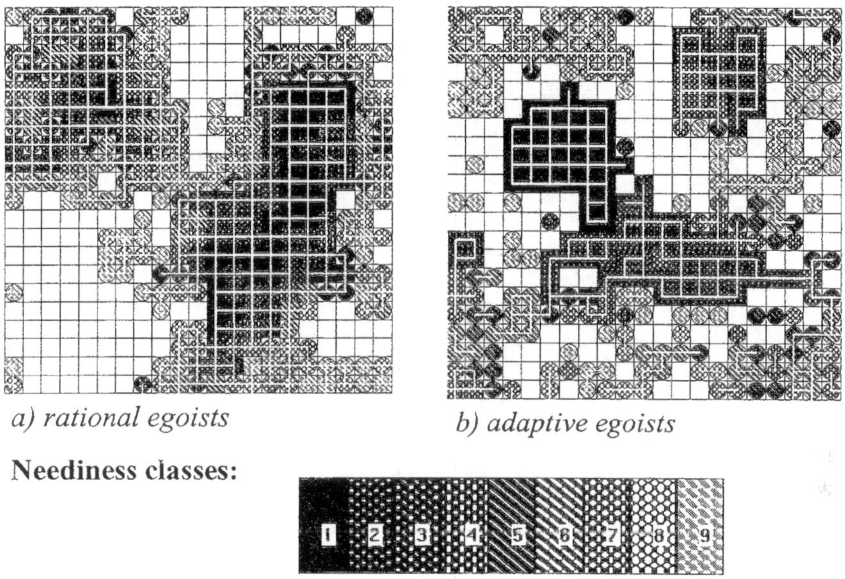

Figure 2: Core-periphery network between rational egoists (a) and class-segregation network between adaptive egoists (b), both after 1000 simulation periods.

neediness Class 2. The further the distance from the core of the network the larger is the degree of neediness of the individuals located there. At the periphery of the network, we almost exclusively find members of the highest neediness Classes 8 and 9. To assess whether this pattern is robust with regard to variation in the behavioural assumptions, we replicate the simulation with the adaptive egoist model, keeping all other parameters constant. For comparison with the simulations of the rational egoist model we assume a learning rate that allows the emergence of stable support relations ($l = 1$)[7]. With this maximum learning rate, the learning algorithm approximates a 'win-stay, lose change' strategy in which propensities are sharply altered following aversive outcomes. Figure 2b shows the solidarity network that typically arises with these assumptions after 1000 simulation periods.

Again solidarity arises despite individuals' self interested behaviour. As in the simulations of Figure 2a, *the emerging network is extensive* in the sense that most members of the population find solidarity partners, even those with high neediness. At the same time, there are clearly fewer solidarity relations in this network than in the network of Figure 2a.[8] Inspection of the distribution of solidarity relations shows that like in the simulations of rational egoists, a *segregation pattern* arose. This time, however, separated clusters of interrelated actors tend to develop. These separated clusters are almost homogeneous with respect to neediness class, indicating a high level of *class segregation*. This segregation pattern entails a highly unequal distribution of help. The less needy actors are in this network, the lower is the neediness of their neighbours, hence the more help they receive relative to the help they need. Conversely, the more needy actors are, the more needy are also their neighbours. As a consequence, those most needy receive the smallest amount of help in proportion to the help they need.

To summarise, the simulations yield three two results:

- Solidarity can arise even between egoistic individuals, but solidarity networks between rational egoists are more extensive than are solidarity networks between adaptive egoists.
- Solidarity networks are characterised by segregation patterns. The level of class segregation is clearly higher between adaptive egoists than it is between rational egoists.

VII. DISCUSSION

To explain the simulation results, we discuss two questions: 1) For which pairings of neediness classes can support relations successfully be established? 2) How attractive are the different partners? We address the rational egoist model and the adaptive egoist model in turn.

The rational egoist model

Figure 3a shows for which pairings of neediness classes support relations can arise according to the rational egoist model. The figure visualises the PD- and COOP-

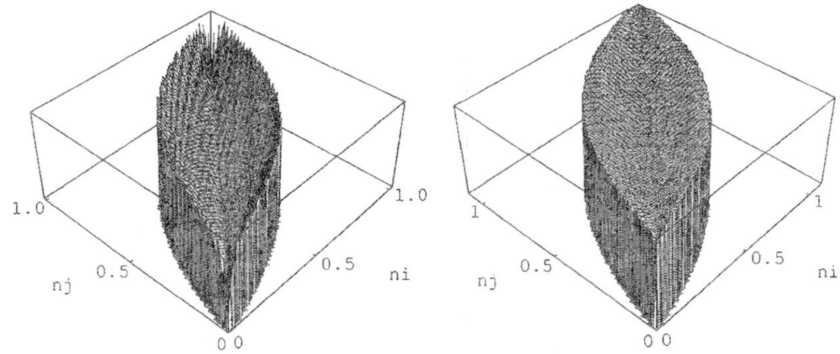

a) Rational egoists: a^*_{ij} *a) adaptive egoists:* a^{**}_{ij}

Figure 3: Threshold continuation probability for the rational egoist model (a) and the adaptive egoist model (b).

condition. The horizontal axes represent neediness levels, i.e. n_i and n_j. The vertical axis indicates the threshold continuation probability α^*_{ij}. The area where no α^*_{ij}-value is plotted shows where the PD-condition (1) is not met. For this range of pairings mutual support is not profitable from at least one player's perspective.

Figure 3a reveals that the following is required for support relations to be *possible at all*:

- The neediness classes *must be sufficiently similar.*
- *Middle neediness* classes have a larger range of potential partners than classes at the *extremes.*

With regard to the *continuation probability*, which is *necessary* for mutual support, we find that

- the *more different* neediness classes there are, the higher is the continuation probability required.

This structure of the PD- and COOP-condition reflects the gains and losses of support of (1). Cooperation is at all possible when for both actors the amount of help they receive is larger or equal to the amount of help they give (PD-condition). This precludes solidarity relations between highly dissimilar neediness classes, because here the actor who is less needy is both capable of giving more help and needs less help than her partner. As a consequence, it is particularly difficult for members of extreme neediness classes to find partners. The larger the difference between neediness classes, the smaller is – from at least one actor's point of view – the amount of help this actor receives relative to the amount of help she gives. Accordingly, these actors need to expect a long duration of a successful support relation to withstand the temptation to unilaterally defect. Therefore the more dissimilar actors are, the larger is α^*_{ij}. The core-periphery pattern, finally, results from the differences in attractiveness of solidarity partners, as they are given by (1). Members of Class 1 could in principle form support relations both with Class 1 and Class 2 members, but they attain a larger payoff with their own kind. As a result, stable cores of mutual choices between members of Class 1 can establish, but only if around them members of the next most attractive Class 2 settle. Members of Class 2, in turn, remain only in positions where they have as many partners as possible of Classes 1 or 2 or of the next most attractive Class 3, and so forth.

The adaptive egoist model

To analyse for which pairings of neediness classes mutual support can arise according to the adaptive egoist model, we compute numerically an estimation of the corresponding COOP-condition. For simplicity we focus on an isolated dyad *ij* playing the support game in consecutive iterations. Broadly, we assess the average number of iterations t^{coop} that is required until the dyad *ij* eventually reaches the absorbing state of full cooperation $(p_{ij}, p_{ji} = 1)$. From t^{coop} we compute the corresponding threshold continuation probability α_{ij}^{**} assuming that the average duration of a support game has to be at least t^{coop} iterations before mutual support can stabilise in *ij*. Figure 3b charts the estimated continuation probability for the learning model.

It is obvious that with respect to the possibility of support relations Figure 3b has the same properties as Figure 3a. This follows from the fact that the PD-condition is a necessary condition for mutual support in both models. In addition, Figure 3b reveals that there is a strong structural similarity between the COOP-conditions. Both models imply that by and large the threshold continuation probability is the larger *the more different are the neediness levels of i and j*. At first glance, the structural similarity of the COOP-conditions seems to contradict the obvious discrepancy between the core-periphery segregation pattern of rational egoists, and the class-segregation structure predicted by the adaptive egoist model. Comparison of the COOP-conditions of the two models reveals, however, that the discrepancy between predicted network structures is a consequence of the fact that across all pairings *ij* the prospects of attaining bilateral cooperation are considerably better for rational egoists than they are for adaptive egoists. Inspection of the COOP-condition of the learning model also shows why solidarity relations in the simulations arose only between members of similar neediness classes. In the simulations we assumed a probability of receiving a migration option of 0.05, resulting in a continuation probability of $\alpha = 0.903$. Figure 3b shows that even the lowest threshold continuation probability for pairings of the same neediness class is slightly higher, with $\alpha_{ij}^{**} = 0.905$. While this demonstrates that the numerical procedure somewhat overestimates the restrictiveness of the COOP-condition for the adaptive egoist model, it also explains why under the conditions of the simulation pairings of at best almost equal neediness succeed in establishing solidarity relations. Successful solidarity relations are established only between similar partners. Accordingly, in their migration decisions adaptive egoists prefer neighbourhoods with as many as possible solidarity partners of neediness classes similar to their own class.

VIII. CONCLUSION

In this paper we analysed solidarity networks using two different micro foundations. In one model, we assumed that actors are rational egoists in a game theoretical sense, in another model we applied a simple adaptive decision mechanism. Both simulations suggest that under the scenario we studied – relatively high stability of relationships – solidarity networks arising between egoistic actors may have the following characteristics:

- Extensive solidarity networks are possible even between egoists, i.e. despite individual opportunism most actors find solidarity partners.
- Solidarity networks are characterised by segregation patterns.

Of course these are tentative conclusions that rely on a number of assumptions about the cost benefit structure of the underlying game, the stability of relations, etc. Future research should explore whether the differences between the two models generalise to a larger range of conditions. For instance, it is of interest to explore whether Hegselmann's (1996a) results about the effects of increasing migration probabilities are robust with respect to variation in micro foundations. So far, we consider it as a strong indication for the robustness of the phenomenon of segregation in support networks that it derives from two decision models that represent almost opposite extremes in terms of the cognitive sophistication of the actors modelled. At the same time, our study demonstrated that using these two micro foundations allows us to assess the range within which the predicted macro phenomena may be affected by properties of individual decision making. In particular we found clearly distinct levels of overall solidarity and different segregation patterns. Solidarity networks between rational egoists were both more extensive and had a lower degree of class segregation than solidarity networks between adaptive egoists.

NOTES

* This research has partly been conducted within a project on modeling the dynamics of social dilemma situations that is financed by the German Science Foundation DFG (Grant no. HE 1412 / 4-1). The authors wish to acknowledge the support of the DFG. Please direct correspondence to andreas.flache@uni-bayreuth.de, or rainer.hegselmann@uni-bayreuth.de.

1. For discussions of rational and adaptive micro foundations see Friedman 1953, Dawes and Thaler 1988, Macy 1990, Hechter 1992, Roth 1992, Macy 1993, Macy and Flache 1995, Flache and Macy 1996, Flache 1996.
2. The support game used here is different from the support game used in Hegselmann (1994 and 1996a). The main difference is that in the game used above *bilateral* help is always possible, though the degree varies for different neediness class combinations. It is therefore possible to assume that both players can observe what the other's strategy has been in the last period in order to react on that. Exactly that is necessary for applying the learning mechanism explained in (4). The support game in Hegselmann (1994) and (1996a) allows in each period only *unilateral* help. As a consequence it becomes a game with *imperfect* information to which the learning mechanism used here can not be applied. Obviously there are different games which model decisive features of what in daily life is simply called *mutual help*.

3. *Tit for tat* is an equilibrium strategy against *tit for tat* iff $\alpha \geq \max\left\{\frac{T-R}{T-P}, \frac{T-R}{R-S}\right\}$. In the support game we have $T - P = R - S$. Therefore the equilibrium condition reduces to (3). Since $R > P$ is always fulfilled in the support game *all* support games can be solved cooperatively by tit for tat strategies, given α is high enough. The latter is not the case for "normal" prisoners' dilemmas iff $T + S > 2R$, a case that can never happen for the support game.
4. This implies the assumption that an equilibrium selection problem is solved.
5. In addition we assume that actors develop a full commitment to either cooperation ($p_{ijt} = 1$) or defection ($p_{ijt} = 0$), once the learning process has driven their propensities sufficiently close to the extremes of the propensity interval. This ensures that bilateral cooperation can stabilise at all, once both players' propensities are large enough.
6. Hegselmann (1994 and 1996a) has studied the effects of stability on predictions of the rational actor model by varying the probability of migration options. We here focus on one scenario only, because we are primarily interested in assessing the effects of varying micro foundations.
7. Macy (1991) has demonstrated that with this learning rate there is a large probability that bilateral cooperation stabilises within a small number of iterations, at least when the R payoffs sufficiently exceed the P payoffs for both players. With a low learning rate (e.g. l=0.1), he showed the emergence of stable bilateral cooperation is practically impossible even with a favourable payoff structure. Too many iterations are required before a dyad reaches the absorbing state of bilateral full cooperation.
8. For reliability, we computed the number of solidarity relations on the basis of 10 simulation runs per model. In the rational egoist model, the average number of solidarity relations was 3.52 (standard deviation 0.06), for the adaptive egoist model we obtained 3.16 (standard deviation 0.08).

REFERENCES

R.R. Bush/F. Mosteller, *Stochastic models for learning*. New York: Wiley 1955.

R.M. Dawes/R.H. Thaler, "Anomalies: Cooperation", in: *Journal of Economic Perspectives*, 2, 1988, pp.187-197.

A. Flache, *The double edge of networks: An analysis of the effect of informal networks on cooperation in social dilemmas*. Amsterdam: Thesis Publishers 1996.

A. Flache, A./M.W. Macy, "The Weakness of Strong Ties: Collective Action Failure in a Highly Cohesive Group", in: *Journal of Mathematical Sociology*, 21, 1996, pp.3-28.

M. Friedman, *Essays in positive economics*. Chicago: University of Chicago Press 1953.

J.W. Friedman, *Game theory with applications to economics*, (2nd ed. 1991). Oxford: Oxford University Press 1986.

M. Hechter, "The Insufficiency of Game Theory for the Solution of Real-World Collective Action Problems", in: *Rationality and Society*, 4, 1992, pp.33-40.

R. Hegselmann, "Zur Selbstorganisation von Solidarnetzwerken unter Ungleichen – Ein Simulationsmodell", in: K. Homann (Ed.), *Wirtschaftsethische Perspektiven I – Theorie, Ordnungsfragen, Internationale Institutionen*. Berlin: Duncker & Humblot 1994, pp.105-129.

R. Hegselmann, "Cellular Automata in the Social Sciences $alpha$ Perspectives, Restrictions, and Artefacts", in: R. Hegselmann/K.G.Troitzsch/U. Mueller (Eds.), *Modeling and simulation in the social sciences from the philosophy of science point of view* (Theory and Decision Library). Dordrecht: Kluwer 1996, pp.209-234.

R. Hegselmann, "Social Dilemmas in Lineland and Flatland", in: W.B.G. Liebrand/D. Messick (Eds.), *Frontiers in social Dilemma research*. Berlin 1996a, pp.337-362.

R. Hegselmann, "Understanding Social Dynamics – The Cellular Automata Approach", in: U. Mueller/N. Gilbert/K. G. Troitzsch/J. Doran (Eds.), *Social science microsimulation*. Berlin 1996b, pp.282-306

M.W. Macy, "Learning Theory and the Logic of Critical Mass", in: *American Sociological Review*, 55, 1990, pp.809-826.

M.W. Macy, "Learning to Cooperate: Stochastic and Tacit Collusion in Social Exchange", in: *American Journal of Sociology*, 97, 1991, pp.808-843.

M.W. Macy, "Backward Looking Social Control", in: *American Sociological Review*, 58, 1993, pp.819-836.

M.W. Macy/A. Flache, "Beyond Rationality in Models of Choice", in: *Annual Review of Sociology*, 21, 1995, pp.73-91.

A. Roth, "Game Theory as Part of Empirical Economics", in: J.D. Hey (Ed.), *The Future of Economics*. Cambridge: Blackwell 1992, pp.107-114.

M. Taylor, *The possibility of cooperation* (revised edition of: *Anarchy and cooperation*, 1976). London: Wiley & Sons 1987.

Institut für Philosophie
Universität Bayreuth
Geschwister Scholl-Platz 3
D–95440 Bayreuth, Germany

KEN BINMORE

EVOLUTIONARY ETHICS*

> He who would understand *baboon* would do more towards metaphysics than John Locke.
> Charles Darwin

I. FOOTNOTE TO PLATO?

Philosophers used to say that all their endeavours were merely a footnote to Plato. In ethics, this is still largely true.

Modern moral philosophers are mostly divided into consequentialists and deontologists. Both see their task as uncovering the nature of an *a priori* Platonic ideal, that consequentialists take to be the Good and deontologists the Right. But neither side doubts that their purpose is to lay down absolute principles that are universally applicable to all human beings.

This approach extinguished the commonsense ideas of sophists like Glaucon in Plato's *Republic*, who held that morality is a matter of social convention. The view was revived again by David Hume [5] and other luminaries of the Scottish Enlightenment in the eighteenth century, but the world continues to prefer the grandiose notions of Immanuel Kant [6], who argued that the only reason that we are sentient is to put into practice the categorical imperative: act only on the maxim that you would will to be a universal law. For example, the most influential ethical work in the twentieth century has perhaps been John Rawls' [9] *Theory of Justice*, in which he seeks to explore the practical implications of operationalizing his version of the categorical imperative. But such an approach denies the obvious fact that what people think is good or right depends on the culture in which they were raised. As the ancient philosopher Xenophanes pointed out: "The gods of the Ethiopians are black and flat-nosed, and the gods of the Thracians are red-haired and blue-eyed."

In recent years, sociobiologists have followed E. O. Wilson in reviving yet again the idea that it is possible to approach ethics as a scientist, by exploring the biological and social facts on which our moral intuitions are based. Such facts are contingent. They would have been otherwise if biological and social history had taken a different course. Moral behaviour in chimpanzees and baboons differs from moral behaviour in humans because their biological history differs from ours. Moral behaviour in other human societies differs from moral behaviour in our society because their social history differs from ours. In ancient Athens, I would perhaps have chased after adolescent boys like Socrates. In antebellum Virginia, I would probably have been ready to keep slaves like Thomas Jefferson.

Such frank relativism is too much for many to swallow. It is painful for an economist to admit, but Adam Smith [10] was the first of many to seek to have it both

W. Leinfellner and E. Köhler (eds.), Game Theory, Experience, Rationality, 277–283.

ways. But those who wish to enter the pulpit to preach that one society is better than another are not entitled to appeal to naturalistic theories of ethics. Even the wishy-washy liberal doctrine that all societies are equally meritorious receives no support from naturalism. There is no culture-free Archimedean standpoint from which to apply a moral lever to the world. If we could liberate ourselves from all cultural prejudices, we would find that morality no longer had any meaning for us.

Of course, as history shows only too well, no holds are barred when one cultural prejudice seeks to sustain itself against another. Sociobiologists are therefore attacked as fascist dinosaurs still peddling the eugenicist filth of the Social Darwinists of Victorian times. In self defence, they have therefore taken to calling themselves behavioural ecologists or evolutionary psychologists. But whatever he calls himself, no modern sociobiologist believes G. E. Moore's [8] claim that evolutionary ethics necessarily maintains that "we ought to move in the direction of evolution simply because it is the direction of evolution". He may reasonably think that his expertise about the way societies work makes it likely that he knows more than others about which reforms are *feasible*. But his relativism forces him to deny that his opinion on what is *optimal* is worth any more than that of the proverbial man on the Clapham omnibus.

II. GAME THEORY

I think that part of the reason that the reaction to sociobiology has been so vitriolic is that it has concentrated its attention in the past largely on biological or genetic evolution. This makes it seem like we are pieces of robotic wetware like ants. I agree that the human universals which constitute the holy grail for traditional moral philosophers are to be found in our genes, but I think that the phenomena that traditionalists actually discuss are almost entirely determined by social or memetic evolution. That is to say, I do not believe that the strategies we use in coordinating with other human beings are hardwired. We learn them as we grow to maturity largely by imitating the behaviour of individuals who are higher in the pecking order.

Maynard Smith's [7] famous *Evolution and the Theory of Games* explains how to adapt the economic discipline of game theory to the study of biological evolution in cases when the behaviour of one animal may affect the fitness of another. Geneticists speak of such problems as being "frequency dependent". When modeling the game of life played by animals in such situations, it is commonplace to look at cases in which only two strategies compete – a normal strategy and a mutant strategy. One then asks whether the normal strategy is evolutionarily stable. That is to say, whether an invasion of the normal population by a small fraction of mutants will eventually result in the extinction of the mutants. The most important requirement for a strategy s to be evolutionarily stable, is that it be a best reply to itself. If there were a better strategy t that could be played against an opponent using s, then a mutant that played t would be able to invade a population of animals all playing s.

A pair of strategies with the property that each is a best reply to the other is called a Nash equilibrium in game theory, after a recent winner of the Nobel Prize for economics. In the biological contexts studied by Maynard Smith and his followers, the models of the game of life employed seldom have many Nash equilibria. But it would be naive to employ similar models in a social context. One may cavalierly use Dawkins's [2] notion of a meme as the social equivalent of a gene, but one then cannot reasonably expect memes to appear one at a time. Whatever meme is currently holding the fort will be under constant bombardment by newly invented memes seeking to become the new king of the castle. To study social evolution seriously, it is therefore necessary to consider games in which the players can choose between large numbers of strategies.

III. RECIPROCAL ALTRUISM

Game theorists think that the simplest class of games that are adequate to study the emergence of human morality are indefinitely repeated games – in which a relatively simple static game is played repeatedly over time by the same players with a random stopping time. Technically, such games have an infinite number of strategies. Their importance lies in the fact that they allow I'll-scratch-your-back-if-you'll-scratch-mine principles to emerge as *equilibrium* behaviour. As Confucius argued, understanding how such reciprocal arrangements operate is the single most important factor in getting to grips with human sociality.

Most people know of the Prisoners' Dilemma. In this famous toy game, each player can either costlessly provide his opponent with a service that is worth $2 to him, or else steal $1 from his pocket. Call these two strategies *dove* and *hawk*. Traditional morality says that each player should choose *dove*, but such behaviour is not evolutionarily stable. In equilibrium, each player chooses *hawk*, because this is a best reply whatever the opponent may choose.

Much effort has been devoted to finding some way around this unpalatable piece of reasoning. It is argued that only stupid people would be caught in such an evolutionary trap. It is said, for example, that really clever people will recognize the validity of Kant's categorical imperative that clearly requires the play of *dove*. But all such arguments are fallacious. If you are trying to maximize your payoff in the game, then no amount of fancy reasoning can evade the simple fact that *hawk* always does better than *dove* however you or your opponent may reason. Of course, people say that the players ought not to try and maximize their own payoff. Instead they should try to do their best for the group of which they are a part. But this is the group selection fallacy of Wynne-Edwards that Richard Dawkins [2] was so anxious to refute in his *Selfish Gene*. Evolution doesn't favour mutations that sacrifice themselves to help their neighbours. It favours mutations that help themselves.

Game theorists believe that all the attention that has been devoted to the Prisoners' Dilemma is largely wasted. If the Prisoners' Dilemma were a suitable paradigm of the human game of life, then we would no have evolved as social animals. For a more suitable paradigm, we need to look at indefinitely repeated games. The famous folk theorem of game theory then applies. This was proved independently by a number of different game theorists in the early fifties. It says that almost any outcome of the static game that gets repeated can be sustained as a Nash equilibrium of the repeated game provided that the players never have reason to think that any particular repetition is likely to be the last. The idea was rediscovered some twenty years later by the biologist Robert Trivers [11], who coined the term reciprocal altruism for the mechanism that makes the theorem work. Axelrod's [1] celebrated *Evolution of Cooperation* provides the most familiar example. It is a Nash equilibrium in the indefinitely repeated Prisoners' Dilemma for both players to use the strategy *tit-for-tat*. This strategy tells the players to begin by playing *dove*, but then to reciprocate whatever the opponent did in the previous round.[1] Both players then cooperate by playing *dove* at every repetition, because any deviation by one player will trigger a punishing response from the opponent that makes the deviation unprofitable.

IV. Group selection

Indefinitely repeated games have large numbers of equilibria. Some of these Pareto-dominate others. In honour of the sociologist Pareto, economists use this jargon to mean that *all* players get a larger payoff at one equilibrium than at another. A Pareto-efficient equilibrium is Pareto-dominated by nothing at all. There is therefore no way that all the players can be made better off at a Pareto-efficient equilibrium. Any movement away from the equilibrium will make at least one player worse off. For example, the equilibrium in which both players use *tit-for-tat* in the indefinitely repeated Prisoners' Dilemma is Pareto-efficient. Since both players receive an income of \$2 per play, it Pareto-dominates the equilibrium in which both players receive an income of \$1 per play as a consequence of always playing *hawk* no matter what.

When a game has many equilibria, one is faced with the *equilibrium selection problem*. How do players go about deciding on which equilibrium to coordinate? Following David Hume, I believe that human morality evolved as a response to this problem.

If this view is correct, it is necessary to abandon the out-of-equilibrium use of *dove* in the one-shot Prisoners' Dilemma as a paradigm for human morality. If a simple example must be adopted, the Driving Game is more appropriate. This game has three equilibria: one in which everybody drives on the left; one in which everybody drives on the right; and one in which players randomize the choice of the side of the road on which to drive. The first and second equilibria are Pareto-efficient and Pareto-dominate the third. Society needs a commonly understood convention that permits the players to coordinate on one of these equilibria.

How do such conventions evolve? This question is easy to answer when one group operates an equilibrium that Pareto-dominates the equilibrium operated by another group. The first group will then prosper as compared with the second. For example, a society in which people always drive on the left will have fewer deaths than a society in which people drive on either side at random. The convention operated by the first society will therefore eventually win out over the convention operated by the second as a consequence of the first society growing in size or founding more colonies.

Such group selection does not contravene the "selfish gene" principle, because the conventions studied require the players to coordinate on an equilibrium. It is therefore always in the interest of each individual player to follow the convention operated in his society, provided that he believes his fellow players will do the same.

But one does not learn anything about the origin of the ideas of justice and fair play that most of us associate with moral behaviour by studying competition between one equilibrium and a Pareto-dominant rival. Such issues only become relevant when a choice has to be made between Pareto-efficient equilibria. Whichever equilibrium is then chosen, somebody will necessarily gain at the expense of somebody else. But how does a society decide who should be lucky or unlucky? One way is to fight it out, but societies that adopt this costly solution will eventually lose out to societies that find their way to a convention that allows the decision to be made quickly and peacefully. The reason is exactly the same as that offered in defence of the claim that Pareto-dominated equilibria will tend to disappear in favour of their more efficient rivals.

The suggestion is therefore that our moral intuitions on justice are derived from the solution that evolution found to the problem of selecting between Pareto-efficient equilibria. Like the Panda's thumb, I believe that this solution required adapting mechanisms that had evolved to fulfil other functions to a new and different purpose.

V. KIN SELECTION

The first mechanism evolved originally to optimize behaviour in the games played within the family. In such games, a player does not optimize by maximizing his own individual fitness, because some of his genes are shared by the relatives with whom he is playing. For example, my sister shares half my genes and hence optimal behaviour for me consists in maximizing a weighted sum of our fitnesses in which her fitness is counted at half the rate of mine. More generally, as Bill Hamilton's [3, 4] rule recognizes, I should maximize a weighted sum of all the fitnesses of my relatives, with weights proportional to their degree of relatedness.

But how do I know what is good for my sister? I need to be able to put myself in her shoes to see things from her point of view – to feel her joy or pain as though it were, in some degree, my own. Not only must I be capable of such empathetic identification, I must be able to quantify my comparison of her plight and mine so as to be sure that her plight is only counted to be half as important as mine. In brief, the mechanics of family life equipped us with a capacity for empathetic identification and interpersonal comparison that we only later learned to use when interacting with strangers.

VI. SHARING FOOD

The second mechanism that I believe needs to be considered in studying the origins of our ideas on justice and fair play, is how our species first learned to share food. Anthropologists commonly see such reciprocal behaviour as lying at the root of our sociality. Its advantage in groups living near the subsistence level lies in in the insurance that such a convention provides for those who may be unlucky in their hunting and gathering. The lucky hunter shares with his unlucky fellow on the understanding that the same service will be reciprocated when their roles are reversed. To operate such an arrangement, the lucky hunter must be able to empathize with his own future self – to put himself in the shoes of the unlucky hunter that he might be tomorrow. If there is an equal chance of occupying either role on any given day, he must then evaluate the share of today's catch that he reserves for himself on the assumption that half the time he will be Mr. Lucky and half the time Mr. Unlucky.

VII. THE ORIGINAL POSITION

Rawls' [9] *Theory of Justice* offers what he calls the device of the original position as an attempt to operationalize the Kantian categorical imperative. To decide what is fair, players imagine themselves behind a veil of ignorance,where they bargain over what compromise to operate without knowing their identities. Each thinks instead that he is equally likely to turn out to be the individual favoured or disadvantaged by the deal. This influential idea has had an enormous impact, but I do not think it is because the Kantian ideas with which it is defended carry much weight. I think it is because we recognize it as a stylized version of a social convention that we *already* use in settling fairness issues. It is a version of the do-as-you-would-be-done-by principle that respects the objection that one should not do unto others as you would have them do unto you because they may have different tastes.

How do I see the device of the original position evolving from the twin mechanisms of empathetic identification within the family and food sharing? Note first that each party to a food-sharing deal must place himself behind a veil of uncertainty where Mr. Lucky sees it as equally likely that he might be himself or a potential future self, Mr. Unlucky. To extend this mechanism to the original position, a player must instead imagine that he is equally likely to be either himself or some other person entirely. That is to say, he must put himself, not in his own future shoes, but in the shoes of another person. But his experience within the family provides him with all the necessary equipment to do this. He must simply empathize with a stranger in the same way that he empathizes with a cousin or an aunt.

But what does he substitute for the degree of relationship that provides the standard for making interpersonal comparisons within the family? It is the different manner in which social evolution solves this problem in different societies that I believe creates the different standards of fairness that we see operating in different societies. It is here therefore that one has to look for an answer to most of the difficulties that beset traditional moral philosophy. I do my best to provide what answers are possible within a simple modeling framework in my two-volume book *Game Theory and the Social Contract* published by MIT Press. The first volume appeared in 1994 with the subtitle *Playing Fair*. The second volume will appear next year with the subtitle *Just Playing*.

NOTES

* I am grateful to the Economic and Social Research Council and to the Leverhulme Foundation for funding this reseach through the Centre for Economic Learning and Social Evolution at University College London.

1. Unfortunately, Axelrod overplayed the virtues of this particular strategy. Robustness studies confirm that cooperative behaviour is very likely to emerge in the indefinitely repeated Prisoners' Dilemma, but the strategies that support such behaviour will not in general be unexploitive, forgiving or simple like *tit-for-tat*.

REFERENCES

[1] R. Axelrod, *The Evolution of Cooperation*. New York: Basic Books 1984.
[2] R. Dawkins, *The Selfish Gene*. Oxford: Oxford University Press 1976.
[3] W. Hamilton, "The Evolution of Altruistic Behavior", in: *American Naturalist*, 97, 1963, pp.354–356.
[4] W. Hamilton, "The Genetic Evolution of Social Behavior, Parts I and II", in: *Journal of Theoretical Biology*, 7, 1964, pp.1–52.
[5] D. Hume, *A Treatise of Human Nature*. 2nd edition. Oxford: Clarendon Press 1978 (Edited by L. A. Selby-Bigge, revised by P. Nidditch, first published 1739).
[6] I. Kant, *Groundwork of the Metaphysic of Morals*. New York: Harper Torchbooks 1964 (Translated and analyzed by H. Paton. First published 1785).
[7] J. Maynard Smith, *Evolution and the Theory of Games*. Cambridge, Cambridge University Press 1982.
[8] G. E. Moore, *Principia Ethica*. Buffalo, NY: Prometheus Books 1988 (First published 1902).
[9] J. Rawls, *A Theory of Justice*. Oxford: Oxford University Press 1972.
[10] A. Smith, *The Theory of Moral Sentiments*. Oxford: Clarendon Press 1975 (Edited by D. Raphael and A. Macfie, first published 1759).
[11] R. Trivers, "The Evolution of Reciprocal Altruism", in: *Quarterly Review of Biology*, 46, 1971, pp.35–56.

Economics Department
University College London
Gower Street
London WC1E 6BT
UK

JOHN C. HARSANYI

A PREFERENCE-BASED THEORY OF WELL-BEING AND A RULE-UTILITARIAN THEORY OF MORALITY

I. INTRODUCTION

Ethics deals with two basic problems. One is what to do to have a good life from our own *personal* point of view, which I shall call the problem of *personal well-being*. The other is what to do to have a good life from a *moral* point of view, which I shall call the problem of *morality*.

PERSONAL WELL-BEING

II. THE CONCEPT OF UTILITY

We owe the concept of a person's *utility level*, interpreted as a measure of his or her[1] well-being, to the great British utilitarian philosophers and economists of the 18th and 19th centuries, from Hume and Bentham to John Stuart Mill and Jevons. These utilitarian thinkers were *hedonists* and regarded people's utility levels as measures both of their *well-being* and of their *happiness* at the same time. Accordingly, they tried to explain economic behavior as maximization of one's happiness or utility.

In contrast, since Pareto and especially since Hicks (1939), modern economics has explained people's behavior in terms of their *preferences* (assuming that the latter satisfy the usual consistency axioms). A person's utility function is regarded merely as a convenient mathematical representation of his preferences. It is still assumed that rational behavior involves maximization of one's utility function. But as the latter is itself defined in terms of one's preferences, it is these preferences rather than utility maximization as such that explain people's behavior.

Indeed, strictly speaking, modern economics could be restated, admittedly at a cost of some inconvenience, in terms of people's preferences alone, without any reference to utility at all. But most economists do retain the utility concept as a convenient mathematical representation of people's preferences.

Economists were attracted to this approach for two main reasons. One was that it enabled them to dispense with the outdated hedonistic psychology assumed by the earlier approach. The other was that happiness is a psychological concept

W. Leinfellner and E. Köhler (eds.), Game Theory, Experience, Rationality, 285–300.

based on *introspection*, whereas people's preferences are expressed by their choice behavior and by their verbal statements, both open to *public observation*. Yet, for the purposes of ethics, it seems to me, this utility concept used in economics must be revised in two important ways. First, besides the usual economic variables, we must include also many *noneconomic* variables as arguments of a person's utility function, such as his state of health, his social position, his relationships with other people, and so on. Second, we must take account of the fact that a person's actual preferences may be based *on incorrect or on incomplete information* and, therefore, may fail to indicate his true preferences at a deeper level.

In view of this, I propose to distinguish between a person's *actual preferences* as expressed by his choice behavior and by his verbal statements, and his *informed preferences*,[2] defined as the *hypothetical* preferences he *would* entertain *if* he had all the relevant information and had made proper use of this information.

For example, suppose that Frank has a serious case of pneumonia and asks to be treated by medication B because he does not know that, according to the best medical information now available, medication A would be much more effective against his illness. Clearly, in terms of his *actual* preferences, Frank would prefer medication B to A. Yet, we obviously have good reasons to assume that, in terms of his *informed* preferences, he *would* prefer A to B. In other words, he *would* choose A over B *if* he knew the relevant medical facts.

I shall use the term *informed preferences* also for describing those actual preferences of a person that *agree* with his hypothetical informed preferences just defined. In contrast, I shall use the term *mistaken preferences* for describing those actual preferences of his that *deviate* from his informed preferences.[3]

In principle, a person's *actual* preferences are empirical variables, disclosed by his choice behavior and by his verbal statements. (This statement, however, is subject to some qualifications. For as I shall argue in Section III, we do not always do what we would really *prefer* to do.)

Unlike a person's *actual* preferences, his *informed* preferences are *theoretical* variables whose nature can only be discovered by deeper analysis of the empirical facts. This is so because the latter are defined as the preferences the *would* entertain under certain conditions.

III. DESIRES AND PREFERENCES

In everyday life we use *both* a person's desires and his preferences to describe the motives of his behavior. Yet, if we study the literature then we shall find that economists analyze human motivation almost exclusively in terms of people's *preferences* whereas philosophers do so almost exclusively in terms of people's *desires*. Yet, as we shall see, we shall obtain better insights into human motivation by considering *both* people's desires *and* their preferences.

No doubt, desire is a simpler and more basic notion than preference is. A desire for A involves only *one* object A and refers to a pro-attitude toward this one object. In contrast, a preference for A over B involves *two* objects A and B, and indicates that the person in question assigns a *higher priority* to his pro-attitude toward A than to his pro-attitude toward B. More fundamentally, a person's preferences indicate not only the priorities he assigns to his various desires as such, but also the *relative importance* he assigns to the *objects* of his various desires, i.e. to the various *alternatives* available to him.

Economists concentrate on people's preferences rather than on their desires because a person's utility function can be defined *only* in terms of his *preferences*. We cannot assume that a given person assigns a higher utility to A than to B unless we know that he *prefers* A to B.

On the other hand, we cannot infer his preference for A over B from facts concerning his *desires* as such. Thus, even if we know that at the impulsive level he has a *much stronger desire* for A than for B, it may not follow that he really *prefers* A over B and will actually *choose* A rather than B. For he may have good reasons to *resist* his stronger impulsive desire for A and to opt for B instead.

For instance, he may find that by following his desire for A he would act contrary to his moral principles. Or, quite apart from moral considerations, he may feel that by doing so he would act contrary to his self-interest (by damaging his health, his financial interests, or his social reputation), or would endanger his friendship with a dear friend, an so on.

To put it differently, instead of simply following their strongest desire, people sometimes prefer to *overrule* their *strongest* desire for the sake of a *weaker*, yet in their own view *more important*, desire – and may have very good reasons for doing so.

Of course, to act contrary to our strongest desire will always require some *willpower* and indeed may sometimes require a lot of it. If we can muster enough willpower, then we *can* defy our strongest desire and can pursue our preferred personal and moral goals. Otherwise, we shall end up yielding to our strongest desire, even if we would very much prefer *not* to do so.

For example, nowadays many smokers would clearly prefer to break their smoking habit because they worry about its long-run effects on their health. Yet, many of them are quite unable to do so, or can do so only with great effort.

IV. THE GLOBAL/LOCAL DISTINCTION FOR OUR DESIRES AND OUR PREFERENCES

We can approach this problem also in terms of Griffin's (1986, p.13) distinction between our global and our local desires and in terms of a similar distinction between our global and our local preferences. He defines our *global* desires as "desires to live a certain sort of life" whereas our *local* desires are desires for more specific things.

By the same token, our *global* preferences are preferences between significantly different kinds of life. We may also say that they are preferences between a life consistent with our basic personal and moral objectives and a life inconsistent with them. In contrast, our local preferences are preferences between more specific things.

Our global desires and preferences express our basic attitudes toward our life as a whole. On the other hand, our local desires and preferences are often based on our genuine needs but can also be based on quite transitory impulses, appetites, or emotional arousals of questionable quality.

As I have already argued, we sometimes have good reasons to *resist* some of our local desires and preferences because they may be incompatible with our important personal and moral objectives. To put it differently, the mere fact that some of our local desires or preferences are *incompatible* with our personal and moral objectives indicates that they are *mistaken* desires and preferences, ones contrary to our *informed* desires and preferences.

On the other hand, we should not *impoverish* our own life by adopting unduly puritanical personal and moral objectives that would require us to *repress* some of our very worthwhile local desires and preferences for no good reason.

V. THE THINGS WE DESIRE: SUBJECTIVE PREFERENCES AND OBJECTIVE OUTCOMES

Hedonists assume that all we want are *subjective experiences* (i.e. psychological states) of some kinds, such as happiness, pleasure, and absence of pain. Yet, in actual fact, we do want many other things as well. Many of these are *not* subjective experiences at all but rather are *objective outcomes* in the outside world, such as having money, social status, influence, success, personal accomplishments, and so on.

Some of the objective outcomes we want amount to *other people's* having certain psychological states. For instance, we may want them to be happy. We may want them also to feel love and respect for us.

Moreover, what we want is the *real thing*. We want to have money and success in *actual fact* rather than merely the *subjective belief* that we have these things when this is not really the case. We do not want to live in a fool's paradise based on false illusions – even if these illusions would be subjectively indistinguishable from the real thing.

It is of course easy to understand why natural selection has endowed us with strong reluctance to base our lives on false illusions. Any animal mixing up imaginary food with real food or mixing up imaginary safety from predators with real safety from them would not last very long.

VI. OUR DESIRES AND PREFERENCES – A SECOND LOOK

Parfit (1984, p.499) has suggested that there are only a limited number of basic personal benefits (he calls them "good things") that can improve our lives from our own personal point of view. It seems to me that this is true. But I shall now propose a somewhat expanded list of such personal benefits. Other people might want to add some further items to this list.

My list is like this: Desire for material comfort and for physical and economic security; for freedom to control our own lives; for having good health; for jobs suitable to our personal abilities and personal interests; for further developing our abilities; for deep personal relations in mutual love, in marriage and in true friendship; for having children and for being a good parent; for knowledge and for understanding of the world and of our own place in the world; for enjoyment of beauty in nature and in art; for having access to the ordinary pleasures of human life; for worthwhile accomplishments of some kind; and for making our own behavior consistent with our basic moral values.

Yet, Parfit never asks why *these* particular things are basic personal benefits for us and why *much the same* things are basic personal benefits for *all* of us. The actual reason is, it seems to me, that all human beings have much the same biological and psychological needs and, therefore, have *much the same basic desires* – except for those people who are prevented from benefiting from some of these personal benefits by some physical or psychological handicap. If I am right in suggesting such an extensive uniformity in our basic desires, as I think I am, then this is an interesting *empirical fact* about human nature, which seems to be of some importance for ethics.

On the other hand, even though people's *basic desires* seem to be much the same, their personal *preferences* – their actual preferences but even their informed preferences – are often quite different, presumably due to differences in their life experiences, native temperaments, special abilities, the social influences affecting them, and so on. As a result, people differ in the relative importance they assign to their various basic desires. For instance, some people may give priority to success in their jobs over avoidance of excessive stress for the sake of their health while others may have a contrary priority. People also differ in their preferences between alternative ways of satisfying the same basic desire. For example, they may have opposite preferences between alternative jobs.

A RULE-UTILITARIAN THEORY OF MORALITY

VII. MORALITY AND SOCIAL UTILITY

It is commonly agreed that any society will greatly benefit by having sensible moral norms and moral values.

Yet, utilitarians go one step further than that. They take the view that compliance with our moral values and with our moral norms has a *rational basis* only to the extent to which this really benefits *human beings* - ourselves, other people, and society as a whole.[4] This view is based on the assumption that in ultimate analysis people have only *two* fundamental objectives, ones that all of us will immediately understand as *rational* objectives. One is *their own* well-being. The other is the well-being of *other people.*

Thus, by utilitarian standards, our moral and political values such as individual freedom, equality, justice, fairness, democracy, law and order, and so on have rational justification only in terms of the benefits that we and other people will obtain if these values are widely respected.

Accordingly, the best moral values and moral norms[5] are those likely to produce the *greatest benefits* for society as a whole as judged from an unbiased and impartial point of view. As I tried to show in earlier publications (see, e.g., Harsanyi, 1977, Chap. 4), the impartially considered welfare of society as a whole at any given time can be measured by its *social utility function*, defined as the arithmetic mean of all individuals' utility functions in this society. Thus, the basic principle of utilitarian theory can be expressed also by saying that the *basic rational criterion* for evaluating our moral norms, our moral values, and morality as a whole is their *social utility.*

Under this definition, our social-utility function would be defined in terms of all individuals' utility functions. On the other hand, each individual's utility function itself is defined in terms of his own (informed) preferences. Thus, *all* individual preferences would enter into our social-utility function. Yet, I now propose to argue that in actual fact some individual preferences should be *excluded* from this social-utility function.

First of all, we should exclude people's *malevolent* preferences, i.e. those based on sheer hostility, envy, sadism or malice. For by including these preferences we would imply that it was desirable from a social point of view to *help* malevolent people in satisfying their preferences to *hurt* some innocent people for no good reason - which is obviously not the case.

Indeed, I want to suggest that a much *broader* class of preferences should be excluded. Dworkin (1977, p.234) made a helpful distinction between *personal preferences* and *external preferences*. The former he defined as "preferences for enjoyment of goods and opportunities." The latter he defined as "preferences for assignment of goods and opportunities to others."

My suggestion is to exclude all *external* preferences, even *benevolent* ones, from our social-utility function. First of all, in the absence of special reasons to the contrary, in dealing with other people we should try to treat them in accordance with *their own* (personal) preferences, i.e. in the way *they* want to be treated, rather than in accordance with *other people's* (external) preferences, i.e. in the way *other people* might want them to be treated. In other words, our social policies should be based on people's *personal* preferences and should pay no attention to their *external* preferences.[6]

Secondly, we must exclude not only malevolent, but also *benevolent* external preferences from our social-utility function. This is so because otherwise we would violate the important utilitarian requirement that our social-utility function should assign the *same* positive weight to each individual's interests. For if we did include these preferences then our social-utility function would assign *much greater* weight to those individuals' interests who have many well-wishers and friends than to those individuals' interests who lack such support.

VIII. ACT UTILITARIANISM AND RULE UTILITARIANISM

As is well known, there are two different versions of utilitarianism, called act utilitarianism (AU) and rule utilitarianism (RU).

AU is the view that a *morally right action* is one yielding the highest amount of social utility under the existing conditions.[7]

In contrast, RU first defines the *optimal moral code* of a society as the moral code that would yield the highest amount of social utility – where a moral code refers to the *set* of *all* moral rules accepted[8] by a given society. Then, it defines a morally right action simply as one in compliance with this optimal moral code.

How can a rational utilitarian *choose* between these two versions of utilitarian theory? I submit that he must make his choice in terms of the basic utilitarian criterion of *social utility*.

Choice between AU and RU will be really a choice between two alternative *moral codes*. One is the *act-utilitarian* moral code, requiring everybody on each occasion to choose the action yielding highest amount of social utility. The other is the *rule-utilitarian* moral code. As we have seen, the latter will be defined simply as the *optimal* moral code, yielding the highest amount of social utility among *all* possible moral codes.

Thus, by its very definition, the rule-utilitarian moral code would provide *at least as much* social utility as any other moral code, including the *act-utilitarian* one, would provide. But I shall try to show that, in actual fact, the rule-utilitarian moral code would provide a *much higher* level of social utility than its act-utilitarian counterpart would do.

IX. THE SOCIAL EFFECTS OF AN ACCEPTED MORAL CODE

The most obvious social effects of a moral code will be the *direct benefits* people will enjoy, and the *direct costs* they will have to bear, as a result of their own and other people's compliance with this moral code. I shall call these – the *positive* and the *negative* – compliance effects of this moral code.

Any sensible moral code will have many *positive* compliance effects because it will try to require only actions benefiting the agent or at least benefiting some other people.

Yet, any moral code will also have some *negative* compliance effects. These will include the effort needed to comply with it as well as the guilt feelings and the social penalties induced by possible noncompliance. These negative effects will be particularly important in the case of *very demanding* moral codes. This may make adoption of such moral codes unattractive even if their *positive* compliance effects would be quite appealing. (For an excellent discussion of what I call positive and negative compliance effects, see Brandt, 1979, pp.287–289.)

Apart from the positive and negative compliance effects, another important class of social effects – so far largely neglected in the literature on ethics – are the *expectation effects* of a moral code. People will not only themselves comply with the accepted moral code to some extent but will expect also *other people* to do so. This expectation will give them some *assurance* that their legitimate interests will be respected by most other people. It will also give them *incentives* to engage in some socially useful activities they would not otherwise undertake. Accordingly, I shall divide the expectation effects of a moral code into *assurance effects* and *incentive effects*.

The compliance effects and the expectation effects of individual *moral rules* can be defined in much the same way.

X. EXPECTATION EFFECTS: THE CASE OF PROMISE KEEPING

Suppose that A made an important promise to B, but now finds it very inconvenient to keep it. Under what conditions will he be morally free to break his promise? According to conventional morality, he will be permitted to do this only in rather exceptional cases, such as situations where keeping his promise would cause real hardships to him or to third parties. *Rule-utilitarian* theory will reach much the same conclusion by analysis of the relevant expectation effects. In contrast, *act utilitarianism* will have to take a much more permissive position.

If A were permitted to break his promise, he would obtain some utility *gain* by being released from a burdensome obligation; whereas B would suffer some utility *loss* by losing A's promised service and having his expectations disappointed. The mathematical difference between this gain and this loss I shall call the *balance of local utilities*. But besides A's and B's *personal* gain and loss, we must also consider the general *social* interest in maintaining the *expectation* that promises will be kept. If this expectation were seriously undermined, many people would face the future with *less assurance* and with *weaker incentives* to engage in various forms of socially useful behavior, such as performing services for other people in exchange for promised future rewards or agreeing on future joint activities with others, and so on.

The question, of course, is whether people's belief in promises would be seriously affected if A were permitted to break his promise in this particular case. Under the rule-utilitarian assumptions, the answer would have to be in the affirmative. For a rule utilitarian could give his moral approval of A's breaking his promise on this *one* occasion only if he were willing to propose a moral code *generally* permitting people to break their promises on *all* similar occasions. Yet, such a moral code might easily have significant effects on people's expectations because it would tell them that even the most conscientious members of society, those fully guided by its moral code, would feel morally free to engage in the *general practice* of breaking their promises in the relevant class of situations.

Unlike rule utilitarianism, act utilitarianism is logically precluded from considering the expectation effects (or any other social effects) of alternative *moral codes*. All it can consider are the expectation effects of *individual actions*. Thus, in our example, an act utilitarian could consider only the effects that this *one* act of promise breaking by this *one* person A is likely to have on the credibility of promises. Yet, this one breach of promise is likely to have *negligible* effects.

Thus, an act utilitarian would have to conclude that, by and large, A will be morally free to break his promise if the resulting *balance of local utilities* itself is nonnegative, i.e., if his interest in breaking his promise is at least as strong as B's interest in having his promise kept.

This shows how little a promise would mean in an act-utilitarian society. For in spite of A's promise to B, his moral obligations to B would be determined solely by the balance of local utilities - much in the same way as if he had made no promise to B at all.

To be sure, like other moral rules, the rule that promises must be kept is subject to some *exceptions*. According to rule-utilitarian theory, the best set of permissible exceptions will be one having the *highest social utility* - by being the best compromise among the promisees' interests, the promisors' interests, and society's general interest in maintaining the credibility of promises.

One can show by similar reasoning that, besides promise keeping, some other rules of conventional morality, such as abiding by our contractual obligations, telling the truth in important matters, being grateful to our benefactors, and so on, likewise owe most of their social utility to their socially favorable *expectation effects* - which rule-utilitarian theory is able, whereas act-utilitarian theory is unable, to take fully into account.

XI. INDIVIDUAL RIGHTS AND SPECIAL OBLIGATIONS

Another important advantage of rule utilitarianism is its ability to recognize morally protected individual rights and special obligations[9] that normally take precedence over direct social-utility maximization. By its own inner logic, act utilitarianism cannot do this, because it has to take the position that we must *always* choose the action maximizing social utility – even if this action violates other people's rights or our own special obligations to some other people.

The social utility of morally protected *individual rights* lies in the fact that they greatly increase the personal freedom, independence, and security of the persons holding these rights. Some individual rights also provide important incentives to socially desirable behavior. These beneficial effects of individual rights must be weighed against the *social costs* they entail by restricting the freedom of action of other members of society, required to respect these rights.

As an example of individual *rights*, consider a person's private *property rights* over his car. By the very concept of private property, the owner of the car is morally free to use it whenever he so desires, whereas other people must not use it without the owner's consent (except perhaps in some emergency situations). According to rule-utilitarian theory, such private property rights tend to have considerable social utility because they make economic relationships more secure and more predictable (assurance effects) and because they are an important incentive to hard work, saving, investment, and entrepreneurship (incentive effects).

But act-utilitarian theory cannot recognize such property rights because its inner logic forces it to hold that everybody is morally free to use another person's property, even without the owner's permission, as long as he thinks that he will produce *more* social utility by using this car than the owner would produce by using it.

On the other hand, our *special obligations* arise from the various social roles we play as parents, spouses, neighbors, members of an occupation, etc. They are important because they establish many forms of socially very beneficial *division of labor* in the society. For example, it is desirable to have a division of labor among adults in caring for the children of the community, with each adult being in charge of no more than a rather small group of children so that he or she can easily find out each child's special needs and can develop close emotional ties with each of them.

In our society, this division of labor mainly takes the form of the parents' looking after their own children. In order to protect this special relationship between the parents and their children, both conventional morality and the rule-utilitarian moral code will impose *special obligations* on the parents toward their own children. In particular, it will require the parents to give *special priority* to their own children's needs over those of other people's, whether children or adults.

According to rule-utilitarian theory, this arrangement has considerable social utility because it gives both parents and children a feeling of security (assurance effects) and gives both of them incentives to many forms of socially desirable behavior (incentive effects). But act-utilitarian theory would have to *disapprove* of this arrangement because it must hold that the parents cannot give any special priority to their own children's needs when they might be able to produce a *greater* amount of utility by satisfying the possibly *more urgent* needs of some other people - whether some other children or some adults.

XII. THE UNACCEPTABLY HIGH COMPLIANCE COSTS OF THE ACT-UTILITARIAN MORAL CODE AND THE VALUE OF INDIVIDUAL FREEDOM

All moral codes have to restrict people's individual freedom by requiring them in many situations to respect other people's fundamental interests and society's common interests, even when this is very much against their own personal preferences and interests. Yet, subject to this requirement, most of the conventional moral codes give people *considerable freedom* to pursue their own preferences and interests.

Given the high value most of us assign to individual freedom and, in particular, to freedom from very oppressive moral obligations, the *optimal* moral code of rule-utilitarian theory would have to be very much alike these conventional moral codes in this respect. Indeed, it seems to me, it would have to be simply a more humane, more enlightened, and more consistent - and therefore a *more rational* - version of these conventional moral codes.

In contrast, the act-utilitarian moral code would be *radically different* both from these conventional moral codes and from that of rule-utilitarian theory. It would be a *highly restrictive* moral code, leaving little room for individual choice. For it would require us *all the time* to act in such a way as to maximize social utility, i.e., to give *absolute priority* to society's interests over our personal preferences and interests, and even over our commitment to our family and friends. It would hardly ever permit us a free choice between alternative courses of action. As far as we could comply with such a highly restrictive moral code at all, we could do so only by *extreme* effort and at *unacceptably high* psychological costs.

XIII. RULE UTILITARIANISM AS THE PREFERABLE INTERPRETATION OF UTILITARIAN THEORY

To sum up, we have considered various advantages of the rule-utilitarian over the act-utilitarian approach. These advantages clearly suggest that, other things being equal, a rule-utilitarian society would enjoy a *much higher* level of social utility than an act-utilitarian society would.

For it is better to live in a society whose moral code – unlike the act-utilitarian moral code – respects our individual rights and our special obligations, and recognizes the social value of promise keeping, abiding by our contractual obligations, truth telling, being grateful to our benefactors, and of some other moral rules of conventional morality because of their desirable expectation effects.

Likewise, it is better to live in a society whose moral code allows people a good deal of free choice in their personal lives and does not impose unacceptably burdensome restrictions on people's personal behavior.

XIV. MORAL INTUITIONS

Whereas utilitarian authors base their ethical theories on one rational principle, that of social utility, most nonutilitarian authors base theirs on their *moral intuitions*. Yet, it is easy to see that our moral intuitions are very *unreliable* guides in ethics.

First of all, the very term "moral intuitions" is a misnomer. There is no reason whatever to assume that we have direct intuitive access to moral truths, in the way we have intuitive access to some basic mathematical truths.

Moreover, what people describe as their moral intuitions differ far more from person to person, and are far too dependent on the society and the particular social group they have been born into, for being based on direct intuitive access to some moral truths of objective validity.

Indeed, it is well known that most people born into slave-holding societies, caste societies, or societies badly oppressing women or their racial, ethnic, or religious minorities, almost always insisted that their moral intuitions *fully supported* the social practices of their societies, even the most objectionable ones by our own moral standards.

These facts clearly refute the claim that people's moral intuitions are based on direct access to objective moral truths. What these facts do suggest is rather that people's so-called moral intuitions simply represent moral beliefs and political attitudes, both reasonable ones and unreasonable ones, these people absorbed from their social environment.

Thus, the mere fact that an ethical theory is based on its *author's* moral intuitions - or even that it happens to agree with *our own* moral intuition - tells us very little on its real merits as an ethical theory.

No doubt, our moral intuitions can have a useful *heuristic role* in ethics, even in utilitarian ethics, because they may call our attention to some moral problems we might otherwise have overlooked. But they cannot replace our *rational judgment* on how to resolve these moral problems in the best interest of the people affected. We certainly cannot make our moral intuitions the *final arbiters* of morality, as many nonutilitarian authors have tried to do.

The problem is well illustrated by the sharp differences between the ethical theories of the two leading nonutilitarian philosophers of our age, those of Rawls (1971) and of Nozick (1974). Following his moral intuitions, Rawls arrived at a radically *left-liberal* and strictly *egalitarian* theory; whereas following his own intuitions, Nozick arrived at a radically *right-liberal* and strictly *libertarian* theory.

It has not occurred to either of them to subject his theory and the social policies suggested by his theory to a *rational test* independent of his personal moral intuitions - for instance, by asking whether his theory and the social policies he is advocating would not be *highly detrimental* to our society if adopted by the latter. Had either of them subjected his theory and his social policies to this rational test then he could have hardly avoided the obvious conclusion that his theory and his policies would in fact be highly detrimental to our society.[10]

It seems to me that utilitarians have very good reasons to make *social utility*, i.e. the real interests of human beings, rather than their own moral intuitions, the basic criterion of morality.

XV. RULE-UTILITARIAN THEORY AND CHANGES IN OUR SOCIETY'S MORAL CODE

In Sections IX and X, I discussed how rule-utilitarian theory can help us *theoretically* to understand the social utility of some of our accepted moral rules. Now I propose to show how it can help us *practically* as a rational moral standard in assessing the social utility of proposed *changes* in our society's moral code.

We live in a period of fast social and technological changes, which give rise to proposals to change some of our accepted moral rules from time to time. At least initially, these proposed changes tend to be very controversial. But rule-utilitarian theory gives us a *rational standard* for assessing the social utility or disutility of the proposed changes in terms of the human interests affected.

For instance, when we significantly liberalized our moral and legal rules concerning *divorce* about twenty-five years ago, there was fairly general agreement that this was a desirable reform. But our subsequent experience seems to suggest that this liberalization may have gone *a little too far*, and that our present social attitudes put insufficient pressure on the spouse seeking divorce to give adequate consideration to the other spouse's interests and to their children's interests.

Only time can tell whether this situation will eventually produce another agreed change in our moral and legal rules concerning divorce, this time perhaps taking the form of somewhat better protection for the other parties' interests. In any case, this tendency to replace some of our moral rules from time to time by alternative rules that our society considers to have *greater social utility* is likely to result gradually in significant improvements in our society's moral code.

Of course, opponents of utilitarianism will *reject* social utility as a valid criterion. But they have never proposed any reasonably clear alternative criterion.

XVI. MORALITY AND OTHER VALUES OF HUMAN LIFE

Before summing up my conclusions, let me briefly discuss the role of moral values in human life from a utilitarian point of view.

The basic task of morality is to induce people to help other people in achieving *their own* objectives. Yet, these are in most cases *nonmoral* objectives, such as economic prosperity, a good social position, good health, friendship, love, knowledge, enjoyable aesthetic experiences, and so on.

Thus, in an important sense, morality is primarily a *servant* of many other human values, rather than itself the highest value. Moreover, even though nobody can have a rich and well-balanced life without strong moral commitments, for most of us these moral commitments are not the central focus of our life, but rather it is our work, our family and friends, and our various intellectual, cultural, social, and political interests.

Indeed, from a utilitarian point of view, strong devotion to one's idealistic moral and political objectives is not always an unmixed blessing because it can easily change into socially disastrous moral and political fanaticism. People willing to make great sacrifices for the sake of their moral and political objectives are much too often quite willing to sacrifice also *other people's* interests, and even other people's lives, for the sake of these objectives. After all, by all accounts, Robespierre was a man devoted to the highest moral principles. Yet, perhaps society would have been better off had he been somewhat *less* devoted to them.

XVII. CONCLUSION

To sum up, I proposed a nonhedonistic theory of personal benefits, arguing that the objects of human desires are not only *subjective experiences* but also *objective outcomes*. I suggested a new interpretation of human preferences. I argued that human beings *have much the same* basic desires for personal benefits whereas their personal preferences are often very *different*.

Then, I outlined a rule-utilitarian theory of morality, suggesting that an important factor determining the social utility of our moral rules are their *expectation effects*. I argued that the act-utilitarian moral code would be unable to give proper weight to the socially very important *expectation effects* of various moral rules, and would unacceptably restrict our *individual freedom*.

Finally, I considered the roles of morality and of other values in human life.

NOTES

1. In what follows, for stylistic reasons in such contexts I shall often omit the female pronoun.
2. The term "informed preferences" has been suggested to me by Griffin's (1986, p.14) term "informed desires".
3. The concept of mistaken preferences raises the question of how far society should go in discouraging people from following their mistaken preferences. Should society pursue a *liberal* policy, permitting people "to make their own mistakes", or should it pursue a *paternalistic* policy, preventing any behavior likely to damage the agent's self-interest?

   In my opinion, *active* paternalism, which would try to prevent any behavior contrary to the agent's self-interest, can be justified only if it would cause utterly unacceptable damage to the agent, or would seriously damage innocent bystanders. But in general we are morally free to engage in *passive* paternalism by withholding our help, including our financial help, from what we consider to be self-destructive behavior. We have no right to force an alcoholic not to drink, but we have every right not to buy him another glass of liquor.
4. No doubt, morality must also protect the welfare of *nonhuman* sentient creatures. Yet, in my view, human interests must have some *priority* over nonhuman interests because they are qualitatively different interests. For only human beings have the ability (at least if they have normal use of their mental faculties) to make a *rational assessment* of their own interests. Only human beings seem to have a clear idea of the *future*, and seem to be able to consider their own and other people's *future interests*.
5. Our *moral norms* tell us what we are morally *required* to do. In contrast, our *moral values* tell us what would be morally *desirable* for us to do even if, strictly speaking, we may not be morally required to do it. Both are important elements of our moral life. Our moral norms set the *limits* of morally acceptable behavior, whereas our moral values suggest desirable options available to us *within* these limits.
6. On the question of how this exclusion of external preferences can be implemented mathematically, see Harsanyi, 1986, pp.9-10.
7. Because of the *uncertainty* associated with the future effects of our actions, the term "social utility" actually should be replaced by "expected social utility", referring to the mathematical expectation of social utility. But for stylistic convenience, I shall throughout omit the adjective "expected".
8. Social acceptance of a moral code or moral rule primarily means its wide acceptance as *morally binding*. It may not necessarily mean a high degree of *actual compliance* with it by the society in question.

9. By special obligations I mean obligations based on one's social position (such as being a father, a neighbor, an employee, etc.).
10. Nozick suggested the *inhumane* policy of abolishing all social programs trying to alleviate poverty if financed out of funds collected by general taxation.

On the other hand, Rawls suggested that people making important contributions to the well-being of our society *do not deserve* any moral credit for doing so if their contributions are based on their *special talents* or on their *special efforts* as a result of their *good character*. His argument for this strange doctrine is that people owe their *special talents* to good luck with their genetic endowment, and owe their *good character* to good luck in being born into a favorable social environment. He seems to forget that, even though having special talents is *not* our personal merit, our *use* of these talents for socially desirable purposes and our efforts to *develop* our talents, often by very hard work, *are* to our moral credit. By the same token, to be born in a favorable social environment is *not* our personal merit, yet if we do acquire a good character in *any kind of environment*, this *will always be* to our moral credit.

Rawls's strange doctrine to the contrary is not only *utterly false* but would also do *great harm* to our society by strongly discouraging people with special talents and/or with good character from full use of them for our common benefit, and by giving these people the impression that our society *fails to appreciate* any form of *superior* performance and of *personal excellence*.

REFERENCES

R.B. Brandt, *A Theory of the Good and the Right*. Oxford: Clarendon Press 1979.

R.M. Dworkin, *Taking Rights Seriously*. Cambridge, MA: Harvard University Press 1977.

J. Griffin, *Well-Being*. Oxford: Clarendon Press 1986.

J.C. Harsanyi, *Rational Behavior and Bargaining Equilibrium*. Cambridge: Cambridge University Press 1977.

J.C. Harsanyi, "Individual Utilities and Utilitarian Ethics", in: A. Diekmann and P. Mitter (eds.), *Paradoxical Effects of Social Behavior*. Heidelberg: Physica-Verlag 1986, pp.1-12.

J.R. Hicks, *Value and Capital*. Oxford: Clarendon Press 1939.

R. Nozick, *Anarchy, State, and Utopia*. Oxford: Blackwell 1974.

D. Parfit, *Reasons and Persons*. Oxford: Clarendon Press 1984.

J. Rawls, *A Theory of Justice*. Oxford: Oxford University Press 1971.

Antonio Camacho

Variability, Uncertainty, and the Emergence of Firms and Markets

A new theory to explain the emergence of firms and markets to carry out economic activity was presented in this author's recent book (Camacho, 1996). In this theory, called the *Variability Approach*, the multi-person firm emerges to carry out a production program, agreed upon during the preceding market period. This occurs when the participants in the program need to exchange among themselves additional information, additional to the information already exchanged during the preceding market period. The organization that emerges to facilitate the required additional communication and to carry out the production program is called a *multi-person firm*.

In this paper I shall argue, after elucidating the distinction between variability and uncertainty, that it is variability rather than uncertainty which together with other factors creates a situation requiring a multi-person firm to carry out efficiently a production program.

To make the paper self-contained I shall start in Section I with a brief overview of the variability approach. Then in Section II, I will discuss the distinction between a *variability problem* and an *uncertainty problem*. That it is variability, not uncertainty (although variability may refer to uncertainty), that determines if a multi-person firm is needed to carry out efficiently a production program, is argued in Section III. Among all the theories known to this writer, related to the existence of firms and markets, F. H. Knight's (Knight, 1921) is certainly the one where the role of uncertainty is more important. For this reason, in Section IV, I shall compare the variability approach with Knight's approach. To complete a general view of the field I shall carry out in Section V a similar discussion regarding transaction cost [(Coase, 1937), (Williamsom, 1975,1985)], property rights [(Grossman and Hart, 1986), (Holmstrom and Tirole, 1989), (Hart and Moore, 1990), (Moore, 1992)], and (Alchian and Demsetz, 1972) approaches. The paper is concluded with some final remarks in Section VI.

I. The Variability Approach: Brief Overview

As the name of the approach suggests, the notion of variability is basic in this theory. It is used here to mean that things may and do vary or change. And in the spirit of Hayek (Hayek, 1945, p. 524), who states "The economic problem of society is mainly one of rapid adaptation to the changes of particular circumstances of time and place", it means also that the variations to which it refers call for timely responses that must follow the observation of the variation.

W. Leinfellner and E. Köhler (eds.), Game Theory, Experience, Rationality, 301–314.

According to the variability approach the two fundamental types of organization of a free enterprise economy, *the market* and *the firm*, can be explained as emerging in response to variability. Two types of variability are considered. One type requires, to respond to it, the emergence of markets; the other, the emergence of firms.

To give the reader a feeling for the type of variability that, to respond to it, calls for the emergence of markets, and for the type that, to respond to it, calls for the emergence of firms, I shall discuss bellow, in Subsection I.1, the notions of market activity and market; and in Subsection I.2 those of firm activity and firm.

In real world economies what we will refer to as market activity and firm activity may (and does most of the time) take place simultaneously. In the present model, however, we will make the simplifying assumption that time is partitioned into a sequence of consecutive periods, starting with a market period followed by a consumption-production (from now on production, for short) period, followed by a market period, and so on. During the market periods, only market activity will take place. During the production periods, consumption and firm activity will also occur.

I.1 The Need for Market Activity and the Emergence of Markets

Skills and resources are dispersed throughout society, and the preferences of its members vary, in general, from one to another. To make an efficient use of these dispersed skills and resources in satisfying their preferences, members of society must learn more about those skills and resources and about their preferences. To this end they must engage themselves in exchanges of information and in calculations leading to the determination of who will use what resources, produce what amounts of what commodities, and consume what amounts of what goods. In other words, members of society determine during the market period the production programs to be carried out during the following production period as well as the amounts of each good to be consumed by each of its participants.

We refer to an activity as the one described in the previous paragraph as *market activity*. As an illustration of market activity consider the simple example from international trade of the wine producer from Porto (Portugal), skillful in producing Port wine and not skillful in producing cloth, and the cloth producer from Manchester (England), not skillful in producing wine and skillful in producing cloth. If there is no market activity, each must be self-sufficient. But if market activity is introduced, the wine producer and the cloth producer, after meeting and exchanging information during the market period, realize that each would be better off if the wine producer concentrates on the production of wine, the cloth producer on the production of cloth, and if they exchange cloth for wine. This much is stated in the text books and shows clearly the possible gains that can be derived from engaging in market activity.

What is not stressed in the textbooks or in the economic literature, perhaps because it is so obvious, is that unless variations occur in the tastes, production capabilities or in anything else that may alter the agreement of exchange that our wine producer and our cloth producer reached during their meeting in the first market period, no further market activity is necessary.

More generally we can state that, unless variations occur that alter the appropriateness of the constellation of agreements reached during the first market period, no further market activity is necessary. This observation provides us with the insight that market activity is necessary, and therefore it is designed, to deal with those variations that alter the constellation of agreements that are reached during a market period, that is, the general equilibrium of the market. The set of all such variations constitutes the type of variability that requires market activity.

We will refer to the organizations created to facilitate the carrying out of the market activity as *markets*. These markets are similar to, but more general than both, the Walras auctioneer market, and Hurwicz mechanisms (Hurwicz, 1959). For while the language used in the Walrasian auctioneer model to carry out the exchange of information consists of prices and quantities of the different commodities, and that used in Hurwicz mechanisms consists of sets of input output vectors, in our markets there is no restriction to the language to be used. Also worth noticing is that while "firms", whatever this term means in the Walrasian or Hurwicz models, are given in these models, they are determined in the variability approach in each market period.

I.2 The Need for Firm Activity and the Emergence of Multi-person Firms

As stated at the discussion of Subsection I.1, part of the outcome of the market activity during the period preceding a production period is the determination of the production programs that must be carried out during the given production period.

Some or all of these production programs may involve several individuals that need to communicate among themselves to carry out efficiently the corresponding production programs. The needed communication is in addition to the communication that already took place during the preceding market period. It did not take place during that period because the corresponding information was not available then. The programs that do involve several individuals that need additional communication as just described above in this paragraph, are called *multi-person programs*. The rest of the programs are either *single-person* programs consisting of one person, or programs consisting of several persons who don't need to communicate with each other during the production period to carry out efficiently their corresponding production programs. The latter programs can obviously be decomposed into (independent) one person programs and hence be considered as single-person programs. Thus the production programs to be carried out during a production period are either multi-person programs or single-person programs as defined above.

The activity to carry out a multi-person program consists roughly of the following: make observations by different participants, communicate the results of the observations to the appropriate fellow participants, determine the corresponding coordination adjustments, communicate these adjustments to the pertinent participants so that they introduce them, and carry out the physical production part of the program. We refer to the activity just described as *firm activity*. The organizations which carry out this type of activities are called *multi-person firms*.[1] Thus we can say that, in the *variability approach*, the multi-person firm emerges when there are multi-person programs, determined during the preceding market period, that must be carried out during the present production period.

I have discussed elsewhere, (Camacho, 1996, Chapters 2 and 7) three types of situations where if the proper conditions, regarding variability, coordination requirements, and structure of information are met, a multi-person firm must emerge to carry out efficiently the production activity. One type of situation refers to the production of units of a product consisting of several parts in the production of which advantages could be derived if (1) each part is produced by a different specialist, and (2) these specialists are able to communicate among themselves to respond to variations brought about by the variability conditions. The second type refers to an *indivisible* input whose services are more efficiently used if (1) they are shared in the production of two or more products, and (2) the different groups participating in the production of the different products can (easily) communicate among themselves to coordinate the sharing of the services. The third type involves two activities, one of which is subject to external diseconomies from the other. These external diseconomies can be more efficiently managed if there is appropriate communication between the participants of the two activities.

II. VARIABILITY AND UNCERTAINTY

I shall try to make clear now the distinction between my notions of *variability* and of *uncertainty*. I will proceed by describing first two situations where, in my sense, variability, not uncertainty, exists.

Then I will change appropriately the conditions of the two variability situations discussed and transform them into uncertainty situations. Finally, I will define the distinction in more abstract terms.

Situation 1

Consider an orchard and the orchard-keeper. The orchard-keeper knows that tomorrow it is either going to rain; or to be hot, dry, and sunny; or to be dry and windy.

The best actions, that is the actions that produce the best net results (after discounting for the costs of the actions) when each of the given events obtains, are:

(i) When it is going to rain put fertilizer on the orchard. With rain, the fertilizer penetrates the soil in an homogenous way and produces the best result.

(ii) When it is going to be hot, dry and sunny, irrigate the orchard.

(iii) When it is going to be dry and windy, pick up the fruit of the orchard so that it doesn't get spoiled.

The conditions regarding the orchard-keeper are:

(a) The orchard-keeper doesn't know today what is going to happen tomorrow. But he/she will know early tomorrow what will happen with enough lead time to take the best corresponding action.

(b) Any action that the orchard-keeper can take today he/she can also take early tomorrow, and for any given event, both the outcome and the cost of the action remain the same.

In this situation I would assert that the orchard-keeper doesn't face an uncertainty situation, in the sense of decision making theory, despite the fact that today he/she doesn't know which of the three possible events is going to obtain tomorrow. He/she faces what I call a variability situation and should act as follows. Wait until early tomorrow, find out the event that is going to obtain, and then select and apply the best action given the event that will obtain.

Situation 2

Consider a barber who operates his/her own barber shop alone. Every working day he receives 20 costumers. But he doesn't know in advance who the costumers are going to be any specific day. Assume also that our barber is indifferent regarding whose hair he cuts, and that every potential costumer pays, including tip, the same amount for a hair cut. Assume finally that our barber possesses all the tools necessary to best cut the hair of any possible costumer.

Under these conditions, our barber opens his/her barber shop every working day, waits for the costumers to come, observes the type of head and hair that he/she has, makes a judgement regarding what is the best hair cut for the given costumer, selects the appropriate tools and does the best haircut, given the type of head and hair of the costumer.

Also in this situation we should assert that our barber does not face an uncertainty situation in the sense of decision making, despite the fact that he is uncertain every working day regarding the type of the heads and the corresponding hair of the costumers that he will have to serve that day. The barber, we say, faces a variability situation.

Change now Situation 1 to Situation 1* by assuming that the decision regarding any action, to be feasible, must be made before the event that will obtain or has obtained is known. Such would be the case if there is no lead time to implement an action if the decision to take it is made early the day the event obtains instead of the previous day. Situation 1* is an uncertainty situation. Similarly, we can change variability Situation 2 to to uncertainty Situation 2* by assuming that our barber must choose every day, before the costumers start arriving, the set of tools that he must use that day to cut the hair of his/her costumers, and that how well a given set of tools works depends on the type of head and hair of the costumers coming that day.

With the help of the previous examples we should be able now to characterize more precisely what is a variability situation and what is an uncertainty situation. Consider now a situation where different events may obtain, and assume that to each event corresponds one or several best actions that change from event to event. Assume also that the decision maker knows the function $F(a, E)$ that assigns to each combination of action a and event E the net result of applying action a, given that event E has obtained.

Definition: A decision maker, under the assumptions of the previous paragraph, faces an uncertainty situation if the decisions regarding the action to be taken are decisions that must be made before the event that has obtained or will obtain is known. He/she faces a variability situation if the decision regarding the action to be taken is made knowing the event that has obtained or will obtain.

III. VARIABILITY OR UNCERTAINTY? THE EMERGENCE OF THE MULTI-PERSON FIRM

I shall argue now that it is variability, not uncertainty, which, together with other variables (structure of information, coordination requirements, ...) causes the multi-person firm to emerge.

I will carry out my argument by presenting four examples. In the first, the decision maker faces an uncertainty, not a variability, situation and will be able to respond to it by using the market without having to form a multi-person firm. Thus it is shown that uncertainty is not a sufficient condition for the multi-person firm to emerge.

The second example refers to a situation where, without uncertainty, a set of conditions regarding variability, structure of information, and coordination requirements call for the emergence of a multi-product firm. Therefore it is shown that uncertainty, as defined in the previous section, is not a necessary condition for the multi-person firm to appear.

The third example considers a situation where both uncertainty and variability, in the sense of Section II above, are present and a multi-person firm is needed to carry out efficiently the production program. This example is important because it represents the *diversification* argument, to cope with uncertainty, used in the economics literature (see Teece, 1982) to explain the emergence of the multi-product firm. But, since the second example, which, as we shall see, is the third example less the uncertainty, also calls for the emergence of a multi-product firm, we conclude that it is variability, not uncertainty, which brings about the appearance of the multi-product firm. Diversification, to cope with uncertainty, can be achieved through the market without the formation of a multi-product firm, as the first example will show.

Finally, the fourth example represents the situation where neither uncertainty nor variability are present.

First Example: Uncertainty, No Variability

Consider an individual who possesses two identical plots of land. There are two types of plants, type *A* and type *B*. Type *A* plants grow better and have a greater yield during rainy years (R-years) than during dry years (D-years). Type *B* plants grow better and yield more during D-years than during R-years. The decision to grow type *A* or type *B* plants in a given plot must be made before it is known whether it is going to be an R or D year. To make things simpler suppose that the net yield of a type *A* crop of plants on an R-year equals the net yield of a type *B* crop of plants on a D-year; and conversely that the net yield of a type *A* crop of plants during a D-year equals the net yield of a type *B* crop of plants during a R-year. The probability of any given year to be a R-year is equal to the probability of it to be a D-year that, since we are assuming that every year is either R or D, is equal to ½. The individual tending the type *A* plants doesn't need to communicate at all with the individual tending the type *B* plants. The owner of the two plots of land, we assume, is risk averse and therefore wants to reduce uncertainty. This he/she will achieve by contracting through the market, with a specialist in growing type *A* plants, the production of this type of plants in one of the plots; and with a specialist in producing type *B* plants, the production of these plants in the other plot. This way our plots' owner is able to reduce the uncertainty to zero, without forming a two-person firm, just by using the market.

Second Example: No Uncertainty, Variability

We will consider here a version, adapted to better clarify the points we want to stress, of the well-known Coasean example of "straying cattle which destroy crops growing in neighboring land" (Coase, 1960). Thus we will be considering two production activities: cattle raising and farming. We will assume that each production activity must be carried out within a single firm. Therefore, there will be a rancher raising cattle and a farmer tending to the agricultural activity.

To make things simpler, let us assume that the size of the cattle-raiser's herd is fixed, although the yield of the fixed-size herd changes with how well-fed its members are. Let us assume also, for the same reason, that the farmer's crops are single-product crops, say he/she grows only wheat on all his/her land.

There is no uncertainty in this case, in the sense that nothing that may happen requires a response before the outcome of the event is known. There are conditions regarding variability, structure of information (who knows what and when), and coordination requirements that call, I shall argue, for the emergence of a multi-person firm to carry out efficiently the production program. These conditions are stated below.

Variability

Suppose that, although wheat grows uniformly on all the farmland, it gets infected in different small areas of the farm land at different times throughout the production period. Suppose also that to avoid the spread of the infection throughout the whole field, the infected wheat plants must be eradicated as soon as they are detected, and that a way to eradicate the infected wheat is to move the herd of cattle to the infected area and let them feed on the contaminated plants.

Structure of Information

The rancher observes the variations occurring in the cattle raising activity that require appropriate action to carry out efficiently that activity. The farmer observes any variation that is significant regarding his/her production activity. In particular, I want to emphasize that the farmer is the first to observe the occurrence of an infection on any part of the farm land.

Coordination Requirements

If the procedure adopted to eradicate the infected wheat is to move the herd to the infected area and let the cattle feed on the contaminated plants, coordination is required during the production period between the cattle raising and the wheat production activities. Under this procedure, the coordination requirement consists of moving the herd to the infected area immediately following the observation of an infection.

Imagine now the calculations during the preceding market period leading to the decision that will be adopted. Assume that only the three following alternatives are feasible: (1) Building a fence, separating the cattle land from the farm land. (2) Not building a fence and letting the cattle wander unrestricted over the farm land. (3) Have the farmer inform the cattle raiser whenever an infected area is observed and have him/her move the herd to the infected area to eradicate the infected wheat and at the same time feed the herd; have the cattle raiser restrain the herd to the cattle land whenever no infected area is observed.

If alternative (3) is considered the best, its adoption, according to the variability approach, requires that the farmer communicate with the rancher during the production period and therefore that cattle raising and farming be integrated into a two-person firm.

Third Example: Uncertainty, Variability

Take the second example and introduce uncertainty by making the net yield of both the farm activity and the cattle-raising activity depend on weather conditions. Assume, as in the first example, that there are only two types of weather: R-years and D-years, each occurring with probability ½. Assume also that R-years are good for farming and bad for cattle-raising, D-years are good for cattle-raising and bad for farming; and that the net yield of farming in an R-year is equal to the net yield of cattle-raising in a D-year, and that the net yield of farming in a D-year is equal to the net yield of cattle-raising in an R-year. Assume finally that the farmer and the cattle-raiser are both risk averse.

Clearly, under these assumptions, if there wasn't the possibility of infection of portions of the wheat plants and therefore variability were suppressed, our farmer and our cattle-raiser will meet during the first market period to determine whether to build a fence separating the two lands or to let the cattle wander through the farmland is the efficient solution to the straying cattle problem; and to reach an agreement on how to share its cost.

They will also agree to divide in a given proportion the yield of the two activities combined which, given our assumptions, is constant over the years. This will be accomplished by using only the market, without integrating the two activities into a two-person firm.

But if infections in the wheat crops do occur throughout the years, and if the best way to cope with such infection is to proceed as discussed in the second example, then a two-person firm integrating the farm and cattle-raising activities needs to emerge to carry out efficiently the production activity.

Fourth Example: No Uncertainty, No Variability

In this case we don't need to present a detailed example. The result is that no multi-person firm needs to emerge. And if the no variability includes the type of variability that calls for the existence of markets, there won't be any market beyond the first period.

IV. THE VARIABILITY APPROACH AND THE KNIGHTIAN ENTREPRENEUR

I will briefly argue now that, contrary to what seems to be the case with almost all the present theories of the firm oriented towards concepts, such as transaction costs, property rights, the principal-agent relationship – where there is no place for the Knightian entrepreneur –, in the variability approach there is room for such type of input.

We have not emphasized in the variability approach the role of the entrepreneur. But this doesn't mean that there is no role for such an input, as we will see in what follows. Remember that in the variability approach we start with a market period, to be followed by a production period, to be followed by a market period, and so on. During each market period the production programs to be carried out in the following production period are conceived and formulated. And it is in this role of conceiving and formulating production programs and, when necessary, forming the appropriate multi-person firms, that I would place a Knightian type entrepreneur in the variability approach. This entrepreneur may exist without multi-person firms, operating only by using the market, when there is no need for such organizations. As we know, the multi-person firm will be created when additional communication, additional to the communication that already took place during the preceding market period, is necessary among the participants, during the production period, to carry out efficiently the production program.

This need, as we saw in Section I above, is explained in the variability theory as the result of the presence of variability, coordination requirements, and structure of information. But this need may appear also when a Knightian type entrepreneur comes with the conception, during a market period, of a production program involving several persons and the only way for the entrepreneur to explain what each of the remaining participants (which can be only one) must do is to tell each of them what to do while the execution of the program unfolds during the production period. This explanation of the firm is an example of what Barzel (1987) has referred to as Knight's *"uncertainty" explanation of the firm*, and Gunning (1993) as Knight's *entrepreneurial explanation of the firm*.

Notice that here it is not necessarily an agency problem which brings about the emergence of the multi-person firm. It is the impossibility of transmitting by the entrepreneur to the participants the precise instructions of what to do before the execution of the program starts unfolding. An agency problem can, however, bring about the emergence of a variability approach firm. Such would be the case if we have a production program involving, for instance, two persons: the principal and the agent; the agent has the tendency to shirk; and the lowest cost way of alleviating the shirking problem is by additional communication, during the production period, between the principal and the agent by bringing the agent under direct supervision of the principal (the agent may feel embarrassed if he/she shirks when his/her boss is watching and therefore the mere presence of his/her boss inhibits him/her to shirk). This case can be considered as an example of what Barzel (1987) referred to as Knight's *"moral hazard" explanation of the firm*.

The discussion above, although not exhaustive, clearly indicates that the variability approach can be adapted to easily accommodate both the *uncertainty* or *entrepreneurial*, and the *moral hazard* Knightian explanations of the firm.

V. VARIABILITY AND OTHER EXPLANATIONS OF THE FIRM: BRIEF COMMENTS

Holmstrom and Tirole in their *Theory of the Firm* (1989, pp. 65,66) state, "What is the purpose of firms and what determines their scale and scope? These are two basic questions that a theory of the firm must address. Yet satisfactory answers have proved too difficult to come by."

This comment was made in 1989. Since then, the variability approach has been developed. It answers fully the first basic question mentioned in the quotation, as can be seen from this paper, and at least partially the second (see Camacho, 1996, Chapters 2 and 7). None of the other theories that I am aware of has answered these questions yet. Below I will take two of the most influential theories, *the property rights* and *the transaction costs*, and argue that they focus on situations which, although of interest, are not germane to a theory of the firm. We will end this section indicating the complementarity between the Alchian-Demsetz (1972) and the variability approaches.

These theories, the *property rights* and the *transaction costs* (à la Williamson), focus on cases involving decisions regarding long duration investments that may lead to situations attractive for exercising opportunistic behavior. We can easily imagine examples of such situations that seem to pertain to what we think of as firms. Take, for instance, the training of a worker to operate a particular set of machines, or the locating of an electricity-generating plant adjacent to a coal mine that is going to supply it. But we can as easily imagine also the same type of situations pertaining not to what we think of as firms. Such is the case of the young married couple where one of the spouses invests in (works to help) the acquisition by the other of, say, a Ph.D. degree. It seems to this writer that it is not justifiable to claim as a theory of the firm the discussion of a class of situations, examples of which can as easily be found *within* firms as in environments totally unrelated to what we think of as firms.

Alchian and Demsetz in their 1972 theory of the firm assume the existence of team production that in certain situations increases productivity with regard to separable outputs produced by the same vector of inputs. Then, given this assumption, they develop a theory of the firm by analyzing how incentives within the team should be allocated to secure an efficient organization. They don't explain, however, why the same inputs yield a greater output when working as a team than when organized as independent production units.

The missing explanation comes easily under the variability approach. Under team production the participants on the team communicate with each other during the production period. When organized as independent production units (independent firms) members belonging to different units do not communicate with each other. The greater communication under team production may facilitate a greater division of labor and, under appropriate conditions, a greater productivity.

VI. Conclusion

Above I stated the precise distinction between *variability* and *uncertainty*, and then argued that, in the variability approach, it is variability, not uncertainty, which is the factor that, together with others (coordination requirements, structure of information, advantages from the division of labor), is relevant to explain the emergence of firms.

We compared briefly the variability approach with Knight's 1921 contribution to the theory of the firm presented in his celebrated *Risk, Uncertainty and Profit.* The result of this comparison is the suggestion that the role of the Knightian entrepreneur can easily be incorporated into the variability approach. Such an entrepreneur would operate carrying out market activity during the market periods. It suggests also that the obscure distinction between uncertainty and risk is unnecessary to develop a theory explaining the emergence of firms, or the role of the Knightian or Schumpeterian (Schumpeter, 1934) entrepreneurs.

The transaction cost (à la Williamson) and the property rights approaches were also briefly considered. The conclusion reached was that at least part of the core of problems discussed by those theories, although of importance to a theory of the firm, are not really germane to such a theory. You can carry out practically the whole discussion of these theories without having to consider the production of even a single pin. Almost the only thing you need is what I refer to (Camacho, 1996) as goods indivisible with regard to length, that is goods of long duration, and the occurrence of some appropriate changes in the preferences and the number of participants.

Another distinction worth mentioning is that, at least in the Coasean version of the transaction cost approach, the firm and the market are viewed as substitute "inputs" to produce "transactions", while in the variability approach, firm activity is considered a complement to market activity that comes first in carrying out economic activity. Remember that in the variability approach time is partitioned into a sequence of consecutive periods starting with a market period which is followed by a production period which is followed by a market period and so on. The firm is conceived during a market period as an organizational tool to facilitate the carrying out during the following production period of a production program.

A final remark: The firm of the variability approach is derived by abstracting away incentives and related issues such as opportunistic behavior, and by focusing instead on variability, coordination requirements, information structure, and the need for communication. The focus of the transactions cost approach is the opposite. Certainly the "firm" of the real world is an organization that emerges to cope both with the variability, coordination, information structure, need for communication issues, and with the incentives, and related issues. A synthesis of these approaches may, hopefully, produce a theory that provides a good representation of the firm of the real world.

NOTE

1. From this definition of multi-person firms, it follows that members belonging to the same firm need to communicate directly or indirectly among themselves during the production period, regarding the production program they are carrying out, to execute efficiently the program. Members belonging to different firms don't need such communication. Thus our characterization of firms implies that, during the production periods, there is communication, direct or indirect, among the members of any given firm; there is no communication between members belonging to different firms.

   It has been remarked in some of the conferences and workshops where the variability theory has been presented that, in the real world, members of different firms do talk to each other during a production period when they are participating in a joint production program through subcontracting. I feel that this remark is valid and stresses the fact that the notion of "firm" in the real world, which I view as a *legal* notion, is not the same as the notion of "firm" of the variability approach, which I consider as an *organizational* notion.

   The frequently observed fact of two different "legal" firms communicating with each other, when carrying out under contract a joint program without becoming integrated into *one* "legal" firm, can be interpreted, from the variability approach perspective, as *two* "legal" firms that during a market period discover a program that is advantageous for the two to carry out jointly during the following production period. Since, to carry out efficiently this joint production program, communication is needed between the members of the two "legal" firms, the two "legal" firms, from the variability approach perspective, become one firm although, from the legal point of view, they retain their legal identity and remain as two different "legal" firms. They formalize their relationship of carrying out the joint program through a contract.

   An interesting question for empirical research is to investigate how likely it is that a "legal" firm be also a variability-approach firm and vice versa.

REFERENCES

Armen Alchian and Harold Demsetz, "Production, Information Costs, and Economic Organization", in: *American Economic Review*, 62, no. 5, 1972, pp.777–795.

Yoram Barzel, "Knight's 'Moral Hazard' Theory of Organization", in: *Economic Inquiry*, 25, January, 1987, pp.117–120.

Antonio Camacho, *Division of Labor, Variability, Coordination, and the Theory of Firms and Markets*. Dordrecht, The Netherlands: Kluwer Academic Publishers 1996.

Ronald Coase, "The Nature of the Firm", in: *Economica N. S.*, 4, 1937, pp.386–405.

Ronald Coase,"The Problem of Social Cost", in: *Journal of Law and Economics*, 1960, pp.1–44.

Sanford Grossman and Oliver Hart, "The Costs and Benefits of Ownership: A theory of Vertical and Lateral Integration", in: *Journal of Political Economy*, 94, no. 4, 1986, pp.691–719.

J. Patrick Gunning, "Entrepreneurists and Firmists: Knight vs. the Modern Theory of the Firm", in: *Journal of the History of Economic Thought*, 15, Spring, 1993, pp.31–53.

Friedrich Hayek, "The Use of Knowledge in Society", in: *American Economic Review*, 35, September, 1945, pp.519–530.

Oliver Hart and John Moore, "Property Rights and the Nature of the Firm", in: *Journal of Political Economy*, 98, no.6, 1990, pp.1119–1158.

Bengt Holmstrom and Jean Tirole: "The Theory of the Firm", in: R. Schmalensee and R. Willig, (Eds.), *Handbook of Industrial Organization*. Amsterdam: North-Holland 1989.

Leonid Hurwicz, "Optimality and Informational Efficiency in Resource Allocation Process", in: K. J. Arrow, S. Karlin, and P. Supes (Eds.), *Mathematical Methods in the Social Sciences*. Stanford: Stanford University Press 1959, pp.23–45.

Frank Knight, *Risk, Uncertainty and Profit*. New York: Harper and Row 1921, 1957.

John Moore, "The Firm as a Collection of Assets", in: *European Economic Review*, 36, 1992, pp.493–507.

Joseph Schumpeter, *The Theory of Economic Development*. Cambridge, MA: Harvard University Press 1911, 1934.

David Teece, "Toward an Economic Theory of the Multiproduct Firm", in: *Journal of Economic Behavior and Organization*, 3, 1982, pp.39–69.

Oliver Williamson, *Markets and Hierarchies: Analysis and Antitrust Implications*. New York: Free Press 1975.

Oliver Williamson, *The Economic Institutions of Capitalism: Firms, Markets Relational Contracting*. New York: Free Press 1985.

The University of Illinois at Chicago
Department of Economics (M/C 144)
College of Business Administration
601 South Morgan Street, Room 2103
Chicago, Illinois 60607–7121
USA

JAAKKO HINTIKKA

A GAME THEORY OF LOGIC – A LOGIC OF GAME THEORY

This paper does not deal with the applications of game theory for which this theory was first developed, that is, for modelling economic systems and rational decision making. But I do not want to consider games as abstract mathematical structures, either. I want to concentrate on what it is that makes a game actually playable. This playability means both the possibility of finding and formulating the strategy that a player uses and the feasibility of actually applying the strategy in question in making one's moves.

In general, it seems to me that the philosophical and other theoretical interest of game-theoretical concepts and conceptualizations extends way beyond its economic and decision-theoretical uses. This interest is in effect acknowledged by professional game theorists when they say or imply that game theory is a general theory of rationality. I will not consider the pros and cons of this claim in my paper. I am on the record, however, of saying that a lot of interesting phenomena should be considered, not as rule-governed processes, as the case now is, but as goal-directed ones.[1] This paradigm switch is likely to make the processes in question amenable to game-theoretical conceptualizations.

One class of such applications that I have worked on concern logic and language theory. Such applications are of considerably wider interest to game theorists in that they illustrate some of the general conceptual problems that come up in game theory, especially those related to the nature and genesis of the strategies players use. They are also of interest for logic and language theory. For instance, I have shown there how the semantics of both formal (logical) and natural languages can be based on certain stereotypical games of verification and falsification.[2] Unsurprisingly they are called semantical games. They can be thought of as regimented versions of Wittgensteinian language-games.[3] It is not my purpose to expound or to study them here in any detail. Suffice it to indicate only some of their most interesting aspects, especially those aspects that show how game-theoretical concepts are used in them.

(i) The most crucial concept of game theory is the notion of strategy first introduced explicitly by von Neumann (or was it Borel?).[4] Now in my game-theoretical semantics (GTS) this notion plays a critical role in that the all-important concept of truth of a sentence S is defined as the existence of a winning strategy for the (initial) verifier in the correlated verification game G(S). The falsity of S means that there is a winning strategy for the (initial) falsifier in G(S).

The importance of the game-theoretical idea of strategy for the developments discussed here is illustrated by the history of the subject. The game-theoretical interpretation of quantifiers is not a new idea. It is implicit in mathematicians'

W. Leinfellner and E. Köhler (eds.), Game Theory, Experience, Rationality, 315–323.

time-honored jargon when they use phrases like "given any ϵ, we can find δ such that". It was spelled out in so many words by C.S. Peirce.[5]

(ii) From this definition it follows that the law of excluded middle becomes tantamount to the determinacy of some semantical game. In view of the strength and of the consequent implausibility of determinacy assumptions in general, the law of excluded middle can safely be expected to fail in all sufficiently strong logics. This turns out to be the case, putting the very concept of negation in interesting new light.[6]

(iii) In the semantical games correlated with ordinary first-order logic, determinacy holds. However, we obtain a much stronger basic logic simply by allowing informational independence in the very same semantical games. In this logic, which might be called independence-friendly first-order logic, the law of excluded middle does not hold.[7]

(iv) The game-theoretical truth conditions explained above for first-order sentences (ordinary or independence-friendly) can be integrated into an explicit truth-definition for those languages. What is more, such a truth-definition can be formulated in independence-friendly *first*-order language, including a truth-definition for that very language itself.

Such game-theoretical truth-definitions are applicable in several different situations in which the received (Tarski-type) truth-definitions are not applicable.[8] Such situations may for instance involve violations of compositionality (i.e. violations of semantical context-dependence) or infinitely long plays of a semantical game or restrictions on players' strategies. Not surprisingly, logicians and mathematicians have spontaneously resorted to game-theoretical conceptualizations precisely when traditional truth-definitions become inoperative. Examples are offered by Henkin quantifier sentences, infinitely deep formulas and Diophantine games with strategies restricted to recursive ones.

(v) We can formulate an interesting special case of the general problem of equilibrium selection for semantical games. If there exists a pure winning strategy in a semantical game G(S) connected with S for either player, S can be assigned a unique truth-value, true or false. But if there exists no pure equilibrium, what should we do? One idea is to consider a Nash equilibrium point for such games and to consider the probability of the verifier's win as the generalized truth value of the sentence S when the players are using strategies corresponding to the equilibrium. The problem of how to do this uniquely the same as in semantical games is an instance of the problem of somehow choosing a unique Nash equilibrium point.

(vi) These observations suffice to illustrate the interest of applications of game-theoretical notions to logic and to some extent *vice versa*. These interconnections can be pushed further.

What an interpreted first-order sentence S gives us is an extensive form of a game. The corresponding von Neumann normal form is codified by the second-order sentence S that assumes the existence of what are known as the Skolem functions of S (in a slightly extended sense of the term). They are the functions that codify the (dependent) truth-making choices of values of existentially quantified variables and disjuncts. They express parts of the winning strategy for the verifier in semantical games when such a strategy exists. For instance, consider the game correlated with

(1) $(\forall x)(\exists y)S[x,y]$

(where S has no logical constants). Its truth can be expressed by quantifying over the strategies available to the two players. The expression is therefore a second-order sentence, in fact

(2) $(\exists f)(\forall x)\mathrm{S}[x, f(x)]$

Here (2) can be considered as a game-theoretical second-order translation of (1). Likewise, the second-order translations of (3)-(5) are (6)-(8), respectively.

(3) $(\forall x)(\forall z)(\exists y)(\exists u)\ S[x,y,z,u]$
(4) $(\forall x)(\exists y)(\forall z)(\exists u)\ S[x,y,z,u]$
(5) $(\forall x)(S_1[x]\ \vee\ S_2[x])$

(6) $(\exists f)(\exists g)(\forall x)(\forall z)\ S[x, f(x,z), z, g(x,z)]$
(7) $(\exists f)(\exists g)(\forall x)(\forall z)\ S[x, f(x), z, g(x,z)]$
(8) $(\exists f)((S_1[x]\ \&\ (f(x) = 0))\ \vee\ (S_2[x]\ \&\ (f(x) \neq 0)))$

Informationally independent existential quantifiers require Skolem (strategy) functions with fewer arguments (or constants instead of functions). Thus the von Neumann translation of

(9) $(\forall x)(\forall z)(\exists y/\forall z)(\exists u/\forall x)\ S[x,y,z,u]$

is

(10) $(\exists f)(\exists g)(\forall x)(\forall z)\ S[x, f(x), z, g(z)]$

where / indicates independence. In general, there is no translation of (9)-(10) to ordinary (unslashed) first-order logic. Likewise, the translation of

(11) $(\forall x)(S_1[x]\ (\vee/\forall x)\ S_2[x])$

is

(12) $(\exists b)(\forall x)((S_1[x]\ \&\ (b = 0))\ \vee\ (S_2[x]\ \&\ (b \neq 0)))$

which is equivalent to

(13) $(\forall x)\ S_1[x]\ \vee\ (\forall x)\ S_2[x]$.

In more complex cases there is no ordinary (slash-free) first-order sentence equivalent to the normal form of a slashed formula.

One of the main points I want to make in this paper concerns the relationship of the extensive form of games to their von Neumann normal form. This relationship is, as every game theorist worth her utility matrix knows, a most delicate matter. The first major triumphs of game theory were obtained by means of considerations pertaining to the normal form. Over the years, it became clear that not everything that can be established in the extensive form of a game carries over to the normal form – and vice versa. For instance, not all solution concepts are adequately represented in the normal form.[9] In the last twenty-odd years, mainstream game theorists have worked predominantly with the extensive forms of games. Almost ten years ago Binmore[10] wrote that "Selten's (1975) view, that von Neumann and Morgenstern were mistaken in identifying a game with its normal form, is now widely accepted".

The first observation I want to make here is that the semantical games offer an instructive special case of the problems that arise in connection with the relationship of the extensive form and the von Neumann normal form of games. I believe that in this way valuable light can be thrown on this relationship in general. Without attempting to exhaust the subject, some observations can clarify the subject.

There is a temptation to think that the normal form of a game captures everything there is to be said of it, apart from purely contextual or other pragmatic factors. This temptation manifests itself in the logical context as a temptation to say that the equivalences illustrated by (3)–(8) are really matters of second-order logic, in that the second-order translation gives the real meaning of the given first-order sentence. Not surprisingly, nobody has suggested this in the case of ordinary first-order sentences, but it is the explanation of choice of those who want to deny that independence-friendly quantifiers are "really" first-order.[11] Unfortunately for them, it is the normal (second-order) form that is an abstraction out of the realities of extensive forms of games. Indeed, by showing that the second-order translations of nothing but codifications of certain aspects of the normal forms of semantical games, game-theoretical viewpoint helps us to appreciate the fact that independence-friendly quantifier logic is indeed a genuinely first-order logic, and not second-order logic in sheep's clothing.

For another thing, game-theoretical semantics illustrates vividly the fact that purely logically and mathematically the relationship between first-order sentences and their second-order translations, i.e. the relationship of the extensive form of semantical games to their von Neumann normal form, is quite sophisticated. In order to see this, consider the simplest nontrivial case of a statement that asserts the equivalence of a first-order formula to its second-order ("von Neumann") translation:

(14) $(\forall x)(\exists y)\ S[x,y] \leftrightarrow (\exists f)(\forall x)\ S[x, f(x)]$.

A moment's reflection shows you that this is one of the many forms of the axiom of choice. The equivalence (14) may be a valid principle of logic, as Hilbert thought (and as I believe), but it is not a trivial one. If we are operating with set theory, instead of second-order logic, it has to be expressed by an axiom that is known to be independent of the other usual axioms of set theory.

In more complex cases something even more remarkable happens. The second-order translation T(S) of a first-order sentence S asserts the existence of Skolem functions for S, as illustrated by (2), (6)–(8) and (10). In the simplest case, the equivalence of S with T(S) is a consequence of the axiom of choice. However, if the situation is expressed in set-theoretical terms instead of second-order ones, it can be shown that the equivalences

(15) $S \leftrightarrow \mathrm{T}(S)$

(in their set-theoretical incarnation) are not all provable in axiomatic set theory. What is more, they cannot all be provable in any (first-order) axiomatic set theory.[12] The reason can be seen in the light of game-theoretical semantics. If all the equivalence (15) were provable, a game-theoretical truth-definition for axiomatic set theory could be formulated in axiomatic set theory itself, along the lines indicated above. (Indeed, the equivalence (15) expresses the game-theoretical truth-condition of S.) This would contradict Tarski's famous undefinability result.

In short, in axiomatic set theory it is impossible to prove the equivalence of the extensive form with the normal form for all semantical games. This is a striking example of the problems and puzzles that are associated with the relationship of the extensive and the normal form of games.

There are interesting further problems which are associated with the notion of strategy and which are illustrated by semantical games. For instance, it can be suggested that only recursive strategies are actually playable.[13] For how can you play a game in accordance with a strategy if you do not have a general method of effectively deciding what your next move will be?

This is a very interesting suggestion. One reason for the interest is that this suggestion embodies an unusual and theoretically interesting modification in the definition of semantical games. This kind of restriction does not affect the move-by-move rules of the game in question. Yet it can affect what happens in an infinite game by eliminating certain rows and columns in the game matrix. It can turn a game with a winning strategy for a certain player to a game where no such a winning strategy does not exist. It can turn a determinate game into an indeterminate one.

The remarks just made can be extended to all humanly playable games with pure strategies. Should we perhaps restrict the strategies of human players to recursive ones in all of them? Games with such restrictions have in fact been studied in the literature, beginning with Rabin's 1957 paper.[14] For instance, restriction to recursive strategies in the Diophantine games of number theorists[15] has in fact been utilized in the literature.[16]

This kind of possibility is illustrated by the semantical games on which our logic is thought of as being based in my game-theoretical semantics. If the recursivity requirement is imposed on these semantical games we end up with a kind of constructivistic logic instead of classical logic. Is this constructivistic logic the true game-theoretical semantics of logic?

This suggestion is a very interesting one. However, it is in my judgment not feasible in the form it was just presented. The basic reason is that the notion of recursivity is too narrow to serve as the set of strategies to which we can restrict our attention. There are two different ways in which it can be relaxed. Both of them remain to be examined, even though they are very much in the spirit of game theory.

The first consists in making explicit the epistemic element which there is in game situations.[17] A player can choose a strategy function only if he or she knows that function. This obviously makes a difference to many of the things I have said, in the sense that they cannot be carried over to the epistemic logic I am envisaging. For instance, if it is known that there exists a choice function, it does not follow that any such function is known. Thus, even though

(16) $K(\forall x)(\exists y)\ S[x,y]$

implies of course

(17) $K(\exists f)(\forall x)\ S[x, f(x)]$

it does not imply

(18) $(\exists f)\ K(\forall x)\ S[x, f(x)]$.

which can also be written

(18*) $K(\exists f/K)(\forall x)\ S[x, f(x)]$.

When the intuitionists have criticized the axiom of choice, what they in fact have objected to is not the transition from (16) to (17), but from (16) to (18).

I believe that such explicit use of epistemic notions might be useful in clarifying the foundations of game theory. However, they have not been used there, and I am not in a position to show here how they might be so used.

In a different way, I have argued that the restriction of strategies is a mistaken one.[18] It is tempting only as long as we assume that the game is presented only in the normal form. In actually playing a game in practice we play it, so to speak, in the extensive form. In other words, we typically do not choose our strategies from the normal form matrix, but form them move by move in response partly to other players' moves.

If so, it should arguably be possible that one's eventual strategy function is a noncomputable one in the usual sense of Turing machine computability (recursivity). Even though any finite sequence of moves will be compatible with a recursive strategy, in terms of a suitable complexity theory (for instance Kolmogorov's theory) we can even calculate how likely the sequence in question is a segment of a recursive sequence. In other ways, too, it can be argued that we can make sense of playing a game in accordance with a nonrecursive strategy.

Nevertheless, the strategy one uses must be formed in some regular way on the basis of the intermediate outcome of one's own moves and those of the other players.

Also, psychologically speaking one must be able to learn the strategies of a game on the basis of what happens in repeated plays of the game. This also imposes some restrictions on the strategy sets. What restrictions?

I do not have a definitive answer to such questions. I am mentioning them as challenges to future research. Pending the results of such research, I can meanwhile offer two suggestions.

One suggestion I can make here is to point out that there exists a class of functions which are computable in an extended sense of computability and which might serve as strategy functions more realistically than recursive functions.[19] They are deterministically fixed in that the value of such a function for each given argument is determined by a finite number of deterministic steps. On the other hand, they are learnable in the sense that if one plays the same game repeatedly, changing one's strategy according to deterministic rules, the strategies one uses converge to the strategy defined by the function in question, in the sense that for each initial sequence of argument values the strategy function values are the right ones from some finite stage on.

My suggestion therefore amounts to restricting the strategy functions to functions that are computable in this *extended sense*. In semantical games, this still yields classical logic. At the same time, each move in a game played in accordance to such a function is in a sense determined by a finite number of deterministic steps. Furthermore, such functions are learnable in the sense outlined above.

The precise definition of this extended sense of computable function is presented elsewhere. Here I mention them as a possible tool in clarifying the relationship between the extended form and the normal form in games with infinite game matrices.

My second suggestion or hypothesis also concerns the process of strategy formation. More specifically, it concerns the psychology of the way a human player forms his or her strategies "across the board" in the everyday social and linguistic games in response to other players' moves.

This question deserves a general methodological comment. The process of forming strategies in response to other people's moves might seem to be relatively unimportant for game theory as a whole. Perhaps it is uninteresting for the applications that most game theorists are interested in, for instance economic ones. But in a wider theoretical perspective this matter cannot be overlooked. Game theory is often claimed to be a general theory of rationality. Now one aspect of rationality is the ability to choose the strategies of one's own behavior, not just once and for all, but in response to what happens in the course of a goal-directed interactive process. I will not discuss here the question whether this interactive strategy formation is *sui generis*, independent of the general problem of strategy selection, only when a player's knowledge of other players' utilities and/or beliefs is incomplete or when the optimal strategies are not recursive or if these strategies were too complicated to be dealt with in practice. Even with such restrictions, the process of reactive strategy formation is important in practice and in theory.

There does not seem to be much work on step-by-step strategy construction reported in the literature. One suggestion I want to make here is that language-games of communication might very well offer an instructive case study of spontaneous human strategy formation. Roughly speaking, one's understanding of what another person says is not merely a matter of taking in the literal meaning of what he or she says. It also involves some grasp of what one's interlocutor "aims at" or "has in mind". It is not far-fetched to interpret this second component of language understanding as involving some insight into the other party's strategies. And the right conversational response is not merely to react appropriately, but to modify one's own strategies so as to adapt them to the other person's strategies.

I cannot pursue this suggestion here, beyond registering my belief that part of the payoff of such an approach is likely to be an enhanced grasp of the psychological implications of game-theoretical ideas and concepts.

NOTES

1. See my paper, "Paradigms for Language Theory", in: *Acta Philosophica Fennica*, 49, 1990, pp. 181-209.
2. See Jaakko Hintikka/Gabriel Sandu, "Game-Theoretical Semantics", in: Johan van Benthem/Alice ter Meulen (Eds.), *Handbook of Logic and Language*. Amsterdam: Elsevier 1997, pp.361-410.
3. Cf. here Merrill B. Hintikka/Jaakko Hintikka, *Investigating Wittgenstein*. Oxford: Basil Blackwell 1986.

4. John von Neumann, "Zur Theorie der Gesellschaftsspiele", in: *Mathematische Annalen*, 100, 1928, pp. 295-320; cf. his "Communication on the Borel Notes", in: *Econometrica*, 21, 1953, pp. 124-125.
5. See Risto Hilpinen, "On C.S. Peirce's Theory of Proposition: Peirce as a Precursor of Game-theoretical Semantics", in: E. Freeman (Ed.), *The Relevance of Charles S. Peirce*, La Salle, Illinois: The Hegeler Institute 1983, pp. 264-270.
6. Cf. Jaakko Hintikka, *The Principles of Mathematics Revisited*. Cambridge: Cambridge University Press 1996, ch. 7.
7. See *op. cit.*, Chapters 3-4.
8. See *op. cit.*, Chapters 4 and 6.
9. See Eric van Damme, "Extensive Form games", in: John Eatwell et al. (Eds.), *The New Palgrave: Game Theory*, , New York: W.W. Norton 1989, pp.139-148, especially p. 143, where further information is provided about the relationship between the extensive and the normal form.
10. Ken Binmore, "Modelling Rational Players, Part I", in: *Economics and Philosophy* vol. 3, 1987, pp. 179-214, especially p. 189.
11. This is the most general gambit used to disparage the general theoretical interest of independence-friendly logic. It backfires on its users, however, for independence-friendly logic can be obtained from ordinary first-order logic by merely relaxing the use of parentheses. See my paper "No Scope for Scope?", forthcoming.
12. See here *op. cit.*, Note 6 above, Chapter 8.
13. Here and in the following, see *op. cit.*, Chapter 10.
14. Michael Rabin, "Effective Computability of Winning Strategies", in: M. Dresher et al. (Eds.), *Contributions to the Theory of Games III*. Princeton University Press 1957, pp. 147-157.
15. James P. Jones, "Recursive Undecidability - An Exposition", in: *The American Mathematical Monthly*, 81, 1974, pp. 359-367.
16. Cf. e.g., Yuri V. Matiyasevich, *Hilbert's Tenth Problem*. Cambridge, MA: MIT Press 1993.
17. See *op. cit.*, Chapter 11.
18. *Op. cit.*, Chapter 10.
19. See Jaakko Hintikka/Arto Mutanen, "An Alternative Concept of Computability", forthcoming.

Department of Philosophy
Boston University
745 Commonwealth Avenue
Boston MA 02215
USA

Dean P. Foster
H. Peyton Young*

Learning with Hazy Beliefs

Consider a game that is played infinitely often by a set of boundedly rational players. At the beginning of each period, each agent has a belief about the strategies that his opponents will use in that period. These beliefs are derived by observing the play of the game so far, that is, they are conditioned on the information contained in the history of play together with initial conditions such as prior beliefs. We need not assume, however, that the posterior beliefs are derived via Bayes' rule (though they might be). We consider a much more general framework in which beliefs are determined by an *arbitrary function* that maps initial information and past history to present beliefs (as in Jordan, 1992). We suppose further that players are *myopically rational* in the sense that they choose a best reply given their beliefs in each period.

The question we examine is whether such players can learn to play the stage game over time. By "learn" we mean that everyone becomes increasingly good at predicting the period-by-period strategies that their opponents actually use. Since players always choose best replies given their predictions, learning implies that they come closer and closer to playing a Nash equilibrium of the stage game (in a sense that we shall make precise below). We show that if there is even the tiniest bit of uncertainty ex ante about the others' payoff functions, then for almost all realizations of the game, *no* plausible belief updating process permits myopic players to learn.

We then show that this impossibility theorem can be overcome if we adopt a more flexible approach to the learning process. Namely, suppose that a person's belief at any given time is not fully determined by the history and the priors, but is a random variable whose *distribution* is determined by the history and the priors. Of course, one can develop many hypotheses about how beliefs might be generated randomly; here we shall suggest one model of the belief formation process that seems intuitively plausible and that guarantees learning with probability one in any finite game.

The basic idea is that each player adopts a model of the world (a forecast of how the other players are going to behave) on a *provisional* basis. Unlike a Bayesian player, who commits himself once and for all to a model of the world conditional on the information he receives, our players are only willing to commit themselves to a model temporarily. Every so often they compare the model to empirical data. If it has failed to predict well over the recent course of history, they simply abandon it and adopt a new model. Moreover, even when they are working with a provisional model, they allow themselves to experiment or tinker with small variations of the model. The combination of these two features is sufficient (under rates of adjustment that are game-independent) for

W. Leinfellner and E. Köhler (eds.), Game Theory, Experience, Rationality, 325–335.

the players to learn to forecast their opponents' behavior accurately in *any* finite game. (For detailed proofs, see Foster and Young (1996).)

This result is related to previous work of Kalai and Lehrer (1993) and Nachbar (1997), who postulate that players are Bayesian and fully rational. In other words, each player has a prior over all future contingencies in the repeated game, which they update using Bayes rule. They then choose a strategy (at the beginning of the game) that maximizes their discounted expected utility given their priors. Kalai and Lehrer's principal result is that such players can learn to predict *provided* the priors attach positive probability to every measurable set of histories that has positive probability under the actual repeated game strategies.

Nachbar (1997) argues that this proviso is very strong, and at least in some games is inconsistent with the ways in which thoughtful players might be expected to form their priors. Say that a set of priors (in a two-person repeated game) is *neutral* if the support of each player's prior includes at least one best reply by the opponent to every strategy that is in the support of the opponent's prior. Nachbar shows that if the priors are neutral and their support is sufficiently diverse, then there are many games that rational players cannot learn, including elementary games like matching pennies. In other words, there exist plausible priors that violate the Kalai-Lehrer condition. While Nachbar's result is derived under different assumptions than ours (in particular full rationality and Bayesian learning) the general spirit is the same: namely, it is problematical whether players (boundedly rational or otherwise) can learn to play some types of games under all plausible initial conditions. Indeed, we show that this is true for any deterministic belief formation process, Bayesian or otherwise. But if we allow some randomness in the beliefs, then learning can occur from any initial conditions.

I. LEARNING RULES

Let G be an n-person game with finite action spaces $X_1, X_2, \ldots, X_n$ and payoff functions $u_i: X \to \mathbb{R}$, where $X = \Pi X_i$. A one-period *outcome* is an n-tuple of actions $x \in X$. All actions are publicly observed. Let Δ_i denote the set of probability distributions over X_i, and let Δ denote the product set of mixed strategies: $\Delta = \Pi\Delta_i$. A *belief* of player i will be denoted by $p_{-i} \in \Delta_{-i} = \Pi_{j\neq i}\Delta_j$. A (mixed) *strategy* of i will be denoted by $q_i \in \Delta_i$, and a vector of strategies by $q \in \Delta$. If we are given a vector $q \in \Delta$ and wish to consider all components except i, we shall write $(q)_{-i}$.

Let us first consider the situation if G is played exactly once. Before the game is played, the state can be described by a family of n pairs $\{(p_{-i}, q_i)\}_{1\leq i\leq n}$ where p_{-i} is player i's belief and q_i is i's strategy. *Rationality* says that each player chooses only best replies given his beliefs, that is,

$$\forall\ \mathrm{i}, \qquad q_i(x_i) > 0 \Rightarrow x_i \in \mathrm{BR}(p_{-i}). \tag{R^0}$$

By itself this does not take us very far – some further behavioral assumption is necessary to obtain an equilibrium. This is provided by the idea of prediction: i is a *good predictor* if his beliefs about the others' coincide with their actual plans:

$$\forall \text{ i}, \qquad | p_{-i} - (q)_{-i} | = 0. \qquad (P^0)$$

Any family $\{(p_{-i}, q_i)\}_{1 \le i \le n}$ that satisfies both (R^0) and (P^0) corresponds to a Nash equilibrium of the game; conversely, any Nash equilibrium can be described in this manner.

With these ideas in hand, let us now consider the dynamic case. Let G^∞ be the infinitely repeated game of G. A *history of length* $t \ge 0$ is a sequence of t outcomes drawn from X: $h^t = (x^1, x^2, \ldots, x^t)$. We let h^0 denote the null history. Let H^t be the set of all length-t histories, and let $H = \cup H^t$. A *behavioral strategy* for player i is a function from histories to strategies. Since the strategy also depends on the payoff function, we shall write g_i: $U \times H \rightarrow \Delta_i$, where U is the set of all payoff functions on X. Let Γ_i denote the set of behavioral strategies for i. A *belief* by player i is a probability distribution on $\Pi_{j \neq i} \Gamma_j$. Any such belief can be represented by a function f_i: $U \times H \rightarrow \Delta_{-i}$, where $(f_i(h^t))_j$ is the probability that i assigns to j's actions in period $t+1$, given the history h^t up to that point. Note that the beliefs are built on the assumption that the other players' strategies are independent of one another. This assumption is justified because the strategies are independent in fact.

Suppose now that the players are (myopically) Bayesian. Let 0_i be i's prior at the beginning of the game about the behavioral strategies the others are going to use. As above we can represent π^0_i by a function f^0_i: $U \times H \rightarrow \Delta_{-i}$. Consider the situation at the end of period t, when i has observed the history h^t. What is i's posterior belief about j's behavior in period $t+1$? The answer is $f_i^0(h^t)$ – the product probability distribution that i *originally believed* would describe the others' actions in period $t+1$, conditional on the history h^t actually occurring (see Kalai and Lehrer, 1993). As we shall soon see, the essential point here is not that players use Bayes' rule to update their beliefs, but that their beliefs at each period are *fully determined* by the observed history and the initial conditions.

Following Jordan (1992) we say that a *learning process* (with deterministic beliefs) is a family of 2n functions $\{f_i, g_i\}_{1 \le i \le n}$ such that

$$f_i: U \times H \rightarrow \Delta_{-i} \text{ and } g_i: U \times H \rightarrow \Delta_i. \qquad (D)$$

A *learning path* is an infinite sequence $\{(h^t, (p^t_{-1}, q^t_1), (p^t_{-2}, q^t_2), \ldots, (p^t_{-n}, q^t_n)): t = 1, 2, \}$ where $p^t_{-i} = f_i(h^{t-1})$ is player i's belief at time t and $q^t_i = g_i(h^{t-1})$ is i's strategy at time t. Note that a learning path is not fully observed, only the histories h^t are. For brevity we shall denote a learning path by $\{(h^t, (p^t_{-i}, q^t_i)\}$.

We assume that the players are *myopically rational:* in each period they choose strategies that maximize their expected payoff given their beliefs about the others' strategies in that period. Thus every learning path $\{(h^t, p^t_i, p^t_{-i})\}$ satisfies

$$\forall\, t, \qquad q^t_i(x_i) > 0 \Rightarrow x_i \in \mathrm{BR}_i(p^t_{-i}). \qquad \text{(R)}$$

A simple example of a myopically rational learning process is fictitious play. Here the belief of player i at time t is simply the empirical frequency distribution of actions taken by the others up through time $t - 1$, and i's strategy is a distribution over the set of best replies to these beliefs.

We remark that myopia is not unreasonable under some conditions. For example, even though the game is in fact repeated, the players might not believe that it is going to be repeated. Alternatively the players might change from one period to the next. Under either interpretation, it is rational to be concerned only with the outcome of play in the present period.

Player *i learns to predict* the behavior of another player j if his forecast of j's probability distribution over actions becomes better and better over time. A strong way of stating this condition is to require that for all distinct i and j, $\lim_{t\to\infty} |\, (p^t_{-i})_j - q_j^t \,| = 0$. A somewhat weaker version is that the mean square error in prediction go to zero:

$$\forall\, i, \qquad \lim_{T\to\infty} \sum_{t=1}^{T} (p^t_{-i} - (q^t)_{-i})^2/T = 0. \qquad \text{(P)}$$

Admittedly this is still a strong requirement. For example, it says that if the column player in a two-person game plays action 1 in odd periods and action 2 in even periods, then the row player learns to predict this behavior with near-perfect accuracy in almost all time periods. This is obviously much more demanding than saying that the row player learns to predict the *average* behavior of the column player.

Most learning processes one can think of – including fictitious play – do not lead to good prediction in this sense. In fact, there are many games in which fictitious play does not even successfully predict average behavior (Shapley, 1964; Jordan, 1993; Foster and Young, 1996). One might therefore worry that this definition of prediction is too demanding. We adopt it partly because it has become an established benchmark in the rational learning literature cited above; more importantly, it is achievable under some forms of learning, as we shall show in the second part of the paper. But it is not achievable under myopic Bayesian learning (except in a very restrictive sense) as we are about to argue now.

II. An impossibility theorem

Let G be an n-person game with utility functions $u = (u_1, u_2, \ldots, u_n)$, which we shall suppose (for the moment) are common knowledge. Let p^* be a Nash equilibrium of G, either pure or mixed. Suppose that each player i predicts that p^* will be played in every period, that is, $g_i(h^t) = (p^*)_{-i}$ for every i and every initial history h^t. Suppose further that each player chooses the strategy $f_i(h^t) = p^*_i$. Then obviously (R) and (P) are satisfied. However, they are satisfied in an unsatisfying sense. First, if the game has multiple equilibria, why should the players all predict the same p*? Second, what if the players are slightly uncertain about the others' payoffs? For example, suppose that the payoffs are not common knowledge, but are distributed according to some continuous density that is common knowledge. Specifically, we can think of each payoff function u_i as a point in $\mathbb{R}^{|X|}$, and the vector of payoff functions u as a point in $\mathbb{R}^{n|X|}$. Let μ be a density on $\mathbb{R}^{n|X|}$ that is bounded away from zero on some small open neighborhood B of u^*, where the game u^* has a unique Nash equilibrium that is fully mixed. Then p^* is *not* a Nash equilibrium for almost all realizations of the game, and there is no reason to think that the players would ever be able to forecast the true equilibrium exactly. This does not rule out the possibility, however, that the players could learn to forecast something close to the true equilibrium over time. This is the question that Kalai and Lehrer (1993) and Nachbar (1997) study in a setting where players are rational. Here we show that when players are myopically rational, and beliefs satisfy a certain informational consistency condition, such learning will not occur with μ-probability one.

To motivate the consistency condition, let the payoffs be independently distributed, that is, $\mu = \Pi\mu_i$. Assume for the moment that the players are Bayesian, and that μ is their common prior. Since the distributions μ_i are independent, i's revealed payoff function u_i gives i no new information about the payoffs of anyone else. What then will player i forecast about j's behavior in period 1? It seems implausible that i would condition his forecast on information (namely u_i) that i knows j cannot infer from his current information except through the prior μ. Moreover, i cannot condition his forecast on information (namely u_j, $j \neq i$) that he himself cannot infer except from μ. This means that i's forecast p^1_{-i} depends only on μ. The same reasoning extends to every player in every subsequent period t : i will base his forecast of j's behavior only on the publicly observed data h^{t-1} and the common prior μ. Note that this is a plausibility assumption only: a Bayesian player *could* make his forecast of others' behavior contingent on his own payoff; we are only saying that this is implausible under the assumed informational structure.

A learning process $\{\mu, f_i, g_i\}$ is *informationally consistent* (I) if, whenever the payoffs are independently distributed, the beliefs are independent of the realized value of the payoffs, that is, $f_i(u, h^{t-1}) = f_i(h^{t-1})$ for all initial histories h^{t-1}. Note that this condition is reasonable whether or not players are Bayesians.

THEOREM 1. *Let $\{\mu, f_i, g_i\}$ be a learning process where $\mu = \Pi\mu_i$ is absolutely continuous with respect to Lebesgue measure, and every game in the support of μ has only mixed Nash equilibria. If players are myopically rational and beliefs are informationally consistent, then with μ-probability one not all players are good predictors.*

This result is similar to a theorem of Jordan (1992) in that it relies on a simple counting argument. Suppose the players are myopically rational and their beliefs are informationally consistent. One can show that, except for a subset of games having μ-measure zero, the best replies along any infinite history are uniquely determined for all players. For any such game the path of play is uniquely defined. Along such a path, there is at least one tuple, say x^*, that occurs at least $1/|X|$ of the time. Suppose now that all players are good predictors. Then they must predict x^* with increasing accuracy in almost all periods where x* occurs. Suppose that player i forecasts that the others will play x^*_{-i} with high probability next period. By myopic rationality i's best reply to this forecast is x^*_i. However, there is at least one i such that x^*_i is not a best reply to x^*_{-i}, because by assumption the game has no pure equilibrium. This contradiction shows that not all players can be good predictors.

The simplicity of this result might lead one to think that prediction is too demanding a condition, but this is not the case. There are reasonable learning rules that lead to good prediction not only with μ-probability one but for all games in the support of μ. (In fact for all finite games period.) Before stating this possibility theorem, however, let us ask what happens when myopically rational players succeed in predicting well. The answer is that both beliefs and strategies must be arbitrarily close to the set of Nash equilibria almost all of the time.

PROPOSITION. *Let* G *be a finite game and let Δ^* be the set of its Nash equilibria. Let $\{p^t_{-i}, q^t_i\}$ be an infinite sequence of beliefs and strategies for $1 \leq i \leq n$ that satisfy* (R) *and* (P). For every $\epsilon > 0$ *and every time T, let $\pi(T, \epsilon)$ be the proportion of times $t \leq T$ such that*

$$\exists\, p^* \in \Delta^*, \forall\, i \quad |(p^t_{-i} - (p^*)_{-i}| \leq \epsilon \text{ and } |q^t_i - p^*_i| \leq \epsilon.$$

Then for every $\epsilon > 0$, $\lim_{T \to \infty} \pi(T, \epsilon) = 1$.

III. LEARNING WITH PROVISIONAL BELIEFS

In this section we introduce a new class of learning rules that have all three of the properties posited in Theorem 1 and that work for every finite game (not just generic games). The key idea is that beliefs are not fully determined by the data; they are random variables whose *distribution* is generated by the data. In particular, we posit that a belief is a merely a provisional or temporary model of the world. Periodically an agent tests his provisional model against reality. If it passes the test he keeps it. If it fails he abandons the model and tries a new one (drawn at random from the class of all possible models).

Quite apart from the fact that this approach works, we believe it represents a plausible model of how people react to complex learning environments. The idea can be illustrated by an everyday example. Consider a fisherman who is deciding where to search for the day's catch. He begins with some generalized belief about the areas most likely to contain fish. He then chooses an area that he believes to be promising, anchors his boat, and casts his line about at random to see whether the fish are biting. If they are he continues to fish there, if not he moves on to another area.

This familiar type of search rule has two essential elements. First, the fisherman's beliefs about the likelihood of catching fish at each particular *point* could be described as indeterminate or *hazy*. In the short run, he is not committed to an *exact* model of where the fish are, but to an approximate model. Second, in the long run he is not committed to an approximate model either. If it does not pass a reality test after a period of time, he abandons it in favor of an alternative model. By contrast, a Bayesian is committed to an exact model of the world at all times, and is also committed to revising the model in a specific way given his posterior information. Thus the novel elements of prospecting are constant local revision (which smoothes the best response), plus global revision when the model fails to predict well (which prevents getting locked in to a bad model). A similar story could be told for people prospecting for oil, searching for lost items, and so forth. Note, however, that fishing has the special feature that the environment constantly changes in response to one's own actions, and is therefore more like a game than some other types of search.

We now specialize this idea to games, and show that with appropriately defined parameters (which do not depend on the game), prospecting players learn to play arbitrarily close to Nash equilibrium with probability one in any finite game.

Given is an n-person game with payoff functions $u = (u_1, u_2, \ldots, u_n)$ drawn from a prior distribution $\mu(\cdot)$. Each person knows his own payoff function u_i and may or may not be able to make inferences about the payoffs of the others. In particular, we do not assume that μ is a product distribution. Furthermore, μ may be degenerate and put probability one on a particular payoff function u. Since the learning process does not depend on $\mu(\cdot)$, and works for all games u, we henceforth drop μ from the notation.

A player's "beliefs" at any given time represent a provisional model or forecast of how the other players are going to behave next period given the data observed so far. The key idea is that a player is not committed to any one model *ex ante*; rather, models are formed and discarded according to a random (but plausible) belief formation process. The basic structure of this process is shown in Figure 1.

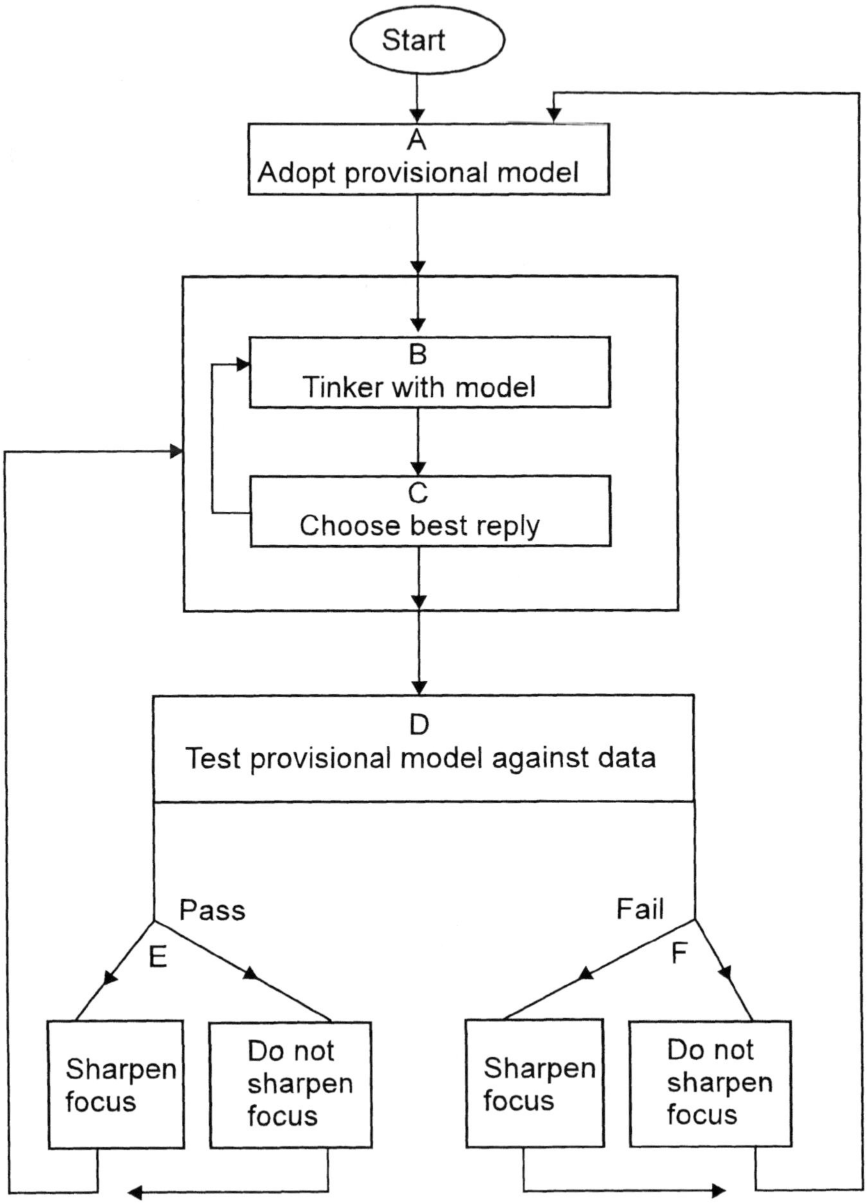
Start
A
Adopt provisional model
B
Tinker with model
C
Choose best reply
D
Test provisional model against data
Pass
Fail
E
F
Sharpen focus
Do not sharpen focus
Sharpen focus
Do not sharpen focus

Figure 1

The transitions between steps are governed by probability distributions that we now specify explicitly. The transition to step A is governed by a density ν_i on $\Delta_{-i} = \Pi_{j \neq i} \Delta_j$ that is continuous and has full support. This represents the probability with which various new models will be adopted. Here we shall assume that ν_i is fixed. In fact, ν_i can be modified to incorporate new information that arises during the learning process, as we mention at the conclusion of the paper.

A new *round r* begins whenever the "focus" is sharpened at the bottom of the flow chart. Denote the successive rounds by $r = 1, 2, 3, \ldots$ Focus is expressed by a parameter σ_r that we initialize at round 1 to $\sigma_1 = 1/2$. Whenever the focus is sharpened we divide σ_r by 2, so that $\sigma_r = 2^{-r}$. In any given path through E or F the probability that a sharpening occurs is $\pi_r = 2^{-(1/\sigma_r)}$

Every time the process enters A we then draw some $p^*_{-i} \in \Delta_{-i}$ via ν_i. The idea is that i's belief is temporarily anchored at p^*_{-i}, which we shall call his *core belief*. His actual belief at each time t is determined by a draw from the distribution $\omega_{i,r}(p_{-i} \mid p^*_{-i})$, where the draws are independent across periods. We assume the following conditions on $\omega_{i,r}$:

i) for every $p^*_{-i} \in \Delta_{-i}$, $\omega_{i,r}(p_{-i} \mid p^*_{-i})$ converges (as $r \to \infty$) to the degenerate distribution with mass one at p^*_{-i};

ii) $0 < \text{var } \omega_{i,r} \leq \sigma_r^2$;

iii) $\omega_{i,r}$ is Lipschitz in the L_1-norm with a constant that is at most $O(1/\sigma_r)$, that is, for some constant L and every $p'_{-i}, p''_{-i} \in \Delta_{-i}$,

$$\|\omega_{i,r}(\cdot|p'_{-i}) - \omega_{i,r}(\cdot|p''_{-i})\| = \int_{\Delta_{-i}} |\omega_{i,r}(p_{-i} \mid p'_{-i}) - \omega_{i,r}(p_{-i} \mid p''_{-i})| \, dp_{-i} < (L/\sigma_r)|p'_{-i} - p''_{-i}|. \tag{9}$$

A canonical example is the uniform distribution on $S(p^*_{-i}) \cap \Delta_{-i}$, where (p^*_{-i}) is the sphere of radius σ_r around p^*_{-i}. Note that we cannot expect the Lipschitz constant to be independent of σ_r, since var $\omega_{i,r} \leq \sigma_r^2$. The bound L/σ_r says that the density does not change too rapidly given that the variance of $\omega_{i,r}$ is at most σ_r^2.

It must be emphasized that player i has definite beliefs at any given time, even though he may not know how he came to hold these beliefs. In each period he therefore chooses a best reply given these beliefs. We shall assume that i's best response function is stationary, that is, he has a fixed best reply mapping $B_i : \Delta_{-i} \to \Delta_i$ that applies at all times t :

$q_i = B_i(p_{-i})$ where $q_i(x_i) > 0$ implies x_i maximizes

$$\sum_{x_{-i} \in X_{-i}} u_i(x_i, x_{-i}) \, p_{-i}(x_{-i}) \tag{8}$$

From the other players' perspective, i's beliefs are random and his actions are described by the probability distribution

$$q_i = \nu_{i,r}(p^*_{-i}) = \int B_i(p_{-i})\, \omega_{i,r}(p_{-i} \mid p^*_{-i})\, dp_{-i}. \qquad (10)$$

Assume that loop B-C is repeated $(1/\sigma_r)^3$ times. Thereafter, the test phase is entered with probability σ_r and the tinkering phase is repeated with probability $1 - \sigma_r$. We summarize all of this as follows.

A. Pick p^*_{-i} at random from Δ_{-i} using the density v_i .

B. Pick p_{-i} at random from Δ_{-i} using the density $\omega_{i,r}(p_{-i} \mid p^*_{-i})$, and choose a best reply according to the (mixed) strategy $B_i(p_{-i})$. Do this for $(1/\sigma_r)^3$ periods in succession, making independent draws from $\omega_{i,r}$ each time.

C. Go to step D with probability σ_r; repeat step B with probability $1 - \sigma_r$.

D. Compute the empirical frequency distribution $Q_j(x_j)$ of actions taken by player j over the last $(1/\sigma_r)^3$ periods, and let $Q_{-i} = \Pi_{j \neq i} Q_j$. If $|Q_{-1} - p^*_{-i}| \leq 2\sigma_r$, go to E; if $|Q_{-i} - p^*_{-i}| > 2\sigma_r$, go to F.

E. With probability $\pi_r = 2^{-(1/\sigma_r)}$ enter the next round, that is, increase r by 1 and divide σ_r by two; with probability $1 - \pi_r$ keep r and σ_r fixed. Go to B.

F. With probability $\pi_r = 2^{-(1/\sigma_r)}$ enter the next round, that is, increase r by 1 and divide σ_r by two; with probability $1 - \pi_r$ keep r and σ_r fixed. Go to A.

When all of the players learn by prospecting, they generate various learning paths $\{h^t, (p^t_{-i}, q^t_i)\}$ with probabilities that depend on the algorithm. We assert that with probability one the players learn to predict along these paths. Hence the process converges to Nash equilibrium in the sense of the Proposition. The reader is referred to Foster and Young (1996) for the detailed proof.

THEOREM 2. *Let* G *be a finite* n*-person game. If all players learn by prospecting, then they are rational and good predictors along almost all realized learning paths. Hence along almost all learning paths both beliefs and strategies come arbitrarily close to Nash equilibrium almost all of the time.*

Obviously there are many variations of this process that will also lead to good prediction. We briefly mention one such modification that brings the approach a bit closer to the spirit if not the letter of Bayesian updating. Suppose that we modify the wide-area search distribution using information that emerges during the learning process. Specifically, let $f_i : H \to \Delta_{-i}$ be a deterministic belief formation process as described in section II (such as ordinary Bayesian updating). Suppose that the prospecting process (for player i) loops to A at time t, and that

h^t is the history of actions up to that time. Let $p^t_{-i} = f_i(h^t)$ and let $\mathbf{1}(p^t_{-i})$ denote the degenerate distribution with mass 1 at $p^t_{-i} \in \Delta_{-i}$. We can think of p^t_{-i} as i's current estimate of the most likely spot at which to anchor his core beliefs. Suppose that player i conducts the wide-area search at time t according to the distribution

$$v^t_i(h^t) = \lambda^t v_i + (1 - \lambda^t)\mathbf{1}(p^t_{-i}),$$

where λ^t is an appropriately chosen weight. It can be shown that if λ^t goes to zero no faster than $1/\log t$, theorem 2 continues to hold. In other words, if players are sufficiently patient in updating their beliefs, and the best reply process is smoothed by local random variation, then myopically rational players can in fact learn to play any finite game.

NOTE

* Dean Foster was supported by the Center for Rationality at Hebrew University, and Peyton Young by the National Science Foundation, grant SBR-9601743.

REFERENCES

Dean Foster/H. Peyton Young, "On the Nonconvergence of Fictitious Play in Coordination Games", Working Paper WP-96-33, International Institute for Applied Systems Analysis, Laxenburg, Austria 1996.

Dean P. Foster/H. Peyton Young, "Learning with Hazy Beliefs," Working Paper, The Hebrew University of Jerusalem 1996.

James S. Jordan, "Bayesian Learning in Games: A NonBayesian Perspective", Preprint, University of Minnesota 1992.

James S. Jordan, "Three Problems in Learning Mixed-Strategy Equilibria", *Games and Economic Behavior*, 5, 1993, pp.368-386.

Ehud Kalai/Ehud Lehrer, "Rational Learning Leads to Nash Equilibrium", *Econometrica*, 61, 1993, pp.1019-1045.

John H. Nachbar, "Prediction, Optimization, and Learning in Repeated Games", in: *Econometrica*, 65, 1997, pp.275-310.

Lloyd Shapley, "Some Topics in Two-Person Games", in *Advances in Game Theory, Annals of Mathematics Studies* vol. 52 (M. Dresher et al., eds.), 1964, pp.1-28.

Dean P. Foster
Department of Statistics
Wharton School
University of Pennsylvania
Philadelphia, PA 19104
USA

H. Peyton Young
Department of Economics
Johns Hopkins University
Baltimore MD 21218-2685
USA

LUCIAN KERN

A GAME-THEORETIC MODEL OF THE WAR IN CHECHNYA

I. INTRODUCTION

The end of the East-West confrontation has by no means put an end to the possibility of war in Central Europe. The outbreak of ethno-religious hostilities in former Yugoslavia made it clear that the previous formation of the Eastern and Western blocks contained a Pandora's box of ethno-religious conflicts which opened up after the end of the Cold War.

In the context of applied game theory, these developments suggest a simple question: Can game theory shed light on the new ethno-religious conflicts and can it provide a meaningful analysis of possibly ensuing wars? This paper seeks to provide a positive answer to the question by indicating the scope for using game theory to understand ethnic hostility in a recent violent conflict, namely the war in Chechnya.

II. THE SEPARATISM GAME

To model the Russian-Chechen conflict - and to explain the Russian military intervention in Chechnya - we begin with an adaptation of Selten's *Chain Store Game* (Selten 1978). In doing so, it is assumed that Russia's decision to intervene militarily in Chechnya is akin to the behaviour of a monopolist trying to deter prospective competitors from entry into a market: if the Chechens win independence, other peoples in the Caucasian and Siberian regions would demand the same.

We therefore recast Selten's game as a *Separatism Game*: Imagine a nation comprising many peoples (for instance Russia), where the peoples in 10 regions (players R1, R2, ..., R10) are considering independence, that is, they successively will, or will not, declare independence (strategies *I* or *nI*). To this move, the central power (player C) can react in each region, either by negotiating/compromising, or by using a strategy of military intervention (strategies *Neg* or *Mil*).

Payoffs are characterized by letters whose values satisfy the following inequalities: $a > 1 > b > 0$ (cf. Wilson 1992, p.320). And as in Selten's game, the worst outcome for both players occurs if C intervenes militarily after a declaration of independence by R1 (like the price war ensuing after the entry

W. Leinfellner and E. Köhler (eds.), Game Theory, Experience, Rationality, 337–347.

into the market of the first competitor and the choice of an aggressive price strategy by the chain store).

Players		Central power	
	Strategies	*Negotiations (Neg)*	*Military intervention (Mil)*
Region 1	*Independence (I)*	***b*/0**	$b-1/-1$
	No independence (nI)	$0/a$	$0/a$

Table 1
Separatism game in normal form (1st round)

(Payoff relations: $a > 1 > b > 0$; iterated dominance equilibrium in bold type.)

Both players will be better off if C chooses the strategy of negotiation after a declaration of independence by R1. C would be even better off if the region drops its claim. But then the region would be worse off than in the case where it chooses independence, given that C should negotiate.

If we look at the game in the normal form (1st round), given in Table 1, it is apparent that C has a weakly dominant strategy: *Neg*. If it is assumed that C is a rational player, strategy *Mil* can be deleted from the strategy set of C. This means that *I* is a dominant strategy for R1 and that the payoff pair ***b*/0** resp. the strategy pair (***I*/*Neg***) is an iterated dominance equilibrium of the game, once we consider the *Separatism game* to be a one-shot game in the normal form (Rasmusen 1994, p.20).

There is a second equilibrium in this game: $0/a$ or (*nI*/*Mil*), but this one is *not* an iterated dominance equilibrium. Thus, the notion of the iterated dominance equilibrium very neatly distinguishes between the two equilibria in favour of ***b*/0**.

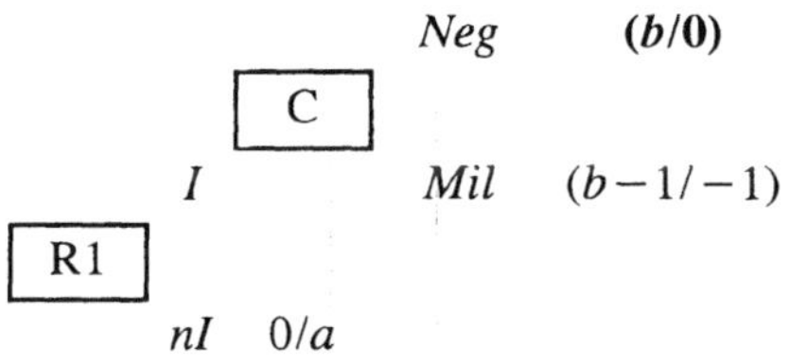

Figure 1
Separatism game in extensive form (1st round)

(Players: C = Central power, R1 = Region 1; strategies: *I* = Independence, *nI* = No independence, *Neg* = Negotiations, *Mil* = Military intervention; payoff relations as in Tab. 1; order of payoffs: R1/C; subgame perfect equilibrium in bold type.)

With respect to the *Separatism game* in the extensive form, shown in Figure 1, it is first to be noted that there are also two equilibria, i.e. ***b*/0** and 0/*a*, but only one of them is subgame perfect. Why? A subgame perfect equilibrium is one which implies a credible threat. Now, 0/*a* is obviously not a credible threat for R1 because, if threatened by the central power with a military intervention, Region 1 will choose independence and thereby induce a loss of −1 to the central power. A threat accompanied by a loss to the threatener is not credible. Thus, the notion of subgame perfectness distinguishes the two equilibria in the same way as the notion of an iterated dominance equilibrium.

The question then is, whether the equilibrium ***b*/0** applies to all ten rounds of the game in its extensive form. Selten (1978) shows that, by using the argument of backward induction, this is indeed the case because, when dealing with the last region in the sequence, a rational central power knows that to intervene militarily will have no further deterrent effect on future regions, as there are none left to deter. Since the central power's rationality and the number of regions (or rounds) are *common knowledge* among the players, everyone is aware of this conclusion, so the last region will declare its independence.

Since the last region is not going to be deterred, it would not be rational for the central power to deter the next-to-last region, which is aware of that and will thus declare its independence. Hence backward induction leads us to conclude that *I* and *Neg* are the equilibrium strategies for the respective players in *all* rounds (Bicchieri 1993, p.193). But this was not what happened in Chechnya.

III. THE DETERRENCE ARGUMENT

Selten also considers an argument which is in conflict with this conclusion: the deterrence argument. This view holds that even if, in the short run, fighting for the central power is costly, it may well be its winning strategy in the long run, since observation of the central power's past behaviour will convince other regions to abstain from independence.

If player C chooses *Mil* in the early rounds and if this strategy is effective, so that the other regions are deterred from declaring their independence, C gets a higher payoff from all rounds than by sticking to *Neg* and the other regions declaring their independence – despite the fact that C will have a zero payoff in the first round.

The intuitive argument of deterrence, on which Selten's paradox relies, can be supported game-theoretically by assuming that, at the outset, there is a small uncertainty in the game, in the form of incomplete information about the central power's type. A central power may be *strong* or *weak*. If the latter is the case, payoffs are as given in Table 1 (and Figure 1), but if it is strong, the payoffs are different, as Figure 2 shows. The reason is that in this case negotiations are less favourable to the strong central power and military intervention is more favourable to it than to the weak central power (Kreps & Wilson 1982, Milgrom & Roberts 1982, Wilson 1992, Osborne & Rubinstein 1994, pp.239-40).

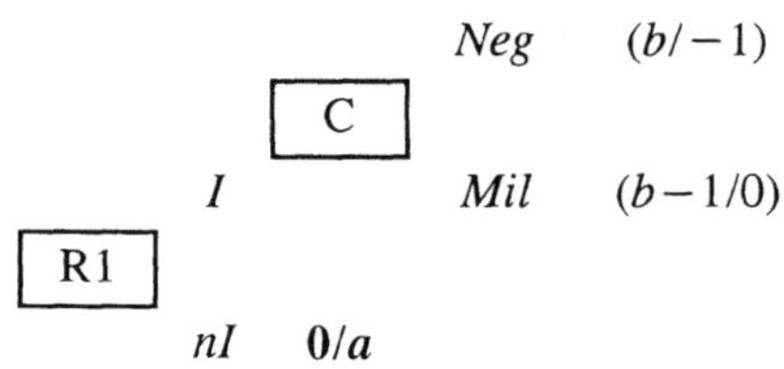

Figure 2
Separatism game with *strong* central power
in extensive form (1st round)

(Players, strategies, payoff relations, and order of payoffs as in Fig. 1; subgame perfect equilibrium in bold type.)

As a consequence the equilibrium strategies change: now *Mil* is a weakly dominant strategy for player C and, deleting *Neg* from the strategy set of C (assuming that C is a rational player) reveals ***nI*** as a dominant strategy for the region. Hence ***0/a*** is the subgame perfect equilibrium of this game, or in other words: a strong central power will always intervene militarily in the face of a declaration of independence, because doing so maximises its single-period payoff and has no adverse long-run consequences.

Therefore a central power playing the deterrent strategy is either a strong central power using its (weakly) dominant strategy or a weak central power trying to discourage independence-seeking regions by mimicking the behaviour expected of a strong central power. In other words, even the weak central power is led to use the deterrent strategy although it is not its (weakly) dominant strategy (Friedman 1986, p.138).

This leaves the regions without a chance of becoming independent. If even a weak central power will use the deterrent strategy, the regions cannot declare their independence if they want to avoid a negative payoff. Thus, the regions are caught in a dilemma: either they have to forego independence or they must take the risk of military intervention - a risk that materialised in the Chechnyan case. So apparently the backward induction argument is inconsistent with events in this case. However, there are also difficulties with the deterrence argument, which, though explaining the Russian military intervention, does not explain Chechnya's decision to declare independence *and* to stick to it in the face of Russian military action.

IV. THE RUSSIAN-CHECHEN CONFLICT AS A MULTIPLE GAME

One view is that the Chechen leaders, by immediately taking up the fight against the Russian aggression (as they understood the intervention), introduced an additional strategy which is not in the strategy set of the game (conceived as a *Separatism* or *Chain Store Game*). In fact, in doing so, they introduced another game, which in the course of time turned out to be a game of military escalation - modelled by Shubik as the *Dollar Auction Game* (Shubik 1971).

Thus we are confronted, not with a simple game, but with a game within a game, or what might be called a "multiple game".[1] As distinguished from a simple game, a multiple game has as at least one of its outcomes, not a pair or an n-tuple of payoffs, but *another* game. Evidently, the game as such is still a *Separatism game*, but the game within it is Shubik's *Dollar Auction Game.*

This is illustrated by the following Table 2, where three of the entries in the payoff matrix are taken over from the *Separatism Game* and in the fourth entry, the payoff matrix of the first round of the *Escalation Game* is inserted with its payoffs. Below this matrix another matrix is shown, representing the payoffs of the second round of escalation.

<table>
<tr><td>Players</td><td colspan="3">Central power</td></tr>
<tr><td></td><td>Strategies</td><td>Negotiat-
ions
(Neg)</td><td>Military
intervention
(Mil)</td></tr>
<tr><td rowspan="2">Region 1</td><td>Independence
(I)</td><td>$b/0$</td><td><table><tr><td></td><td>Neg</td><td>Mil</td></tr><tr><td>Neg</td><td>b/b</td><td>$\mathbf{0/a}$</td></tr><tr><td>Mil</td><td>$a/0$</td><td>$-b/1$</td></tr></table></td></tr>
<tr><td>No Independence
(nI)</td><td>$0/a$</td><td>$0/a$</td></tr>
</table>

↓

	Neg	*Mil*
Neg	b/b	$0/a$
Mil	$\mathbf{a/0}$	$1/b-a$

Table 2
Multiple game of *Escalation within the Separatism game* in normal form (1st and → next round of the escalation)

(Payoff relations as in Tab. 1; changing equilibria in the Escalation in bold type)

The escalating character of Shubik's *Dollar Auction Game* comes about because the rule in this auction is that the highest *and* the second-highest bidder have to pay their bids, but only the highest bidder gets the prize (one dollar). Thus the second-highest bidder will keep loosing until s/he bids higher than the highest bidder (thereby forcing the latter into a loss). This introduces the dynamics of escalation into the auction, driving bids higher and higher, even higher than one dollar, so that losses are certain even for the winner. In this case the bids rise because bidders are trying to reduce their losses.

Extending the original *Separatism Game* with an *Escalation Game* (the extension is shown in Figure 3 in the extensive form) results in a change of the equilibrium compared with Selten's analysis. Instead of the payoff pair ***b*/0** or the strategy pair (***I*/*Neg***) as in the original game, we now have **0/*a*** or (***Neg*/*Mil***) as a new equilibrium in the first round of the escalation.This shift indicates the profitability for the parties to enter the *Escalation Game*. In fact both players will gain from starting this game (if they win in the next step of escalation) or they get at least no less than in the original game (if they do not win in the next step of escalation).

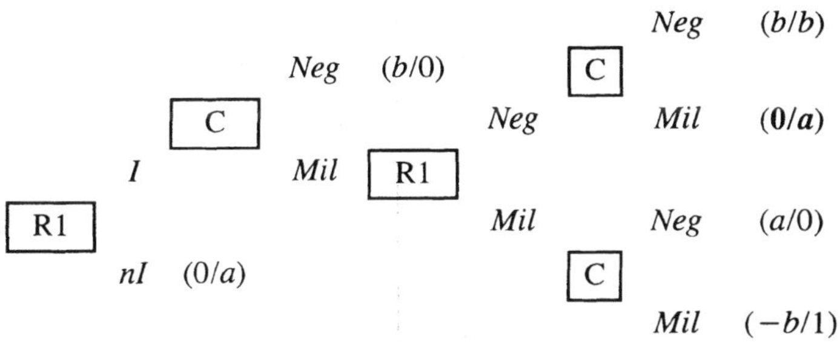

Figure 3
Multiple game of *Escalation within the Separatism Game* in extensive form
(1st round of the *Separatism Game* and 1st round of the *Escalation Game*)

A closer look at the dynamics of the *Escalation Game* reveals that the equilibrium changes from round to round. In the first round, when the central power takes military action, its payoff is highest and the equilibrium is $\mathbf{0/a}$. In the second round, when the first region "outbids" the central power in escalation, the equilibrium changes to $\mathbf{a/0}$ in favour of the region. The change in equilibrium, therefore, successively favours the central power and then the region. If both parties use their corresponding equilibrium strategies, i.e. *Military action*, the escalation will be going on, and, as Shubik notes, "there is no natural bound to the escalation" (Shubik 1982, p.293).

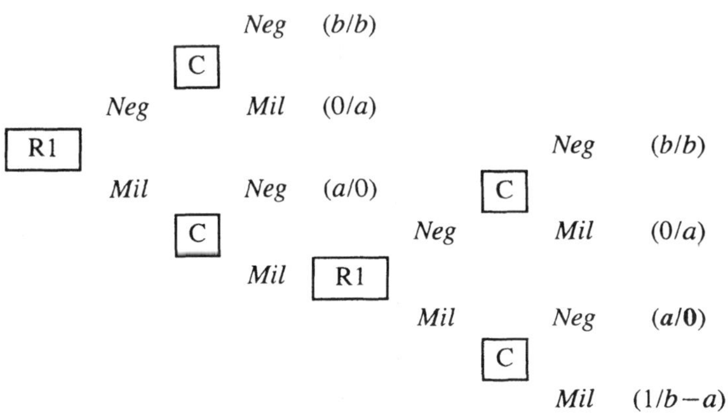

Figure 4
Multiple game of *Escalation within the Separatism Game* in extensive form

(1st and 2nd round of the *Escalation Game*)

V. THE RUSSIAN-CHECHEN SETTLEMENT

Viewing the Russian-Chechen conflict as a multiple game of *Escalation within a Separatism Game* explains much of the dynamics of the conflict, which otherwise is hardly understandable.

First, why has Russia intervened militarily in the first place? As a central power, it had to take into account that other regions besides Chechnya might also demand independence. Hence – following the logic of Selten's *Chain Store Game* – it took an aggressive posture early in the game to deter others. This decision is supported game-theoretically by the Kreps-Wilson model of a strong and weak central power, showing that even a weak central power in this case will have an interest in taking up *Military action*.

Second, why has Chechnya nevertheless not only declared its independence but went on so far as to introduce a military strategy of its own? One answer within the logic of Selten's *Chain Store Game* might be, that Chechnya assumed Russia to be a rational player following the backward induction argument, though this would have meant that Russia opted for strategy *Neg*. But this is not very convincing, because Russia, by openly supporting the Chechen oppositional militia in their fight against the troops of President Dudajew, made clear at an early stage in the game that it would use its military option.

This fact also contradicts a second possible answer, namely that Chechnya might have considered Russia to be a weak central power, who, with the payoffs of Figure 1 in mind and because of its weakness, would not use the military strategy. Chechnya chose *Independence* because this was the strategy that opened the way to the *Escalation Game*, within which it hoped to win, using its military strategy.

Third, why has the spiral of escalation continued for more than two years despite massive losses on both sides? Because each side hoped to justify its losses by winning in the next round. But this presupposes – as the *Escalation Game* shows – that the other side gives in, which – for the same reason – it most probably will not do. This explains the long duration of wars of independence, of which colonial wars are a clear example. Losses must be very high indeed before there is a chance of reaching a negotiated settlement.

In the Chechen case, such a settlement, originally initiated by the former general Alexander Lebed, was reached in 1996. It brought about the end of the war, a minimal form of independence for Chechnya and the complete withdrawal of Russian troops from Chechen soil by the end of 1996. But this settlement is not included in the game-theoretic model presented so far. On the contrary, our model would predict a continuing military escalation. Hence the question arises: what is missing with our model?

Surely the point is that there is nothing wrong with the model as such, only that it is not sufficiently developed to reflect the possibility of a nascent peaceful settlement. It needs, therefore, to be supplemented by yet another game. For this purpose we go back to a game called *War of Attrition*, which is a species of the *Chicken game*, stretched out over time. The respective strategies in this case of a (supposed) military confrontation for both parties are: *Cease* the confrontation or *Continue*. The payoff structure of the game is shown in the southeastern box of Table 4. As Rasmusen (1994, p. 73) describes it, both players start with the choice of *Continue*, and the game ends when one player picks *Cease*.

Players		Central power	
		Negotiations (Neg)	*Military action (Mil)*
Region 1	*Negotiations (Neg)*	b/b	$\mathbf{0/a}$
	Military action (Mil)	$a/0$	*Ce* *Co* *Ce* b/b $\mathbf{0/a}$ *Co* $\mathbf{a/0}$ $-a'/-a'$

Table 3
Multiple game of *War of Attrition within the Escalation game* in normal form

(Strategies in the *War of Attrition: Ce* = Cease confrontation, *Co* = Continue confrontation; payoff relations: $a' > a > 1 > b > 0$; equilibria in the *Escalation* and *War of Attrition* are in bold type.)

As is easily seen, the payoffs for three of the four strategy pairs are the same as in the *Escalation game*, reflecting the fact that, in this game as well, if one of the parties stops the confrontation, he gets a zero payoff and the other player earns a reward for outlasting him. The difference lies in the fourth strategy pair. If both parties continue the confrontation, both get a negative payoff – not just one of them. That is to say that in this game – but not in the *Escalation game* – no positive value is attached to winning the next step in the escalation, because losses nevertheless have been incurred.

If we now combine the *Escalation Game* with the *War of Attrition*, such that one of the outcomes of the *Escalation Game* is not a pair of payoffs but the *War of Attrition* – as is shown in Table 4 – we have a new multiple game which is the *War of Attrition within the Escalation Game*. This game may explain how a peaceful settlement may arise from a military confrontation.

A decisive prerequisite for the transition of the *Escalation* to the *War of Attrition* is the growing awareness of both parties that winning the next step of escalation does not necessarily have a positive reward, even if the "positive reward" only implies being able to justify the losses. In characterising the payoffs of the strategy pair (*Co/Co*) by negative amounts of money, even the notion of "winning" in the *War of Attrition* becomes obsolete. If there is such an awarenes of losses by both parties, then the *Escalation Game* almost inevitably changes to the *War of Attrition*. And this is what must have occured in the Russian-Chechen confrontation.

Of course, there is no guarantee that the transfer to the *War of Attrition* immediately brings about the end of the confrontation. Although there are two equilibria which imply the end of the confrontation, both players may fail to reach any of them if they insist that it is the other player who should give in (thereby trying to secure a positive reward for themselves).

But there is another equilibrium if one looks at the game not in its normal form but in its time-dependent version. As Fudenberg and Tirole have shown (1991, p. 120), if both players use the strategy *Cease at time* t *if the other player hasn't ceased before then* with a positive probability p, this leads to a symmetric, mixed strategy equilibrium. Hence in the long run an equilibrium ending the confrontation will eventually be reached.

The analysis – if it is correct so far – contains a bitter (and paradoxical) lesson: Although both parties will lose more and more, the longer the escalation continues, they have no interest in an early peaceful settlement of the conflict, because in the short run they hope to gain by using their military options. Thus they get caught in a spiral of escalation which is difficult to halt.

In order for it to be stopped, the players have to change to yet another game: the *War of Attrition*. This is possible only if the players become aware of the fact that even winning in the next step of the escalation leads to losses. It is this change in the assessment of the players which opens up the way to a peaceful settlement.

NOTE

1. Games of this type do not have a standard designation yet. They are called *games within games* by Shubik (1982), *nested games* by Tsebelis (1990) or *meta-games* by Laden (1991). Nevertheless they are used – more and more in recent years – for the analysis of political and economic processes and even for the reconstruction of theories of justice (Laden 1991). For examples cf. besides the authors mentiones – Garrett & Weingast (1993) and Heckathorn & Maser (1987).

REFERENCES

Cristina Bicchieri, *Rationality and Coordination*. Cambridge: Cambridge University Press 1993

James W. Friedman, *Game Theory with Applications to Economics*. Oxford: Oxford University Press 1986.

Drew Fudenberg/Jean Tirole, *Game Theory*. Cambridge, Mass.: MIT Press 1991.

Geoffrey Garrett/Barry R. Weingast, "Ideas, Interests, and Institutions: Constructing the European Community's Internal Market", in: Judith Goldstein/Robert O. Keohane (Eds.), *Ideas and Foreign Policy*. Ithaca-London: Cornell University Press 1993, pp.173-206.

Douglas D. Heckathorn/Steven M. Maser, "Bargaining and Constitutional Contracts", in: *American Journal of Political Science*, vol. 31, 1987, pp.142-168.

David M. Kreps/Robert Wilson, "Reputation and Imperfect Information", in: *Journal of Economic Theory*, vol. 27, 1982, pp.253-279.

Anthony Laden, "Games, Fairness, and Rawls's *A Theory of Justice*", in: *Philosophy and Public Affairs*, vol. 20, 1991, pp.189-222.

Paul Milgrom/John Roberts, "Predation, Reputation, and Entry Deterrence", in: *Journal of Economic Theory*, vol. 27, 1992, pp.280-312.

Martin J. Osborne/Ariel Rubinstein, *A Course in Game Theory*, Cambridge, Mass.: MIT Press 1994.

Eric Rasmusen, *Games and Information. An Introduction to Game Theory*, 2nd ed. Oxford: Basil Blackwell 1994.

Reinhard Selten, "The Chain Store Paradox", in: *Theory and Decision*, vol. 9, 1978, pp.127-159.

Martin Shubik, "The Dollar Auction Game: A Paradox in Noncooperative Behavior and Escalation", in: *Journal of Conflict Resolution*, vol. 15, 1971, pp.109-111.

Martin Shubik, *Game Theory in the Social Sciences*. Cambridge, Mass.: MIT Press 1982.

George Tsebelis, *Nested Games. Rational Choice in Comparative Politics*. Berkeley-Los Angeles: University of California Press 1990.

Robert Wilson, "Strategic Models of Entry Deterrence", in: Robert J. Aumann/Sergiu Hart (Eds.), *Handbook of Game Theory*, vol. 1. Amsterdam: North-Holland 1992, pp.305-329.

Geschwister Scholl-Institut
für Politische Wissenschaft
Universität München
Oettingenstr. 67
D-80538 München
Germany

DENNIS C. MUELLER

INFORMATION AND UNCERTAINTY: POWER, PROFITS AND MORALITY

Many important issues in economic and political science can be adequately analyzed under the assumption that individuals are certain about the consequences of their actions, possess perfect information. Many cannot, however, and models that incorporate uncertainty and asymmetric information have become increasingly popular over the last fifty years or so.

This is of course also roughly the life of game theory. The existence of uncertainty is central to the theory of games, and game theory's spectacular rise in popularity in recent years owes much to a growing appreciation of the importance of uncertainty in decisionmaking.

The existence of uncertainty also plays a central role in theories of power and profits, and in what is labeled moral and immoral behavior. The purpose of this note is to illustrate the roles of information and uncertainty in each case, and to discuss the links among them.

I. INFORMATION AND MORALITY

Mores are of two types: they prescribe actions that make others better off, or they proscribe actions that make the actor better off at someone else's expense. The golden rule and good Samaritan actions are examples of the first type of mores, the various "thou shalt nots" of the 10 Commandments are examples of the second. Moral codes do not typically *require* that one sacrifice one's life or health or give away all of one's possessions to benefit others. Indeed, suicide is also prohibited in many moral codes. This feature of morality can make the morality of an action hinge on the information content of a situation, as the following example shows.

M is walking alone through a forest in a foreign land. She sees a stranger approaching. He is carrying a large knife and has a sinister appearance. *M* feels frightened. She is carrying a large walking stick and believes that she could knock the stranger unconscious with a single sudden blow. She would not wish to do so, however, if she thought that the man would pass without harming her.

M faces the strategy choices and payoffs in Matrix 1.

W. Leinfellner and E. Köhler (eds.), Game Theory, Experience, Rationality, 349–357.

Matrix 1

	Stranger would attack M	Stranger would not attack M
M Strikes stranger first	6	3
M Does not strike stranger	0	10

M has been brought up properly and experiences remorse if she strikes the stranger. She would experience more remorse, if she struck the stranger and he would not have attacked her, than if she would successfully deliver a preemptive blow. There is no dominant strategy for *M*. Her optimal action depends on her estimation of what the stranger will do. If a bird were to chirp in *M*'s ear that the stranger would definitely not attack her, *M* would experience the highest utility not striking the stranger. If the bird revealed that the stranger would attack, she would optimally deliver the first blow.

In neither case would *M*'s action raise any ethical issues under conventional ethical systems. If the stranger would not strike *M*, and she knows that, she is happier not striking him given her upbringing, and deserves no applause for ethical restraint. If the stranger would attack and kill her, she is merely defending herself by striking first and most ethical systems will condone this act of self-defense. In the absence of uncertainty the situation raises no moral issues.

If *M* is uncertain about what the stranger will do, her optimal action depends on her estimation of the likelihood of the stranger's attacking. If she believes there is a 50/50 chance that he will attack, her optimal strategy is not to strike. If she guesses that there is a 0.53 or greater probability that the stranger will attack, *M* maximizes her expected utility by striking first.

Now if it is correct to say that *M* did not act immorally in striking the stranger when she knew he would attack her, it would seem that she does not act immorally when she strikes the first blow because she judges the probability of the stranger's attacking to be sufficiently high. If *M* does not act immorally when she strikes first with the knowledge that the stranger will attack with probability 1.0, then surely she does not act immorally, if she believes the probability to be 0.9999. If striking when the probability is 0.9999 is not immoral, then striking when the probability is 0.9998 cannot be, and so on, so long as this probability is greater than the critical value 0.53. To argue otherwise would seem to make moral behavior hinge on the presence of perfect certainty, and would rule out all actions by *M*, however prudent, which would injure the stranger, if she were the least bit uncertain about what he would do. Yet, it would also seem that *M* acts immorally if she strikes first when the probability of the stranger attacking is fairly low, *even if in doing so she maximizes her expected utility*. The *golden rule* would seem to require that M not strike, *if the probability of the stranger's attacking is not high*, i.e., that she take a reasonable chance on the stranger's not attacking her. But to make this argument for the payoffs in Matrix 1 with a probability greater than the critical 0.53 is to postulate some probability between 0.53 and 1.0 at which M's choice, although still rational and optimal, becomes immoral. This in turn makes the morality of M's actions depend purely on her subjective probability calculations, and the weights that she places on the various potential outcomes.

My goal in this paper is not to resolve the semantic issue of whether M's behavior can be rational but immoral in this situation, however tempting that may be, but to explore the role of morality, or moral principles, in situations involving uncertainty. *If M* has been taught to follow the golden rule, *and* she is uncertain about what the stranger will do, she may be more likely not to strike first than if she has not been taught the golden rule. One can think of this change in behavior as coming about, either because the utility payoffs M experiences in the upper row of Matrix 1 are both reduced by her acceptance of the golden rule, or that she imposes a separate set of *ethical preferences* on this choice as argued by Harsanyi (1955), Margolis (1982), and Etzioni (1988). Although I favor the first approach (Mueller, 1986, 1992), this is a question of methodological style rather than substance. Under either interpretation, M's acceptance of the golden rule can shift her choice, given uncertainty about the stranger's actions, toward the nonviolent, do not strike alternative.

The addition of principles like the golden rule can play an additional role in bringing about the nonviolent choice for M. If the stranger has been taught the golden rule, *and M knows that he has been taught this rule*, she may lower her estimate of the probability of his attacking and this would shift her decision toward not to strike. If we change the example so that M is walking in the park near the town where she grew up, and she recognizes the man as being from her town but does not know him personally, then the mere fact that the man is from her town may inform M that he has been taught the golden rule and is unlikely to attack her. Knowledge that others have been taught certain ethical principles can shift one's probability calculations to induce nonviolent choices in situations like Matrix 1.

Many "moral dilemmas" take the form of prisoners' dilemmas. Here, unlike the situation in the previous example, a dominant strategy exists in a one shot game and we expect rational actors to choose the noncooperative strategy. In a multiperiod, repeated game context, however, a rational person may be willing to cooperate if the other player cooperates in return. Knowledge that the other player has been taught the golden rule may increase an individual's expectation that her cooperative behavior will be matched in the games, and thus lead the individual to play the cooperative strategy.

From these examples we draw the following conclusions. Moral codes generally prohibit actions that make one individual better off at the expense of another. They do not, however, typically require that one knowingly forfeit one's life, impoverish oneself, or in some other way harm oneself for the benefit of others. Whether a particular action violates an ethical norm, therefore, may depend upon what the person against whom one acts does. The moral code may prohibit *M* from doing *x*, if *S* does *y* but condone *M*'s doing *x*, if *S* were to do *z*. Before *M* knows whether she should do *x*, she must know whether *S* will do *y* or *z*.

This property of moral systems can make the morality of *M*'s doing *x* depend on her perceptions of what *S* will do, and thus upon her subjective probability calculations of what *S* will do. Knowledge that *S* adheres to a particular moral code can thus change *M*'s calculations of what *S* will do, and thus change the actions of *M*, and our perceptions of the morality of such actions.

II. Information and Power

The word "power" probably connotes even more diverse images than the word "morality" and has led to as much discussion and perhaps confusion. We can, for example, think of *physical power*. Big brother gets little sister to give him her candy by twisting her arm. *Procedural power:* The chairman has the authority to determine whether *A* or *B* gets to speak, and thereby can influence the outcome of the meeting. *Market power* is the ability to raise price. The advantages of possessing each of these forms of power (and still others) are obvious and need no discussion.

Of more interest is the kind of power that allows an individual to obtain an advantage by getting another person to do something *voluntarily* that benefits the first individual and makes the second worse off.[1] Big brother gets sister to give him her candy *without* twisting her arm. The ability to exercise this form of power will be shown to arise in situations involving uncertainty and accrues to those individuals who possess information that reduces uncertainty.[2]

To see the roles of information and uncertainty play in the creation of power consider the arm-twisting, brother-sister example. If big brother twists little sister's arm she drops the candy. Brother gets candy and sister gets sore arm. Brother is better off and sister is worse off. We have a simple exercise of physical power, but not an exercise of power as defined here, since sister did not voluntarily give brother the candy.

Now suppose that brother does not immediately twist sister's arm but only threatens to twist it, unless she gives him the candy. The actions of brother, B, and sister, S, and payoffs to sister are given in Matrix 2.

Matrix 2

		S: Gives candy	S: Does not give candy
	Does not twist arm	a 5	d 10
B	Twists arm	b 0	c 1

If sister gives brother the candy and he twists her arm anyway, he adds injury to insult and sister gets her lowest payoff. She is better off if he does not twist her arm, and best off if she does not give him the candy *and* he does not twist her arm.

In this situation, *B* cannot exert power over *S* under our definition of power unless *S* is uncertain about what *B* will do. If she knows *B* is bluffing, he will not twist her arm even if she does not give him the candy, she will not give him the candy. She will also not voluntarily give him the candy, if she knows that he will twist her arm regardless of what she does. Even in the case where he will not twist her arm if she voluntarily gives him the candy, but will twist it if she does not, he will not be exercising power as we have defined it, if she knows that this is what he will do, e.g. because he is programmed to act in this way. If she is certain that this will be his behavior, then she knows that the only options for her are squares *a* and *c*. The choice is essentially hers. She chooses square *a* because it promises a higher payoff than *c*, and he behaves as he has been programmed to do. *B* only can exercise power over his sister, if he threatens to twist her arm if she does not give him the candy and she is uncertain about what he will do. If she then gives him the candy, when he would not in fact have twisted her arm if she had not given it to him, she will have opted for square *a* when she could have gotten square *d*. Her brother would have exercised the power he possessed given his sister's uncertainty and his knowledge of what he would do. He will have induced his sister to act so as to make him better off and herself worse off.

Although information and uncertainty play their respective roles in generating power according to our definition of power, the uncertainty that gives rise to the power in this example is generated by *B*'s possession of physical power as well as knowledge of his own actions. The reader may question whether information and uncertainty alone really generate power of the sort we have defined.

To see that they do, note first that uncertainty without information does not generate power. If *B* chooses an action by flipping a coin, and *S* knows he does so, she will not give him the candy and he exercises no power.

As a second illustration, let us increase the amount of uncertainty in the situation and give *S* some useful information. She has hidden the candy in her room, and only she knows where it is. Now *B* can only demand that she tell him where it is. She may do so, or she may lie. While he is hunting for it, she may run to their mother and complain. The additional uncertainty and information that *S* has may allow her to keep her candy *and* avoid a twisted arm.

Dictators like Stalin possess procedural/physical power in that they can have whomever they please executed. They typically convert this procedural power into the more general notion of power discussed here by unpredictably executing people who disobey their orders, or commit other acts of which they disapprove.

Secrecy is coveted by every bureaucrat, because it enhances his power. Perhaps no one illustrated better the link between information, uncertainty and power than that unhappy bureaucrat Franz Kafka.[3]

III. Information and Profit

One of the classic studies of profits is Frank Knight's *Risk, Uncertainty and Profit* (1921). In this book Knight emphasized that economic profits exist only because of uncertainty. Knight did not place any weight on information or information gathering, however. Instead, he depicted entrepreneurs as risk takers or uncertainty bearers. In a market economy, individuals with proclivities to bear risks (uncertainty) were self-selected to play entrepreneurial roles.

Information is featured in the other major contribution to profit theory of the early 20th century, Joseph Schumpeter's theory. Schumpeter (1934, ch.4) saw profits as arising from innovations, and so the relevant information in Schumpeter's theory was about new products, new production techniques, new organizational structures, and the like. Although Schumpeter played down the role of uncertainty in generating profits, uncertainty clearly is paramount in the introduction of any new product or process. Thus, in Schumpeter's theory uncertainty can be seen as a kind of entry barrier that prevents firms from introducing new products, production techniques and the like. Those entrepreneurs who possess the information that a particular product will sell, a particular invention will work, etc. overcome this entry barrier and enter (create) the industry. Their rewards are the profits of the innovator.

Information also figures in other prominent theories, most notably in Coase's (1937) theory of the firm in which the entrepreneurial role is to have (gather) information about situations in which hierarchical organizations are more efficient than markets in organizing economic activity.[4]

IV. DISCUSSION

Much of economics examines individual and aggregate behavior under conditions of perfect certainty. Consumers choose consumption bundles, managers choose factor input bundles to maximize utility and profits and the invisible hand guides everything to a Pareto optimal outcome. So wonderful are the workings of the invisible hand that even the introduction of uncertainty does not disturb its beneficence. Entrepreneurs can profit from information in a world of uncertainty and so are moved to gather information. In the process uncertainty and the profits it fosters are eliminated.

As the dismal science is so often apt to do, economics takes actions which other social sciences and moral philosophers frown upon, namely information gathering and use in the presence of uncertainty, and converts it into a virtue. Uncertainty in bureaucratic and political settings creates the potential for the exercise of power, power that goes to the holders of information. This information-induced power allows those individuals who possess information to benefit at the expense of those who do not have it, by getting the nonholders of information to undertake actions that they would not otherwise undertake.

An important category of actions prohibited by moral codes are those that allow one individual to benefit at the expense of another. Thus, these prohibited actions have precisely the same properties as those actions in which individuals exercise power in a nonmarket setting. Not surprisingly, therefore, having political power and exercising that power often is regarded as antisocial or immoral behavior.

Indeed, such is even the case in certain market settings. The Schumpeterian entrepreneur who has information that no one else has creates a monopoly and is able to charge a price above costs, thus benefiting at the expense of consumers. Monopolistic pricing is typically regarded as antisocial behavior and is illegal in many countries in certain contexts. It is tolerated in Schumpeter's (1934) capitalism, because it is thought to be the "price" society must pay to induce the entrepreneur to gather information and introduce the innovation.

Information about corporate actions like mergers that will lead to future stock price movements can also generate "profits" for its holders, and leads individuals to acquire and trade on this information. This insider trading is typically frowned upon and in many countries is illegal. Jack Hirshleifer (1971) has pointed to a wide class of information gathering activities that merely transfer wealth from those who do not have the information to those who do, and thus are in the aggregate welfare reducing.

Thus, in a wide variety of contexts we see that the possession and use of information that makes the possessor better off than nonpossessors is regarded as antisocial, immoral and often illegal. To avoid the social costs of this sort of behavior, societies have traditionally conditioned their members not to exploit their informational advantages in certain situations. Big brother is spanked if he twists his sister's arm, *and* is spanked even if he is caught only *threatening* to twist her arm if she does not give him her candy. Churches promise those who advantage themselves by lying and cheating, eternal damnation. This conditioning process is designed to induce individuals not to exploit the informational advantages that they have. Habits are conditioned that lead people unconsciously to avoid exploiting their informational advantages. Big brother does not even consider twisting his sister's arm or threatening to do so.

Once we recognize a possible role for operant conditioning in human behavior, questions of morality become more difficult, for both good and bad behavior can be conditioned. Suppose big brother has been beaten regularly by his alcoholic father. Big brother is often mean and sadistic. He gets pleasure out of twisting his sister's arm even if she does not have any candy. Such behavior might be better classified as pathological rather than immoral, and big brother should be placed in the care of a psychologist rather than a priest. Big brother only commits an immoral act if he has not been conditioned to commit violent acts, but threatens to commit one. Only if he is capable of consciously weighing whether to threaten or not, and to twist or not, does it seem legitimate to say that he acts immorally, if he takes advantage of his position. To condemn big brother's actions, we must assume that he possesses the "free will" to choose which action to take.

We have come up against a difficult philosophical and psychological question, and one which is related to the discussion of probability information at the beginning of the paper. The sight of the man with the knife approaching acts a a stimulus that sets off certain reactions, probability calculations, and eventually a behavioral response. What these are depends not only on the characteristics of the approaching man, but also on how our heroine *M* has been conditioned to behave in such situations. Both her perceptions of what the man will do, and her response to those perceptions are influenced by her past experience. A strong form version of behavioral psychology would argue that *all* actions are conditioned responses to stimuli, and thus all are essentially amoral. To speak meaningfully of an action's being immoral, one must posit the existence of sufficient free will for the actor to choose what action to take. Since *some* actions are clearly conditioned responses to stimuli, questions of morality hinge upon the existence of a sphere of behavior over which operant conditioning does not govern, and conscious choices are possible. The size of this domain defines the potential scope of moral philosophy. Interestingly, the psychologist-social engineer confronts the same difficult problem as the moral philosopher. How can one create incentives and institutions that induce people to gather information that reduces uncertainty and thereby improves the welfare of everyone, and yet avoid the gathering and use of information that merely makes its users better off at the expense of others?

NOTES

1. This is Robert Dahl's (1957, p.80) definition of political power.
2. I have previously discussed this concept of power in Mueller (1980).
3. A more systematic illustration is presented in three detailed case studies of French bureaucracies by Michel Crozier (1964).
4. For further discussion of the role of information and uncertainty in producing profit, see Mueller (1976).

REFERENCES

Ronald H. Coase, "The Nature of the Firm", in: *Economica*, 4, 1937, pp.386-405.

M. Crozier, *The Bureaucratic Phenomenon*, Chicago: Chicago University Press 1964.

R.A. Dahl, "The Concept of Power", in: *Behavioral Science*, 2, 1957, pp.201-15, reprinted in Bell, E. and Wagner R.H. (eds.), *Political Power*, New York: Free Press 1969.

Amitai Etzioni, *The Moral Dimension*, New York: Free Press 1988.

John Harsanyi, "Cardinal Welfare, Individualistic Ethics, and Interpersonal Comparisons of Utility", in: *Journal of Political Economy*, 63, 1955, pp.309-21.

Jack Hirshleifer, "The Private and Social Value of Information and the Reward to Incentive Activity", in: *American Economic Review*, 61, 1971, pp.561-74.

F.H. Knight, *Risk, Uncertainty and Profit*, New York: Harper and Row 1965 (1st. ed. 1921).

Herbert Margolis, *Selfishness, Altruism and Rationality*, Cambridge: Cambridge University Press 1982.

Dennis C. Mueller, "Information, Mobility and Profit", in: *Kyklos* 29, Fasc.3, 1976, pp.419-48.

Dennis C. Mueller, "Power and Profit in Hierarchical Organisations", *Statsvetenskapling Tidskrift* (The Swedish Journal of Political Science), N.5, 1980, pp.293-302, reprinted in Manfred J. Holler (ed.), *Power, Voting and Voting Power*. Würzburg: Physica-Verlag 1981, pp.65-77.

Dennis C. Mueller, "Rational Egoism Versus Adaptive Egoism as Fundamental Postulate for a Descriptive Theory of Human Behavior," in: *Public Choice*, 51, 1986, pp.3-23.

Dennis C. Mueller, "On the Foundations of Social Science Research", in: *Analyse & Kritik*, 14, 1992, pp.195-220.

Joseph A. Schumpeter, *The Theory of Economic Development*, New York: Oxford University Press 1934.

Center for Management Studies
University of Vienna
Bruennerstr. 72
A-1210 Vienna
Austria

ARNE NAESS

THE SPIRIT OF THE VIENNA CIRCLE DEVOTED TO QUESTIONS OF *LEBENS- AND WELTAUFFASSUNG*[*]

I. ATTITUDE TOWARDS RESEARCH AND COOPERATION WITHIN THE VIENNA CIRCLE

The history of the Vienna Circle is bound up with what was called the *Wissenschaftliche Weltauffassung*. But with the requirements of the members when it came to deciding whether a sentence expressed scientific knowledge or not, the basic sentences expressing a *Lebens- und Weltauffassung* would scarcely qualify as such, nor would hypotheses about a scientific world view. The *Wissenschaftlichkeit* of *physicalism, logical behaviorism, logical syntax, unity of science*, were hypothetical at best, and in my opinion should not be identified with the total philosophical enterprise of the Vienna Circle. To its strictly speaking *philosophical* enterprise I attribute a certain kind of scientific or research *attitude* and *clarity* as much as any set of philosophical opinion of a substantial sort.

As I see it the members *excelled* in clarity, and in fairness and collaboration in *debates*. These characteristics they, like myself, found most effective and ethically acceptable among ardent "scientific" researchers. And, which was unique in Europe, the admirable way of communication and collaboration was applied to philosophical problems - collaboration by very different personalities with widely different backgrounds.

I talked about the above-mentioned aspect of the Vienna Circle a couple of years ago.[1] Here I shall only repeat the excellent technique of offering different formulations of a view in order to arrive at clarification of disagreements and agreements. Typical question in a debate: "Would you accept that one could express the view you have by the sentence S_2 rather than S_1?" In my work within the realm of what I call total views, comprising the classical questions of *Lebens- und Weltauffassung*, I have been heavily influenced by what I personally experienced in the Schlick Seminar in 1934 and 1935.

The term "total" ought to have a bad reputation in politics. But I am speaking of views, explicit, *but mostly implicit*, *both* normative - how do you think you *ought* to live? *and* descriptive - how are the basic features of the conditions of your life *and* the life in general in what you think is reality?

This kind of sentence was used when somebody thought he disagreed with a speaker, but saw a *possibility* of agreement if the speaker could accept a different formulation, S_2.

In trying to express fragments of total views in systematic form we need to make clear which statements are meant to be declarative and which normative. In order always to make this clear I place an exclamation mark at the end of sentences intended to express a "norm", in the sense of announcements that something ought to be, should be etc., and imperatives of different kinds, *including rules*. A system with at least one sentence with an exclamation mark,

W. Leinfellner and E. Köhler (eds.), Game Theory, Experience, Rationality, 359–367.

I call normative. Thus, a small, central part of my total view is expressed, mainly for didactic reasons, containing only one philosophical sentence with an exclamation mark: the one-word sentence "Self-realization!" From the basic philosophical normative sentence "Self-realization!" *plus* a host of declarative sentences, conveniently called "hypotheses", I derive new norms.

One of the reasons that the Vienna Circle tended to dislike what I call total views is that they associated them with claims to absolute certainty and to dogmatic truths. But to hold a total view is compatible with being a sceptic, at least a sceptic of the Pyrrhonic kind, *maintaining* an 'epoché' in relation to every absolute certainty or dogma.[2] It is normal to retain certain basic attitudes year after year, but to change some opinions practically daily.

Why shouldn't the analytic clarity and the research attitude be applied in what I always have taken seriously, namely most general orientations, and their manifestations in conflicts?

II. THE STATUS OF BASIC VIEWS *ON LIFE AND WORLD UNTIL NOW*

Until this century *Lebens- und Weltauffassung* belonged to the central part of academic philosophy, and certainly to "philosophy" in a more general sense. Diogenes "in the barrel" expressed himself mainly through non-verbal behavior. He did not offer long, difficult arguments in favor of his *Lebens- und Weltauffassung*. This is one of the reasons why he is not accorded as many pages in textbooks of philosophy as some other classical Greek philosophers. He is scorned as an academic philosopher, but nobody – and this is important for the present argumentation – contests that he was a philosopher.

In the great cultures of the East, in India and China, philosophy of life, of society and cosmos, had an important function, and if we look up the names of the main figures, until recently we are reminded of great teachers how to live, and how not to live. Science did not play an important role.

What is the status of *Lebens- und Weltauffassung* today? The question comprises both academic and non-academic philosophy. My answer is not uncommon and not original: the concern in academic philosophy is feeble, *sehr schwach*, but among people and in social debates, rather strong and getting stronger.

III. SOURCES OF PRESENT INCREASING INTEREST

The vital concerns which stimulate life- and worldview reflections today, especially in the *materially* richest countries are widely recognized as important. Here is a short list:

The movement from *Gemeinschaft* to *Gesellschaft*: Ferdinand Tönnies published his famous book *contrasting Gemeinschaft and Gesellschaft*, community and society about 100 years ago. We live in a community when we have a positive feeling of having much together, being much together, doing much together, occasionally helping each other. Now we, in the rich countries, live in a mere society when there is a system of rules binding people together in a more mechanical way: a common forum of decisions affecting competing group interests. I might call the societies pressure democracies, with great power of pressure groups protecting vested interest and a police, rather than a father, to trace limits.

One may question whether we in the earlier 20th century had much *Gemeinschaft*. In Norway we had a class society, but more than 80 per cent lived in rural communities with a high degree of community. The important point is the direction of development seems to be towards less and less community: *Autonomy* of the individuals, even within families. One should not rely on help, one needs no assistance. Rely on psychiatrists and other functions of the state. Autonomy is clearly different from former ideals of individualism.

Another new term is "entrepreneurial culture". Society must be organized in such a way that entrepreneurs get better conditions for the realization of their ideas. What kind of ideas? Any non-criminal designs, it seems, but with emphasis on economic growth and competition.

In close connection with the concern about the disappearance or decrease of community, there is of course concern about mindless, physical violence: violence seemingly lacking motivation. What we conceive of as fellow human beings are not always seen as such, but mere objects. And concern about criminality in general: the USA is leading the way with more than 1 million people in prison, a kind of place generally recognized as not highly conducive to change of lifestyle.

Then we have also concerns of a philosophical kind, about mass media and the global force of advertisements now estimated 'to cost' about a trillion US dollars globally. Mass media is ethically, if not neutral, at least presssing for the least possible non-interference by any norms of an ethical and educational kind, and pressing for increased dependence upon markets.

Many feel that there is a resulting threat of *Gleichschaltung*, a decrease of deepness in cultural differences, an increase of standardization in spite of a multiplicity of tiny subcultures, especially in rich, great cities.

I now go back to the question of advertising. The private producers of goods and services on the market very naturally try to increase their sales and cannot be supposed to distinguish needs from desires, a philosophically important distinction. The great thing is to create what people *feel* to be needs. It is now recognized that growing material so-called needs increase ecological unsustainability. Because I have worked more than 20 years in this area of problems I shall go into that in some detail.

In 1972 the Club of Rome published their important *Limits to Growth*. In 1992, two of them, Meadows and Jørgen Randers, wrote the sequel *Beyond the Limits*. An even more important book showing with a great number of diagrams that curves that should go up actually go down, curves that should go down, go up.

The turn from increasing to decreasing unsustainability is not yet generally recognized to require in the rich countries a formidable change of average lifestyle. The rich countries must try to inhibit the increase of the material standard of living occuring in Southeast Asia, China and other parts of the globe. Even if the rich nations succeed in this, the impact on the ecosystems by the say, 800 million richest people may surely reduce the richness (abundance) and diversity of life on Earth.

There is a philosophical situation which is new this last part of our century. We have to take into consideration five warnings:

(1) Ecological unsustainability is increasing.
(2) People in the rich countries live (on an average) in a way that they cannot seriously wish that others, who would also like to live that way, actually realize *their* wishes.
(3) Decreasing the growth of unsustainability implies decreasing material standards of living in the rich countries.
(4) Most of the cost in work and money necessary to change from increasing the decreasing unsustainability must be furnished by the rich countries.
(5) Any delay increases the cost exponentially.

In 1988 the *Worldwatch Institute* roughly estimated the cost of a change from increasing to decreasing unsustainability: about 17% of the sum of global *military* investment at the time, 149 (American) *billion* dollars a year. Today a rough estimate may have reached the sum of 200 or 250 *billion*. It is a vast sum, but not overwhelming.

There is a concept introduced by Immanuel Kant that deserves close inspection today. It is found in one of his early works and very few seem to be aware of its existence. It is a concept of a *beautiful* action as essentially different from a moral action. An action is moral if it is *only* motivated by the respect for the moral law. If you do exactly what is required by that law, but out of inclination (*Neigung*) then you do not act morally. (This was, incidentally, made fun of by Schiller). But Kant then says that if you act in a way that satisfies the law, but out of inclination, then you act *beautifully*.

Who would not like to act beautifully? From the point of view of ethics, as many of us see the situation today, the way to proceed to overcome the great evils is not through preaching but to find ways to describe the goals and the means in such a way that they are *attractive*. Difficult, but not impossible.

The five warnings are of a character that requires philosophical reflection along predominantly new lines.

The worldview, the *Weltauffassung*, centers on our world, conceived as our planet *Earth*, not cosmos. How do we personally experience our participation in changing the life conditions on the planet, the rapid decrease of biodiversity and animal abundance and habitats, of the increase of the domination of humans over increasing areas of the planet? How do we experience the present social and political incapability to meet the challenge?

What kind of ontology, epistemology and ethics are conducive to strengthening the motivation of large-scale measures to ensure a change from increasing to decreasing unsustainability?

Also from questions essentially having to do with our *Weltauffassung*, we are led to questions of *Lebensanschauung*: in the rich nations how to apply slogans like 'rich life, simple means', that is, means requiring less use of energy, less polluting, less waste etc. In the poor countries, how to increase the material standard of living, but avoiding development in the direction adopted by the rich countries. In short, how to implement the increase 'leapfrogging' the way leading to the rich countries' unsustainability.

Of the many contributions to what sometimes are called ecophilosophy and ecosophy I might mention the concept of "the ecological self". Already Aristotle (in the West) declared humans to be "social animals" or better, "social living beings". Humans are, from one very special point of view, highly mobile parts of the surface and lower atmosphere of our planet. The only parts which somehow perceive all other parts and appreciate them.

An extremely important, more or less spontaneous process is that of *identification* with other forms of life, that is a perception of animals as in some ways like us, that one can do something for their own sake, that they, like us, have needs and interests. Philosophy of life has now a new social and political impact. Those who strongly appreciate life on the planet and strongly identify with animals and plants, whether they are in a narrow sense useful or not, are active to change lifestyles, institutions and politics in the direction of sustainability. Important also is that they look for a gain in quality of life whenever from every step towards responsible ecological policy and are willing at any time to give up much of the typical rich countries' goodies.

Green politics, with a capital G, is only one point of so-called "ecosophies", total views in part inspired by work in favor of overcoming the ecological crisis. One may of course be an activist in green political contexts without referring to, or being motivated by, a philosophy of life or worldview. Those who have, I call "supporters of the deep ecology movement". Deepness is in this context defined in terms of chains of premise/conclusion relations. The supporters are in part motivated by their ultimate premises, their ultimate norms and their descriptive views of the world. In short they involve their *Lebens- und Weltauffassung*.

But why had not conditions led previously in this century to a philosophical awakening? One factor is a focus on language rather than on life, society and the world.

The detrimental focus on language started explosively with Wittgenstein's *Tractatus logico-philosophicus* and logical empiricism. According to Rudolf Carnap, human systematic knowledge was, roughly speaking, of two kinds, the scientific and the syntactic. Questions of philosophy were in part reduced to logical syntax, in part dismissed as cognitively meaningless, but emotionally important. With *Tractatus*, questions of value, of normativity, of ethics and therefore also of politics were shoved into a sphere of mystery. There is no place for philosophical *research*. In the 1940s the anti-research attitude was consolidated with Wittgenstein's *Philosophical Investigations* and the ordinary language movement. Empirical *research* on language was shunned or deemed philosophically unnecessary. Intelligent reflection and intuitions concerning one's own language should be enough to solve, or dissolve traditional philosophical problems. This trend is, if no longer dominant, very much alive at the close of the century. Philosophy professors still talk in terms of "getting the fly out of the bottle", that is, through considerations of language to set people free from warring about what was considered to be great, universal philosophical problems in the West and in the East for more than a couple of thousand years.

Less serious, but not without influence, is a quasiphilosophical trend which tries without much success to undermine belief in value priorities, search for truth, scorns any broad and deep movement to come nearer to a solution of great problems of peace, of unacceptable, desperate poverty and of oppression and of the ecological crisis. Instead of systematic approaches, we should limit ourselves to small narratives and cultural conversations!

So far I have talked about *Lebens- und Weltauffassung* in general terms. Now I shall exemplify what I mean by pointing to Spinoza. This thinker was venerated by members of the Vienna Circle, but the tendency was to consider his problems, in so far as they were real, to belong to psychology and the social sciences. I call my total view, if I can be said to have one – "Spinozistic". I refer to a class of total views comprising variations of interpretations, elaborations and reconstructions of Spinoza's system as formulated (mainly) in his *Ethics*. The term "reconstruction" I introduce in order to characterize interpretations of the text which Spinoza himself in part would presumably reject, but which the text admits.

It is my contention that the way I work shows the influence of the spirit of the Vienna Circle.

One may even say that this spirit is traceable in many ways, and that it provokes negative feelings of the same sort as we experienced in the 1930s: the intrusion of tools such as symbols used in symbolic logic into humanistic studies, taking definition and deduction seriously, and related 'hard' ways of thinking. One of the characteristic traits of my study is the following: I take Spinoza's definitions and definition-like ("definitoid") sentences seriously. If a sentence is a definition in a certain narrow sense it implies that you may substitute the *definiens* expression for the *definendum* expression, and *vice versa* in the relevant text (in my case, the *Ethics* of Spinoza), without changing the intended meaning. The style of the text may be heavily damaged because the *definiens* expression may be complicated, perhaps consisting of 100 words, whereas the *definendum* expression (a sentence or a term) may consist of only one or two. But that is irrelevant. Spinoza uses about a dozen sentences which may imply either complete or not complete substitutability. Thus, if he says that something, x, is *the same as* something else, *y*, he may roughly mean what I would declare by saying that *x* and *y* *denote* the same (but that "*x*" and "*y*" do not necessarily *connote* the same). There would scarcely be complete substitutability. Such substitutability *may* sometimes be complete because he in other places in his texts uses stronger expressions to identify the particular *x* and *y*.

Noting the occurrence of the definitoid sentences, I work with a list of 243 definitoid sentences. Some are slightly reconstructed. They are standardized and said to announce "equivalences" among *centrally important* terms in the *Ethics*.

Through the use of the equivalences I leave the words "Deus" ("God") and "substantia" ("substance") out in my version of parts of the *Ethics*. At the time of Spinoza one could scarcely dispense with those words, but today we can, and *the perfectly immanent God* of Spinoza invites us to eliminate the words which for most people denote transcendental entities. Speaking about entities, I use Occam's razor and the second "definition" (*definitio*) in Part 1 of the *Ethics* to eliminate the word "essence". Incidentally, that "definition", a modern logically conscious reader would tend to say are two definitions connected with "or" (*vel*):

"II. I say that to the essence of anything pertains that, which being given the thing itself is necessarily posited, and, being taken away, the thing is necessarily taken away; or, in other words, that without which the thing can neither be nor be conceived, and which in its turn cannot be nor be conceived without the thing."

What Spinoza calls *definitio* is transformed in this case into two equivalence sentences implying universal substitutability in the text.

Using Spinoza's definitions in this way, the *Ethics* can be 'reconstructed', not violating the text at any point. Spinoza would personally object to some of the reformulations, I presume. It would depend in part upon what stage of development of his life we refer to. In his last years he would perhaps tolerate more of the changes than in his ardent youth.

The intricate web of the about 50 *most* central terms and their equivalences can be surveyed in its fullness through the use of symbolic logic. I use only propositional logic and predicate logic. I do not see how we can conveniently keep more than a thousand relations in mind without such an instrument. But to those who are not acquainted with logic, the symbols tend to arouse negative feelings.

Of the terms which are connected with others through equivalence-relations, I may mention the following: *acquiescentia* (5 relations), *aeternitas* (5), *affectus* (19), *amor* (10), *anima* (8), *bonum* (7), *causa* (21), *conatus* (16), *libertas* (3), *potentia* (32), *ratio* (30), *virtus* (18), *volitio* (3). About 100 terms in the *Ethics* are connected with one or more definition-like sentences. It is often said about certain philosophers that their vision is fundamentally simple and "one". The equivalences testify to that. "To be" in itself is equivalent to "to be able to be conceived through itself", which is equivalent to "freedom", which is equivalent to "to be self-caused", and so on. But the kind and intimacy of the equivalences show great variation.

Two years after the publishing of my book presenting "The Structure of a Central Part of Spinoza's *Ethics*", the 300-year centeniary of Spinoza's death was commemorated in Amsterdam. Here he was laid to rest in the Nieuwe Church. I felt it to be a vindication of the acceptability of the sort of analytical approach to his supreme work that I was invited to deliver the speech in that Church. Even if no other Spinoza scholars have so far taken up work of my kind, I believe in a great future of analytical clarity in life- and worldview philosophy. Announce a lecture on Spinoza's immanent concept of God and you may expect 20 students to come, but 200 may show up. Logical clarity and empirical work have a place *within* the framework of creative metaphysics.

Spinoza's theorems about the favorable function of active emotions on the way to higher degrees of freedom are in principle testable empirical hypotheses and at the same time genuine parts of his metaphysical system. This was not denied by Vienna Circle members. It is a grave misunderstanding that the text of the *Ethics* is somehow deductive, or at least meant to be deductive. Of its five parts, Part 3 and the lengthy Part 4 are full of empirical hypotheses. If the interest in them had been great enough, the work could have started to try to devise tests.

The limited empirical attitude of some of the members of the Circle showed itself in discussing whether Alfred Tarski's work on the concept of truth furnished an *adequate* definition of truth. The conclusion Rudolf Carnap and others reached seemed to be that *logical* analysis showed it to be adequate. My view was that adequacy implied agreement with at least one way the words "true", "false", "wahr", "falsch" etc. *actually* have been used in speech and in texts. Only empirical research could, as I saw it, establish the *limits of the domain* of actual occurrences of the words within which the Tarski definition was adequate. I accept that there is such a domain and that it is an important demain, but that there are several usages, even *within scientific texts*, which are outside the domain. This conclusion rests on the analysis of several hundred occurrences of "true", "false", "wahr", "falsch" and closely related words.

I mention this because empirical research on usages of a term indicates that an extended use of what might be called the Mach-Poincaré-Duhem theorem is warranted. Roughly speaking, it says that given a set of observations there are indefinitely many mutually incompatible hypotheses which can be made to cover those observations satisfactorily. Studying occurrences of verbal utterances indefinitely many mutually incompatible sets of rules may be made to cover these occurrences. One cannot simply "see" which rules the occurrences of the word "true" obey in ordinary speech or in scientific publications. It helps that the word belongs to one's mother tongue, but acquaintance is not the same as knowledge. The partial *adequacy* of Tarski's definition can only be corroborated through *research* which in part is empirical. Neither Tarski nor Carnap found it necessary to stimulate *empirical* research of this kind (as a genuine part of philosophical work).

In my reconstruction of a central part of the *Ethics* I use about 40 predicates, for instance

L(x)	x is free (*liber*), partially or totally, adequately or inadequately
Rat(xy)	x acts rationally in relation to y
NPC(xy)	x belongs to that without which y cannot be conceived
Mel(x)	x is in a state of melancholy

The "Mel(x)" could easily be omitted, but certain considerations of terminological symmetry made it natural to introduce it.

As could be predicted, when people with humanistic or literary background saw my text with its strange symbols, they mostly closed it in fear or with disgust. (This is at least my suspicion). But ever since I was 17 years old I read Spinoza in Latin and acquainted myself with some symbolic logic. My total view, if there is any, clearly belongs to the class of Spinozisms. Consequently in my life a reconstruction, including the symbols, is significant in my practical life.[3]

NOTES

* Arne Naess' contribution was read to the symposium "Wissenschaft als Kultur - Österreichs Beitrag zur Moderne", organized by the *Institute Vienna Circle* that constituted a part of the "Focus on Austria" of the 1995 Frankfurt book fair. The German contributions were published in Friedrich Stadler (ed.), *Wissenschaft als Kultur. Österreichs Beitrag zur Moderne*, Wien–New York: Springer 1997 (vol. 6 of the *Veröffentlichungen des Instituts Wiener Kreis*).

1. "Logical Empiricism and the Uniqueness of the Schlick Seminar: A Personal Experience with Consequences" in: F. Stadler (ed.), *Scientific Philosophy and its Developments*, Dordrecht: Kluwer Academic Publishers 1993.
2. More about this in my *Scepticism*, University of Oslo Press and Routledge & Kegan Paul, 1968 and 1969, but out of print.
3. What is said in this article about total views and Spinozistic views is elaborated in various publications. A sample:
   A. Naess, *Freedom, Emotion and Self-subsistence. The Structure of a Central Part of Spinoza's* Ethics, University of Oslo Press, 1975.
   A. Naess, *Ecology, Community and Lifestyle*, Cambridge University Press, 1989.

University of Oslo
Centre for Development and the Environment
P.O. 1116 Blindern
N-0317 Oslo
Norway

Christoph Hofinger
Oliver Holle
Markus Knell

"Sicher ist sicher"

I. Introduction and Background

As a side event to the conference "Game Theory, Experience, Rationality", the Institute for System Simulation of Socio-economic Processes (SYSIS) organized a computer tournament. It was based on the well-known game-theoretical model "Tragedy of the Commons". This model – in many respects similar to the even more famous "Prisoners' Dilemma" – illustrates the instability of "good" equilibria in games, where each player can gain by deviating from the cooperative strategy.

The term "Tragedy of the Commons" was coined by G. Hardin[1], who used it to describe the behavior of a group of villagers. They have to decide how many cows of their herd they will put on their private pasture and how many they will transfer to a village-owned common grazing ground. Each individual farmer has a strong incentive to increase his stock of cows on the commons, as long as the rate of growth there is higher than the revenue that is possible on the private pasture. Widespread yielding to this incentive may well lead to overcrowding and overutilization of the common resource. A similar dilemma is present in many problems of today, and so the "Tragedy of the Commons" metaphor was successively applied to a wide variety of areas, including: fisheries, irrigation systems, groundwater basins, global warming etc.[2]

Nevertheless there are some limitations to the statement that defection is a dominant strategy under all circumstances. First, the *Folk Theorem* has shown that cooperation can be realized in repeated games by incorporation of mutual punishments (trigger strategies) into the strategies of all players (given that the discount rates are sufficiently low or the continuation probability of the game is sufficiently high). Second, it was argued that in a heterogeneous environment of players (strategies), a "nice" strategy may well outperform a strategy that strictly plays the Nash-equilibrium, since the losses it takes against the latter will be more than counterbalanced by the gains realized in encounters with other cooperative strategies.

The validity of this argument was first shown by R. Axelrod[3] in a series of computer tournaments where each submitted strategy played a repeated Prisoners' Dilemma game against each of the others ("round-robin tournament"). The well-known result was that the simplest strategy, Tit-for-Tat, won the tournament and also succeeded in an "ecological" and an "evolutionary" setting. (Tit-for-Tat inventor Anatol Rapoport gave a speech at the very conference held by Axelrod to discuss the performance of the various strategies submitted. Although he did not submit a specific strategy for the tournament, the spirit of his famous

W. Leinfellner and E. Köhler (eds.), Game Theory, Experience, Rationality, 369–378.

strategy lived in several of the algorithms submitted, one even being a direct relative called "Tit-for-Tat Survivor".)

The computer tournament organized at the conference resembled earlier tournaments held by R. Axelrod, but there were three important differences. First, the "Tragedy of the Commons" model had a broader strategy space than the Prisoners' Dilemma. The players had not only to decide whether to cooperate or defect, but could gradually change their level of cooperation between 0 and 100 (corresponding to the number of cows sent on the commons). The second difference was that the players competed in groups of three and not, as in the classical tournament, in a paired setting with two competitors.

As a third difference, the possibility of a complete destruction of the commons was introduced. This captures the essence of many real world "tragedies" where overexploitation can lead to a total collapse of the commons. This present threat was thought to further strengthen the payoff of cooperation.[4] The aim of the tournament was to find out which strategies can be considered successful under these new conditions and whether they resemble the rules stated by R. Axelrod: "1. Don't be envious. 2. Don't be the first to defect. 3. Reciprocate both cooperation and defection. 4. Don't be too clever." (ibd., p. 110).

II. Details of the Model and the Tournament

The setting chosen for this tournament was the following:
All strategies played against all other strategies including themselves in groups of three. All possible groups were built.

- The number of rounds for each single game was 25, where each possible triplet of strategies played 10 such games of 25 rounds. This was done to reduce the influence of chance (the probability of destruction) on the results.
- Each round started with the same symmetric endowment of 100 cows per player.
- The private interest rate was 20%.
- The commons function was given by the polynomial: Output on Commons = 22.5 Cows on Commons - 0.25 (Cows on Commons)^2. A single cow on the commons thus gave a revenue of 22.25 (i.e. a growth rate of 2125%), whereas all available 300 cows on the commons gave a growth rate of −5300%. The social optimum and the Nash equilibrium in this setting were 40 cows and 60 cows per player, respectively.
- The probability of a complete destruction of the commons (i.e. the game stops before the 25th round is reached) was assumed to rise proportionally with the number of cows on the commons and was given by: p = max(0; (cows on the commons − 100)/500). The players could thus set the probability of a resource destruction to zero by limiting the number of the total cows on the commons to 100, i.e. not putting more than 33 cows out per player.

From the very beginning of the conference, SYSIS had organized an information desk in the lobby of the CIU where interested participants could get detailed written information about the settings and rules of the tournament, including hints and examples about how to submit strategies. In addition, on two computers a *Visual Basic* version of the "Tragedy of the Commons" game was installed, which provided participants with a further possibility to get familiar with the details of the particular model.

From the public presentation of the tournament on Wednesday morning to Friday evening conference participants had the opportunity to submit strategies in a "flow chart" format. The strategies were programmed and tested against each other during a "night session", such that the results of the tournament could be presented on Saturday morning. To create a sufficient incentive to participate some exclusive prizes were promised to the best 3 strategies. Following the metaphor of the model the winner was presented with a cow-like tea-pot, the follower with a cow-like milk-can and the third-placed with a cow-like mug. In the event, 13 strategies were submitted.

III. THE PARTICIPATING STRATEGIES

Unless otherwise indicated, the strategies start with cooperation, i.e. the social optimum of 40. The Basic code of the algorithms is listed in the appendix.

1. **PAVLOV:** In each round, this well-known strategy compares the decisions of the other players with its own moves. If they behaved in the same way as he did, PAVLOV cooperates, otherwise he defects.
2. **STEP BY STEP:** In the first five rounds this strategy plays cooperatively (the social optimum of 40 cows), the next five rounds it raises its entries, "step by step", up to the Nash equilibrium. During the last 15 rounds it raises its entries even further, up to the break-even value of 80.
3. **RANDOM WALK WITH DRIFT:** After playing cooperatively in the first round, this strategy then adds a random term of $+/-10$ to its previous entry. Additionally, each round the entry is raised by one cow. However, the range of the strategy is limited by a lower bound of 20.
4. **MEAN AVERAGE:** This strategy checks whether the two other strategies play cooperatively, in the sense that they stay within the range where the destruction risk is absent. If so, it also cooperates in the sense of playing one third of the riskless value (=33). If not, the strategy places the average of the other two players entries in the previous round.
5. **MEASURE FOR MEASURE JR.:** This strategy compares the total number of cows on the commons of the previous round with the social optimum. If it was less than or equal to this amount, then the strategy sets the cooperative value of 40, otherwise it sets zero.

6. **TIT FOR TAT SURVIVOR:** Just as in Tit for Tat, this strategy checks whether the two other strategies played cooperatively in the last round. The difference from the original Tit-for-Tat strategy lies in the definition of its own entry. Playing cooperatively is defined here as a gradual entry between the risk-free amount (33) and the actual social optimum (40). The later in thc game, the closer this value is to the latter. The same is true for the non-cooperative value, which even increases hyperbolically.
7. **DEFECTOR:** This desperado strategy starts off by playing almost all cows (99). From then on it continues playing Nash, except for the case where one of its opponents is a "clone" of itself.
8. **OPAQUE:** This defines a ratio that compares its own last-round payoff with the ones for the other players. From then on it adds to the cooperative value the difference between the other players' decision and their socially optimal behavior, weighted by the ratio determined above.
9. **LUCKY:** Checks whether the others stay below the social optimum. If so, the strategy fills up the remaining amount up to the total social optimum for the commons. If the others defect, the strategy fills up the remaining amount up to the break-even value (80), or plays the higher value of the other two strategies if it exceeds break-even.
10. **BAD MOVES**: This strategy starts with the best riskless value of 34, from then on it compares the decision of the other two strategies with this value, but increases that benchmark each round by twice the number of rounds played. The same happens for its own entry. Cooperation starts at 33, defection at 50, both are increasing by one in each round.
11. **PAT:** In the second round, a complicated classifier mechanism determines what the strategy will play (40, 60 or 80). For the next 10 rounds the strategy keeps playing cooperatively. After the twelfth round, it checks each round whether more than 160 cows were on the commons. If so, it also defects by playing Nash, if not it remains cooperative.
12. **RISK AVERSE:** This risk averse strategy simply sets the riskless value of 33, irrespective of the other players' moves.
13. **SMOOTHED BEST REPLY:** This strategy starts with playing the mean of Nash and social optimum. In the second round it sets half of the difference of 216 minus the sum of the entries of the two other players. From then on it continues in a similar way, with the difference that it increases its contribution by the number of rounds. It also takes the weighted decisions of the last two rounds into account.

IV. THE RESULTS OF THE TOURNAMENT

All strategies were rather inadequate, which hardly came as a surprise considering the small amount of time available for designing and submitting. What did surprise us though was that the simplest and shortest strategy, *Risk Averse* (submitted by Mark Machina), took the first place in the tournament performed.[5] Obviously this strategy is "nice" in the sense that it never defects first, as it in fact never defects at all. Where in the original Prisoners' Dilemma setup without possibility of destruction, a certain amount of punishment was necessary in order to succeed, in the present environment it seems to be optimal to be forgiving and answer an offense by "offering the other cheek".

Final results		
Rank	Name	Points
1.	Risk Averse	1313343
2.	Mean Average	1248865
3.	Tit for Tat Survivor	1214765
4.	Opaque	1198368
5.	Pavlov	1185500
6.	Measure for Measure	1090051
7.	Pat	1082058
8.	Bad Moves	1046749
9.	Lucky	1032194
10.	Step by Step	945660
11.	Smoothed Best Reply	874162
12.	Defector	863766
13.	Random Walk	823552

Table 1: Final results of the tournament

In fact the introduction of the possible destruction of the common resource does not seem to constitute a minor change, as can be seen by looking at the first three positions of the tournament. All three strategies take the threat of destruction into consideration and are, in one way or another, centering their replies around the "risk-free" value of 33. On the lower end of the ranking one can find strategies that either contain strong random elements (*Random Walk*) or strong tendencies towards defection (*Defector*, *Step by Step*). Again this further confirms similar results of many other tournaments.

NOTES

1. Garrett Hardin, "The Tragedy of the Commons", in: *Nature*, 1968.
2. Cf. e.g. Elinor Ostrom, *Governing the Commons: The Evolution of Institutions for Collective Action*. New York: 1990; Elinor Ostrom, Roy Gardner and James Walker, *Rules, Games and Common-Pool Resources*. Michigan: 1994.
3. Robert Axelrod, *The Evolution of Cooperation*. New York: 1984.
4. At a conference held at the Institute for Advanced Studies, Vienna, in March, 1996, a computer tournament was already performed with a similar setting except the possibility of a complete destruction of the common resource. The winning strategies there were all more or less similar (and often based on) Tit-for-Tat. A mimeo with the results of this tournament is available from the authors.
5. All in all there were 455 single games per tournament. Many thanks to Werner Zahnt of SYSIS who was responsible for the program that made it possible to obtain the results in the shortest time possible.

Institute for Advanced Studies
Stumpergasse 56
A-1060 Wien
Austria

APPENDIX: THE ALGORITHMS OF THE PARTICIPATING STRATEGIES

Legend:

r	Round counter
index	Player counter
Cows (index, r)	Cows owned by player (=100)
allCowsCommons (r)	Total cows on the commons
Choice (index, "b")	Risk free value per player (=33)
Choice (index, "c")	Social optimum per player (=40)
Choice (index, "d")	Nash per player (=60)
Lowerbound	Highest total number of total cows on the commons without destruction risk (=100)
TragedyPoint(r)	Amount of total cows on the commons where yields fall under the yields when no cows are sent to the commons (=240)

```
Function Defector(index As Integer) As Integer
   If r = 1 Then
      Defector = Cows(index, r) - 1
      Exit Function
   End If
   If AllCowsCommons(r - 1) = NofPlayers * Cows(index, r - 1) -
   NofPlayers Then
      Defector = Choice(index, "b")
   Else
      Defector = Choice(index, "d")
   End If
End Function

Function Opaque(index As Integer) As Integer
   Dim a As Single, Summe As Single, i As Integer
   If r = 1 Then
      Opaque = Choice(index, "c")
      Exit Function
   End If
   Summe = 0
   For i = 1 To NofPlayers
      If i <> index Then
        Summe = Summe + Yield(index, r - 1)
      End If
   Next i
   a = Yield(index, r - 1) / Summe
   Summe = 0
   For i = 1 To NofPlayers
      If i <> index Then
        Summe = Summe + Commons(index, r - 1)
      End If
   Next i
```

```
   Opaque = Choice(index, "c") + a * (Summe - (NofPlayers - 1) * Ch
   oice(index, "c"))
End Function

Function Lucky(index As Integer) As Integer
   Dim Summe As Integer, Highest As Single, i As Integer
   If r = 1 Then
      Lucky = Choice(index, "c")
      Exit Function
   End If
   Summe = 0
   For i = 1 To NofPlayers
      If i <> index Then
        Summe = Summe + Commons(index, r - 1)
      End If
   Next i
   If r = NofRounds Then
      If TragedyPoint(r) <= NofPlayers * TragedyPoint(r - 1) - Summe
      Then
        Lucky = TragedyPoint(r)
      Else
        Lucky = NofPlayers * TragedyPoint(r - 1) - Summe
      End If
   Exit Function
   End If
   If Summe <= (NofPlayers - 1) * Choice(index, "c") Then
      Lucky = NofPlayers * Choice(index, "c") - Summe
   Else
      For i = 1 To NofPlayers
        If i <> index Then
          If Commons(i, r - 1) > Highest Then
            Highest = Commons(i, r - 1)
          End If
        End If
      Next i
      If NofPlayers * TragedyPoint(r - 1) - Summe <= Highest Then
        Lucky = NofPlayers * TragedyPoint(r - 1) - Summe
      Else
        Lucky = Highest
      End If
   End If
End Function

Function BadMoves(index) As Integer
   If r = 1 Then
      BadMoves = Choice(index, "b") + 1
      Exit Function
   End If
   If AllCowsCommons(r - 1) - Commons(index, r - 1) >
   CInt((NofPlayers - 1) * Choice(index, "b") + 2 * r) Then
      BadMoves = (Choice(index, "d") + Choice(index, "c")) / 2 + r
   Else
      BadMoves = Choice(index, "b") + r
   End If
End Function

Function RandomWalk(index) As Integer
   Dim Z As Single
   If r = 1 Then
      Commons(index, 0) = Choice(index, "c")
   End If
   Randomize Timer
   Z = CInt((Rnd - 0.5) * 20)
   RandomWalk = Commons(index, r - 1) + r + Z
   If RandomWalk <= LowerBound - (NofPlayers - 1) * Choice(index,
"c")     Then
```

```
      RandomWalk = LowerBound - (NofPlayers - 1) * Choice(index,
      "c")
   ElseIf RandomWalk > Cows(index, r) Then
      RandomWalk = Cows(index, r)
   End If
End Function

Function MeanAverage(index) As Integer
   If r = 1 Then
      MeanAverage = Choice(index, "b")
      Exit Function
   End If
   If AllCowsCommons(r - 1) - Commons(index, r - 1) <=
      CInt((NofPlayers - 1) * Choice(index, "b") + 1) Then
         MeanAverage = Choice(index, "b")
   Else
      MeanAverage = min((AllCowsCommons(r - 1) - Commons(index, r -
        1)) / (NofPlayers - 1), Choice(index, "d"))
   End If
End Function

Function Measure(index) As Integer
   If r = 1 Or r = NofRounds Then
      Measure = Choice(index, "c")
      Exit Function
   End If
   If AllCowsCommons(r - 1) <= NofPlayers * Choice(index, "c") Then
      Measure = Choice(index, "c")
   Else
      Measure = 0
   End If
End Function

Function TfTSurvivor(index) As Integer
   If r = 1 Then
      TfTSurvivor = Choice(index, "b") + 2 + r / (NofRounds /
      (Choice(index, "c") - (Choice(index, "b") + 2)))
      Exit Function
   End If
   If AllCowsCommons(r - 1) - Commons(index, r - 1) <= (NofPlayers
   - 1) * Choice(index, "c") Then
      TfTSurvivor = Choice(index, "b") + 2 + r / (NofRounds /
      (Choice(index, "c") - (Choice(index, "b") + 2)))
   Else
      TfTSurvivor = (Choice(index, "d") + Choice(index, "c") * 3) /
      4 + r / (NofRounds / (Choice(index, "c") - (Choice(index, "b")
      + 2))) + r ^ 2 / NofRounds
   End If
End Function

Function BestReplyS(index) As Integer
   If r - 1 Then
      BestReplyS = (Choice(index, "d") + Choice(index, "c")) / 2
      Exit Function
   End If
   If r = 2 Then
      BestReplyS = (2.16 * Cows(index, r) - (AllCowsCommons(r - 1) -
      Commons(index, r - 1))) / 2
      Exit Function
   End If
   BestReplyS = (2.15 * Cows(index, r) + r - ((AllCowsCommons(r -
   1) - Commons(index, r - 1)) * 2 + (AllCowsCommons(r - 2) - Com-
   mons(index, r - 2))) / 3) / 2
End Function

Function Pat(index) As Integer
   If r = 1 Then
```

```
      Pat = (Choice(index, "d"))
      Exit Function
   End If
   ElseIf r = 2 Then
      If AllCowsCommons(r - 1) > NofPlayers * Choice(index, "c")
      Then
        If AllCowsCommons(r - 1) - Commons(index, r - 1) >
        Choice(index, "c") * (NofPlayers - 1) Then
           Pat = Choice(index, "c")
           Else
             Pat = Choice(index, "d")
             End If
      Else
        Pat = TragedyPoint(r)
      End If
   End If
   ElseIf r < NofRounds / 2 Then
      Pat = Choice(index, "c")
   Else
      If AllCowsCommons(r - 1) > (NofPlayers - 1) * TragedyPoint(r)
      Then
        Pat = Choice(index, "c")
      Else
        Pat = Choice(index, "d")
      End If
   End If
End Function

Function RiskAverse() As Integer
   Riskaverse = 33
End Function

Function StepByStep(index As Integer) As Integer
   Select Case r
      Case Is <= CInt(NofRounds / 5): StepByStep = Choice(index,
      "c")
        Case CInt(NofRounds / 5) + 1 To CInt(NofRounds * 2 / 5):
        StepByStep = Choice(index, "d")
      Case CInt(NofRounds * 2 / 5) + 1 To NofRounds: StepByStep =
      TragedyPoint(r)
End Function

Function Pavlov(index As Integer) As Integer
   If r = 1 Then
      Pavlov = Choice(index, "c")
      Exit Function
   End If
   If Commons(index, r - 1) <= Choice(index, "c") And
   AllCowsCommons(r - 1) - Commons(index, r - 1) <= (NofPlayers -
   1) * Choice(index, "c") Then
      Pavlov = Choice(index, "c")
   ElseIf Commons(index, r - 1) > Choice(index, "c") And
   AllCowsCommons(r - 1) - Commons(index, r - 1) > (NofPlayers - 1)
   * Choice(index, "c") Then
      Pavlov = Choice(index, "c")
   Else
      Pavlov = Choice(index, "d")
   End If
End Function
```

GERHARD BENETKA

SCIENTIFIC EMIGRATION

MITCHELL G. ASH, *Gestalt Psychology in German Culture 1890-1967. Holism and the Quest for Objectivity* (Cambridge Studies in the History of Psychology, vol. 5). New York: Cambridge University Press 1995.

MITCHELL G. ASH/ALFONS SÖLLNER (EDS.), *Forced Migration and Scientific Change. Emigré German-Speaking Scientists and Scholars after 1933*. Washington, D.C.: Cambridge University Press 1996.

CHRISTIAN FLECK, "Autochtone Provinzialisierung. Universität und Wissenschaftspolitik nach dem Ende nationalsozialistischer Herrschaft in Österreich", in: *Österreichische Zeitschrift für Geschichtswissenschaften*, 7, 1, 1996, pp.67-92.

ELSE FRENKEL-BRUNSWIK, *Studien zur autoritären Persönlichkeit*. Edited and introduced by Dietmar Paier (Bibliothek sozialwissenschaftlicher Emigranten, vol. 3). Graz: Nausner & Nausner 1996.

I. CULTURAL EXODUS

Since the eighties a new, young generation of scholars has been studying the emigration of German-language scientists after 1933 and 1934/38, respectively. The material that has been collected since then by German and Austrian as well as American, English and Israeli colleagues reveals the contours of a "more differentiated, in some respects more modest picture" of the "intellectual exodus" in Europe as opposed to the past. Given the fact that researchers no longer only focus on the fate of prominent emigrants or on those emigrants who were particularly successful in exile and have now turned to the life and work of those whose careers were less spectacular, the simplistic equation of "losses here", "gains there" has proven inadequate for understanding the "scientific transfer" imposed by fascism. "To inquire only about losses and gains", as the two editors of the volume *Forced Migration and Scientific Change. Emigré German-Speaking Scientists and Scholars after 1933* write, "presupposes a static view of science and of culture, as though the émigrés brought with them finished bits of knowledge, which they then inserted like building-stones into already established cultural constructs elsewhere." (Ash/Söllner, p.4)

In opposition to such an "internalistic" view which – given its fixation on the products of scientific research – has completely lost sight of the process of their production, an explicitly "socio-historical perspective" has evolved and become accepted over the past two decades. This view deduces its problems mainly from a confrontation of cultural, scientific and discipline-specific conditions of aca-

W. Leinfellner and E. Köhler (eds.), Game Theory, Experience, Rationality, 379–386.

demic socialization in the "countries of origin" with the socio-cultural context of academic training, research and practical professional work in the "host countries". It must be examined in what way scientists who were driven away from certain settings that were decisive for the development of their knowledge and their skills could manage in a *differently* structured environment. Ash and Söllner maintain that "the cultures of the so-called host countries" should not be seen as "fixed entities" to which the emigrants had to become adapted in a one-sided way. On the contrary: taking a socio-historical position implies that these scientific structures are themselves to be seen as dynamic and flexible, as "embedded in cultural settings that are themselves fluid enough to change". (ibid., pp.11-12)

It is easy to see how this approach leads to questions that can only be answered beyond the opposition of "loss" and "gain". Wasn't it only forced emigration that made a glorious career possible for a number of emigrants? What, for instance, would have become of Paul Felix Lazarsfeld if he had not been forced to emigrate in 1934? Or – to stick to pyschology – of someone like Egon Brunswik? Can he who precisely in the months prior to the "Anschluss" was finally able to obtain a fixed position at the University of California even be counted among the group of forced émigrés? Psychology then was no longer a "German" discipline; it had begun to shift its focus to the "New World". Didn't fascism only accelerate what had already long been underway: the internationalization (or "denationalization") of science and thus the concentration of research where there were adequate opportunities and means? Especially the example of Egon Brunswik's "voluntary" emigration shows with the necessary clarity that an inclusion of the "forced emigration" of the thirties and forties in the general context of a science and technology transfer typical for the development of 20th century science over and beyond all cultural boundaries cannot serve to legitimate "historization" tendencies of National Socialism. On June 9, 1938 Brunswik was finally able to marry the woman to whom he had long been engaged. The marriage ceremony took place on the boat that had brought the Jewish woman Else Frenkel to America on her flight from the Nazis. The now Mrs. Frenkel-Brunswik was thus able to benefit from the preferential quota for refugees whose relatives had a permanent position in the United States. (Paier, p.33)

II. EXPELLED AND FORGOTTEN

Recent research on Nazi science had compellingly shown, as Ash and Söllner write in the introduction to the volume they edited, that "science that replaced [the] work [of the émigrés] cannot be dismissed simply as Nazified ideology disguised as science" (Ash/Söllner, p.7). The authors use the wealth of available findings to corroborate one of the basic theses of modern (e-)migration research also from the vantage point of the history of science in Nazi Germany: "It now appears exaggerated to speak in any simple or general sense of an 'exodus of reason' or even modernity as such from Nazi Germany after 1933." (ibid., p.8)

If one goes by the titles of Austrian contributions to emigration research one gets the impression that researchers in this country have trouble giving up expression such as "exiled reason" or "cultural exodus".[1] It could be that this reflects a certain backwardness. Christian Fleck discusses at great lengths and describes the shortcomings of, and gaps in, Austrian research in the text he contributed to *Forced Migration and Scientific Change*. (Fleck in Ash/Söllner, pp.203-207). At the beginning of the nineties (the articles in the volume reviewed here are based on lectures already held in 1991), Fleck could still complain about a "lack of research grants for study visits" (especially in the U.S.A.) as one of the reasons for lacking basic research. Today, by contrast, Austrian social-democratic crisis management is hardly able to provide any adequate funding for emigration research. Thus the necessary basic research will still not be available in the near future.

Yet Austrian research is not as backward as some authors who apparently are only familiar with the situation and the literature by hearsay make it out to be.[2] This adherence to statements which American and German historians of science sought to eliminate with good arguments can also be explained by certain specific features of the Austrian development of science before 1938 and after 1945.

III. THE FAILED RETURN

In July of 1946 Austria's diplomatic representative in Washington D.C. sent the Federal Chancellery a concept titled "Memorandum on the Reconstruction of Austrian Universities". The memorandum had been written by the "Austrian University League of America", an organization that had been founded not even a year before by emigrated Austrian scientists as well as American scientists with a particular interest in Austria for biographical or scientific reasons. It also compiled lists with the names of politically irreproachable, non-emigrant scientists and – in particular – emigrated Austrian scientists who were willing to return to their home country. In the most comprehensive of these lists one finds more than 300 persons named, about two thirds of whom were doctors. In an article that was published last year in the *Österreichische Zeitschrift für Geschichtswissenschaften* Christian Fleck evaluated the material stored in various Austrian archives on the activities of the Austrian University League. In his analysis of Austrian university policy in the post-1945 years, he concludes the following with regard to the social sciences:

> The BMU (Austrian Federal Ministry of Education) responded with complete ignorance to these lists [of the Austrian University League of America] which had made it possible to select on the average for every position not filled in 1945 from among less than four candidates. No contact was sought with none of the candidates listed with whom the BMU was not already in contact with. Austrian universities thus presumedly lost professors of the stature of Ernst Gombrich, Hans Herma, Robert Kann, Felix Kaufmann, Ernst Kris, Lise Meitner, Oscar Morgenstern, Hans Motz, Else Pappenheim, Max Perutz, Karl

Popper, Erich Schiff, Alfred Schütz, Erich Voegelin, Friedrich Waismann, Abraham Wald, Emanuel Winternitz, Katharina Wolf, to name above all those scientists who became famous *later*. (Fleck, 1996, pp.86-87)

That the ministries practiced a politics of obstruction and sometimes even open prevention with regard to the remigration of (social) scientists willing to return to Austria is one of the specificities which must be revealed in connection with the development of the sciences in the 2nd Republic.[3] In the social sciences the number of returnees was, according to Fleck, in relation to Germany, negligible.

According to estimates, about one-third of German social scientists returned to East or West Germany. [...] In the Austrian case, it was mainly the proponents of the specific brand of Catholicism that the Ständestaat had fostered who returned home after 1945: Dobretsberger, Mokre, Wilhelm Koppers, and Robert Heine-Geldern. Members of the left and Jews were not welcome and, as in previous decades, were excluded from the universities. (Fleck in Ash/Söllner, p.209)

However, at the same time it must be taken into account, as Fleck states, that the overwhelming majority of the social scientists who emigrated from Austria did not at all consider returning to their home country (ibid., p.220). Surprisingly many of them relatively quickly succeeded in becoming established in their new institutional and professional setting. Fleck sees the reasons for this successful acculturation of the Austrians as being first the structural conditions that were typical for the scientific culture of Vienna in the inter-war period, i.e., innovative social scientific research only took place outside of or only on the fringes of the university - more or less poorly paid but at least completely independent of traditional academic disciplinary divisions and boundaries. Already then flexibility and interdisciplinarity were something natural for the majority of those working in the social sciences.

However there is another aspect that Fleck sees as relevant in this context: Only an infinitely small minority of the emigrants included in his sample were from old established Viennese families. Most of them had later moved to Vienna, and quite a few of them from the eastern areas of the Danube monarchy. As opposed to their German colleagues in exile, many of the emigrants from Austria and their families, respectively, had already had to work through the experiences of migration and cultural adaptation years before, which made it easier for them to reorient themselves in the "host countries" (ibid., p.220). Whether this original thesis, which is also particularly interesting in a socio-psychological sense, is tenable must be proven by further research.

Independent of this, it can be said that intellectual life in Austria after 1945 (for decades) still stood under the sway of clerical-conservative and in part also German national ideologies from before 1938 as a consequence of the failed return of emigrants and of the failed denazification of the universities. Vienna had become provincial in science. Paul Lazersfeld described the impression that he had of Austria towards the end of the fifties in a letter to the Ford Foundation: "As to the Austrian situation at large, I find it as depressing as before. No brains, no initiative, no collaboration. Someone should make a study to find out how a country can be intellectually so dead, and at the same time have such wonderful music festivals." (cited in Fleck 1996, p.92)

IV. NATIONAL STYLES OF SCIENCE

By thematizing the social context of science it is possible, from a comparative perspective, to unearth national or also local specificities of research practice. Mitchell Ash's well-written book, both rich in ideas and insights, on *Gestalt Psychology in German Culture 1890-1967* is very rewarding reading against this background of interests.

Ash interprets Gestalt theory as an attempt to establish holistic thought *within* the natural sciences, and this in a time in which intellectual life in Germany was characterized by a wide-spread rejection of scientistic concepts. He presents his very detailed study as being indebted to a "contextualist scholarship in the history of science" (p.4). The point of departure for the argumentation he develops in his book is the idea "that scientists, like other professionals, belong simultaneously to multiple sociocultural groupings, each of which defines itself by constructing its own discursive practical universe." For Gestalt theory three such "groupings" were relevant:

> The working group in a single laboratory or the scientific school encompassing one or more such groups; the discipline or the subdiscipline, encompassing parallel or rival groups of scientists competing for position in both epistemic or conceptual and institutional space; and the broader cultures and societies in which disciplines and laboratories in turn locate themselves. (Ash, 1995, p.5)

In his book Ash succeeds is showing how the Gestalt theoretists succeed in developing a specific way of overcoming the "crisis of psychology" (itself being nothing but the discipline-specific reflection of an omnipresent crisis of the sciences since World War I), to expand beyond the narrower disciplinary confines of psychology and to subsequently assert itself against competing movements within and without academic psychology. This was possible because of Gestalt theory's special way of responding to the issues resulting from the imperatives of action and roles simultaneously given in various social contexts.

If one tries to reconstruct the interpenetration of the analytical units elaborated by Ash from below as it were, that is, proceeding from the social setting in which the Gestalt theorists did their research work, this automatically leads to a discussion of the understanding of the experimental method as was typical for the German-language tradition of psychology. (Such a discussion can only be conducted detached from diverging developments in the U.S.A.) Under the chapter headed "Methodological commitments" Ash then gives a precise description of the course of Gestalt psychological experiments:

> Described in this way, Berlin school research style seems like a radical departure from the ideals of repeatability and impersonal objectivity central to modern science. The discrepancy disappears when one realizes that the Gestalt theorists located objectivity not in any impersonal procedure or in the use of any particular apparatus, but in the *phenomena themselves*. Sought were phenomena, or invariant relations among phenomena, that were so impressive that they automatically acquired an aura of objectivity, despite being artifacts created under laboratory conditions and then referred back to nature. The Gestalt theorists searched for invariant principles of order and meaning to inhere in phenomena as experienced under particular stimulus conditions, not correlational or other contingent functional relationships between independent and dependent variables. With this emphasis on inherent order and meaning (Sinn), Gestalt experimentation expressed fundamental opposition to technological conceptions of mind. (Ibid., p.222)

This passage actually describes a "pattern of investigative practice"[4] typical not just of the Gestalt theory school but of German psychology in general. Since Wundt, laboratory experimentation meant bringing forth psychic phenomena which first take place in every (normal, adult) mind in the same way and second normally remain unnoticed in everyday life. The approach which had gradually become dominant in the U.S.A. since the end of World War I could be briefly described as a "systematic variation of conditions for verifying an hypothesis". Up until the fifties, this approach remained just as foreign to many generations of German psychologists as the inference-statistical methods corresponding to this style of research.

In his contribution to the volume edited by him and Söllner, Ash shows to what extent the ambivalent reception of Gestalt theory, in particular, in the United States before and after 1933 can be explained by basic differences in style of research. Moreover, the case studies presented on Wolfgang Köhler, Kurt Lewin and Hedda Bolgar clearly show that the acculturation of emigrated scientists cannot be simply described as a "process of submission to the discursive norms and scientific or professional practices that prevailed in a host culture". On the contrary: Emigrated psychologists – in any case most of those who were able to establish themselves in their host countries adhered, according to Mitchell Ash (in Ash/Söllner, p.138), more or less strongly to working and research styles acquired earlier: "The émigrés mobilized biographical, conceptual, and methodological resources and reconstructed them to do a job, often opposing rather than passively accepting prevailing research and professional styles."

V. Acculturation by Opposition

In the introduction to the volume of selected writings of Else Frenkel-Brunswik edited by him, Dietmar Paier presents a detailed and very precise biographical case study of scientific acculturation. He speaks of a "constructive re-orientation" effected by Frenkel-Brunswik on the basis of existing knowledge and the willingness to quickly study recent trends in American (personality and social) psychology. One could assume that the United States offered the perfect conditions for Charlotte Bühler's former student and assistant to become quickly established. Nowhere else in the German language-speaking countries had the developments in American psychology been followed more attentively than at the Vienna Institute for Psychology where in the twenties and thirties Karl and Charlotte Bühler developed a style of research that, given its organization with a rigid division of labor, was modern even by US standards and in the elitarian and isolated German world of science downright unusual.

Notwithstanding all the interdisciplinarity and all efforts to bring about a synthesis of diverging theoretical trends, and in spite of Karl Bühler's methodic pluralism that was also philosophically founded in his *Krise der Psychologie* (Crisis of Psychology)[5], psychoanalysis remained excluded from teaching and research at the Vienna institute. Even though Freud's theory was officially deemed unscientific by the discipline, some of the closest co-workers underwent analysis in the thirties. Else Frenkel was one of them. As all of the others, she saw her preoccupation with psychoanalysis as a private matter. For her scientific work within the Bühler school - Else Frenkel was Charlotte Bühler's first assistent in the field of psychological biographical research - Freudian theory in any case did not play a role.

The integration of psychoanalysis in Else Frenkel's scientific-psychological thinking only took place in the United States and this in institutional setting not exactly typical for American psychology as a whole. At the University of Berkeley where Else Frenkel-Brunswik had found employment, even if a temporary one, the psychologists

> were either very interested in psychoanalysis, like, for instance, Edward C. Tolman and Egon Brunswik, or, as Daniel J. Levinson and R. Nevitt Sanford, were analyzed themselves or came, in the case of Erik Erikson, from psychoanalysis to begin with. As opposed to Vienna, the connections between psychoanalysis and academic psychology, between analysts in San Francisco and and psychologists in Berkeley were intensive and a minimum of communication structures between both disciplines already existed. (p.36)

Frenkel-Brunswik's preoccupation with Freud was influenced by her efforts to clarify the scientific and theoretical status of psychoanalysis. This discussion was conducted on the ground of Logical Positivism. For the philosophy of the Vienna Circle something similar as for psychoanalysis applies. Even though Frenkel-Brunswik had been in contact with representatives of the "Scientific World View" since long before 1938 a direct influence of logical-positivist thinking on her scientific work can only be proven in exile in the United States. "As can be seen in Else Frenkel-Brunswik's example," Dietmar Paier writes, "the forced new orientation often encouraged innovation. Theories and insights that were of peripheral importance for the scientists before emigrating could move to the foreground in a changed scientific context." (ibid., p.56)

That such syntheses were developed and presented on the outside and in part as critique of the "mainstream" is also true for Else Frenkel-Brunswik. One only needs to read the sharp reply that Hans Jürgen Eysenck had written in response to her contribution to a symposium on the European tradition of personality psychology at the 14th International Congress for Psychology in Montreal 1954.[6]

"Acculturation by opposition" was, for Mitchell Ash, a paradigmatic pattern for the emigration of psychologists. The critical position vis-a-vis the main movements of contemporary American psychology had the effect that innovative contributions of German-origin emigrants did not play any role in the "Americanization" of German psychology at the end of the fifties and the beginning of the sixties. It is thus no coincidence that "in neither in the main library of the university [in Vienna] nor in any of the institute libraries one finds the *Frenkel-Brunswik-Papers*; indeed, there is not even a copy at Austrian libraries; only in the Sigmund-Freud Haus is there a copy but it is not registered anywhere."[7]

NOTES

1. Cf. Friedrich Stadler (ed.), *Vertriebene Vernunft I. Emigration und Exil österreichischer Wissenschaft 1930-1940*. Vienna: Jugend und Volk 1987; Friedrich Stadler (ed.), *Vertriebene Vernunft II. Emigration und Exil österreichischer Wissenschaft*. Vienna: Jugend und Volk 1988; Friedrich Stadler/Peter Weibel (eds.), *Vertreibung der Vernunft/ The Cultural Exodus from Austria*. Vienna: Springer 1995 (2nd revised and enlarged edition; 1st edition Vienna: Löcker 1993).
2. Cf. Edith Kurzweil, *Freud und die Freudianer*. Stuttgart: Verlag Internationale Psychoanalyse 1993. Ms. Kurzweil had the following text passage (note 12, pp. 27-28) printed (obviously completely ignorant of the subject she felt obliged to speak about): "The organizers of the conference 'Die vertriebene Vernunft' invited historians and psychoanalysts to study the Hitler-era but this took place in response to questions related to the election of Kurt Waldheim as state president and not motivated by a general interest in knowledge."
3. Cf. also Christian Fleck, *Rückkehr unerwünscht. Der Weg der österreichischen Sozialforschung ins Exil*. In: Friedrich Stadler 1987, *op.cit.*, pp. 182-213.
4. Cf. Kurt Danziger, "Social Context and Investigative Practice in Early Twentieth-Century Psychology", in: Mitchell G. Ash/William R. Woodward (eds.), *Psychology in Twentieth-Century Thought and Society*, Cambridge, Mass.: Cambridge University Press 1984, pp.13-33.
5. Karl Bühler, *Die Krise der Psychologie*. Jena: G. Fischer, 1927.
6. Else Frenkel-Brunswik, "Perspektiven der psychoanalytischen Theorie", in: Helmut von Bracken/Henry P. David (eds.), *Perspektiven der Persönlichkeitstheorie*. Berne: Huber 1959, pp. 138-152; Hans Jürgen Eysenck, "Charakterologie, Schichtentheorie und Psychoanalyse: eine kritische Betrachtung", in: *ibid.*, pp. 248-256.
7. Kurt Rudolf Fischer, "Egon Brunswik und Else Frenkel-Brunswik", in: Friedrich Stadler (ed.) 1988, *op.cit.*, pp. 306-316. "Frenkel-Brunswik-Papers" refers to: Else Frenkel-Brunswik, *Selected Papers*, edited by Nanette Heiman and Joan Grant. New York: International University Press 1974.

c/o Institut Wiener Kreis
Museumstraße 5/2/19
A-1070 Vienna
Austria

KARL H. MÜLLER

REFLECTIONS ON THE AUSTRIAN KNOWLEDGE BASES

F. STADLER (ED.), *Phänomenologie und logischer Empirismus. Zentenarium Felix Kaufmann*. Wien: Springer 1996.

F. STADLER (ED.), *Wissenschaft als Kultur. Österreichs Beitrag zur Moderne*. Wien: Springer 1996.

It would be relatively easy to produce two small reviews on two volumes, unified by one and the same editor (Friedrich Stadler), but diverse in their topics, the one being devoted to the centenary of Felix Kaufmann, the other to Austria's contribution to modernity. At first sight, it seems impossible to unite both reviews into a single article going beyond a superficial composition of two heterogeneous topics. However, a "common theme" can be established which may serve as the indispensable unifying background. This common background theme can be found in the concept of "knowledge bases" which stands, consequently, in the title of the following review.

I. "SCIENTIFIC KNOWLEDGE BASES"

For the following review it will be helpful to point to some very elementary features of "scientific knowledge bases" in general and "regional knowledge bases" in particular.[1]

Scientific Programs: The most essential elements for science-oriented "knowledge bases" consist in scientific programs which can be defined as *any* type of *encoded* scientific output (article, research report, book, measurement, etc.) which fulfils the core standards for scientific procedures and which meets, moreover, the basic requirements for specific scientific "genres" (like articles, reviews, books, etc.)

Scientific Knowledge Bases: The *total* sum of available and accessible scientific *programs* at a specific point in time may be termed as "science centered knowledge base".

Regional Knowledge Bases: Persons^R or groups^R within a particular well-defined region R (..., city, sub-national unit, nation, group of nations, ...) can be characterized as essential providers and producers of a specific *regional* knowledge base (..., city, sub-national unit, nation, group of nations ...). Moreover, regional knowledge bases consist of *two* types of "building blocks", an explicit set of programs and an implicit domain of competencies and performances.

W. Leinfellner and E. Köhler (eds.), Game Theory, Experience, Rationality, 387–395.

Thus, the "Vienna knowledge base" at time *t* must be considered as the sum total of *explicitly* encoded programs as well as the totality of *implicitly* encoded practices and performances. Finally, it is highly relevant to stress the fact that regional "knowledge bases" can acquire a specific status within the "*global* knowledge bases" with respect to the centrality or the marginality of their production and utilization contexts.

Following the last hint, the preliminary characterizations of "science-oriented regional knowledge bases" are sufficient to differentiate between *three regional* science-centered knowledge bases within the *global* scientific system. Following a regional distinction, put forward *in extenso* by Immanuel Wallerstein for economic network formations and nation building at the *P*-levels (Wallerstein 1974, 1979, 1991), an analogous distinction can be drawn for the *G*-Levels and, thus, for the knowledge bases, too.

Core Scientific Knowledge Bases: In the *first* instance, knowledge production within core regions is highly distribution-oriented, setting the standards for the "state of the art" within specific fields of inquiry *elsewhere*, too. Judged from the point of view of an "intellectual balance of international exchanges", the core knowledge base is *diffusion* driven, exhibiting a *global* diffusion potential but being highly selective, in turn, with respect to knowledge bases and programs elsewehere.

Semiperipheral Scientific Knowledge Bases: For the *second* type, a genuine *mixture* between core features and peripheral features can be recorded, since semiperipheral knowledge bases show areas of high *global* competence with a correspondingly high diffusion potential as well as research fields with predominantly receptory features only.

Peripheral Scientific Knowledge Bases: The *third* type, finally, is mainly *reception* driven, exemplifying a high reception potential but being only marginally reproduced and recombined in other regions. Once again, seen from an "international balance of international exchanges", the peripheral knowledge base is characterized by a *local* diffusion potential only, although it is able, albeit with a certain time lag, to *reproduce* the state of the art-standards set in core or semiperipheral knowledge bases.

These preliminary introductory remarks have established a somewhat unusual, but unifying reference frame for the reviews below. Thus, after a brief description of the basic structure and the content of the volume on Felix Kaufmann as well as on "Science as Culture", the final two sections of this article will be centered around the question whether some prominent features of the Austrian "knowledge bases", past and present, can be identified which are *reflected* in *both* volumes.

II. THE VOLUME ON FELIX KAUFMANN (1895 - 1949)

As this highly recommendable volume makes it abundantly clear, Felix Kaufmann was part and parcel of the Viennese science culture of the First Republic which, above all, was characterized by a rich variety of different, internationally well-established schools like the "Austrian school of economics" (Ludwig von Mises, Friedrich Hayek, Oskar Morgenstern, Fritz Machlup, Joseph A. Schumpeter and others), the "Vienna Circle" (Moritz Schlick, Rudolf Carnap, Philipp Frank, Otto Neurath and others) or the "Austro-Marxists" (Helene and Otto Bauer, Friedrich Adler, Max Adler and others) ... Moreover, Hans Kelsen and his "positivist" approach in the field of philosophy of law must be considered as a vital ingredient, especially in the case of Felix Kaufmann, too.

Felix Kaufmann, however, tried to move not only between these local Viennese traditions, but also had close ties to a highly popular philosophical approach throughout the twenties and thirties in the German-speaking world, namely the phenomenological program which had been developed and refined by Edmund Husserl, Martin Heidegger and, especially important for Vienna, by Alfred Schütz.

In view of this rich intellectual milieu, the volume on Felix Kaufmann presents two introductory chapters which can be considered as highly interesting and illuminative. Hans Georg Zilian offers a short summary of the biography of Kaufmann and his major intellectual contributions, and Bruno Kohlberg gives a lucid interpretation of Kaufmann as the mediator between phenomenology on the one hand and the "neo-positivist" outlook of the Vienna Circle on the other. These two articles alone make it a must for anyone interested in the history of the Vienna Circle or in the intellectual diversity of "fin de république"-Vienna to buy the volume on Felix Kaufmann.

What follows, is a combination of four articles by Liliana Albertazzi, Jesús Padilla-Gálvez, Herbert Hrachovec and Hans-Joachim Dahms, which are devoted to special segments of Kaufmann's works and which are centered, suprisingly, on the mathematical contributions of Kaufmann (Gálvez and Hrachovec) and on two methodological issues, namely on the nature of language and logic (Albertazzi) and on "physicalism" (Dahms).

The final round of articles is *not* linked to Kaufmann anymore, consisting of a contribution on Alfred Schütz concerning the problem of relevance (Elisabeth List), and of a comparison between Wittgenstein and Husserl (Jaakko Hintikka).

While the bibliography of Kaufmann, together with the two overview articles at the beginning, make the entire volume a genuinely interesting one, a critical remark has to be added on the uneven composition of the small volume, since two articles bear *no* relationship whatsoever to Felix Kaufmann and the remaining four articles on Kaufmann are concentrated on special issues relatively far from his *central* interests in scientific methodology, law and economics.

III. "SCIENCE AS CULTURE. AUSTRIA'S CONTRIBUTION TO MODERNITY"

The second volume here under review has a highly ambitious goal, trying to assess and evaluate not the impact of a single person or even a single "school of thought", but of the *totality* of Austrian schools and research traditions in the past as well as the *current* state of the "Austrian mind". In doing so, the volume has been separated into four major areas:

- Philosophy, Philosophy of Science, Mathematics, Constructivism
- Art (Music) and Architecture
- Complex and Dynamical Systems
- "Science Meets Women", "Science Meets the Public"

Moreover, these four sections stand in different contexts, the *first* one predominantly is devoted to the period of "Ludwig Wittgenstein and the Vienna Circle" (Heinz von Foerster, Rudolf Haller, Kurt R. Fischer, Heinz Zemanek), the *second* one studies on the *roots* and the *historic* pre-conditions for the emergence of a highly diversified musical culture in Austria (Herta Blaukopf, Kurt Blaukopf, Rudolf Flotzinger) as well as of different "generative styles" in architecture (Achleitner), the *third* one describes the current status of dynamical systems theory (Feichtinger, Oeser, Pichler, Sigmund/Nowak, Zeilinger), including its "science of science" implications (Nowotny), while the *fourth* one is linked to the under-represented role – both historically and contemporary – of female intellectual traditions and, finally, on the present interactions between science and media or, to be more precise, a *specific* radio program (Ö1) in Austria.

As these articles stand, they make interesting reading for a heterogeneous set of readers with interests in the Austrian philosophical tradition, the current state of dynamical systems theory or the changing roles of "transdisciplinary research". At the same time, the cognitive diversity which results from four groups of *very* different topics, leads to a "Caveat lector", *not* to expect a single volume on Austrian's contribution to modernity, past or present. Such a goal would have required four separate volumes at least and would go well beyond the capacity of a single editor. Consequently, the present volume should and must be viewed as a first and, thus, necessary step for a more systematic and more in-depth analysis of the changing role of Austria's intellectual traditions and research programs within the *global* development of science, culture and the arts.

IV. REFLECTIONS ON THE KNOWLEDGE BASES OF THE FIRST REPUBLIC

It would have been a challenging task to work out a sketch of the Austrian "knowledge base", say in 1933 or 1934, since many world-famous scientists like Kurt Gödel, Paul K. Lazarsfeld, Oskar Morgenstern, Karl R. Popper or Abraham Wald were still in relatively low positions and not particularly well known "outside a small circle of friends". Moreover, universities *outside* Vienna and, to a large degree, *inside* Vienna, too, were *predominantly* on the periphery, relying on *local* traditions only – the case of Othmar Spann and his Austro-fascist type of "universalism" – or on the *non*-innovative *reproduction* of available science programs only. Nevertheless, three essential features can and should be discussed which characterize the Vienna knowledge base at the time when Felix Kaufmann was an active member of the Viennese "scientific community".

First, despite an overwhelming amount of primarily peripheral research, the *Vienna* knowledge base exhibited apparently a "critical mass", sufficient for the emergence and the maintenance of a highly creative output across various disciplines, from mathematics, logic and philosophy to the natural sciences and, finally, to the social sciences. In other words, Vienna in the 1920s or early 1930s was full of small units and discussion groups which had built up a very active recursive cognitive network in which new results have been communicated quickly and innovative responses and reactions have been immediate which, in turn, had led to appropriate reactions in other segments of the network – and so *ad finitum* of the years 1933/34/38 ... Turning more specifically to Felix Kaufmann, one can immediately see the *innovative* responses *resulting* from the recombination of results from the *Vienna* knowledge base in philosophy (Vienna Circle), in philosophy of law (Hans Kelsen) or in mathematics (Hans Hahn, Carl Menger, Felix Waismann, Kurt Gödel) with Husserl's "phenomenology". Moreover, Rudolf Carnap or, to a lesser extent, Otto Neurath *reacted* to the types of "phenomenological criticism" put forward by Felix Kaufmann and held the recursive cognitive network in motion.

Second, a surprising feature of the Vienna knowledge base lies, on the one hand, in the *heterogeneity* of places and locations in which "new knowledge" had been produced and, on the other hand, in the *non*-professional character of the innovative network groups. Since the traditional places, namely the universities in Vienna, formed only a relatively *small* part of the innovation-assembly, other locations like the Chambers, those of Commerce and those for Workers, Museums, notably the "Gesellschafts- und Wirtschaftsmuseum", "Volkshochschulen" (People's Universities) and, above all, private households, convening, among others, scientific circles and discussion groups, figured prominently as substitutes for, or extensions of the universities and research institutions. Moreover, the professional careers of many local innovators were situated *outside* the normal areas of "knowledge producers" or "knowledge providers". Ludwig Mises, Karl Polanyi, Otto Neurath or Karl R. Popper were to be found in positions far away from the occupational profile for "normal science". Once again, Felix Kaufmann exhibited this pattern, too, since his professional career in Vienna consisted of a management post in the Anglo-Iranian Oil Company which, like managerial jobs elsewehere, is *not* likely to be associated with the production of a highly sophisticated scientific oeuvre in the field of methodology, economics or philosophy of law...

Third, the Vienna knowledge base has been constructed in a very unlikely format, viewed from its "encoding dimension". Here, one can observe a distinctive one–many relation in the sense that single contributors to the Vienna knowledge base made, on the average, significant contributions in major scientific fields which are normally considered *widely* separated. Otto Neurath, probably one of the most ingenious members of the Viennese network, produced interesting and substantive "programs" in areas like algebra, ancient economic history, architecture, planning theory, psychology, social science methodology, visual communication. Likewise, Ludwig von Mises was able to participate substantially in fields like organization theory, planning theory, social science methodology *very* broadly conceived, theory of money, etc. And Felix Kaufmann, too, turned out to be a highly diversified "knowledge producer", proliferating a still stimulating cognitive input in areas like logic, mathematics, philosophy of education, philosophy of law or social science methodology (encompassing economics).

In this manner, some basic traits especially of the Vienna knowledge bases during the 1920s and early 1930s could be identified which have found their proper reflection in the writings of Felix Kaufmann.

V. REFLECTIONS ON THE CURRENT KNOWLEDGE BASES IN AUSTRIA

Starting with a fairly recent article from the *Scientific American* (Gibbs 1995), the contemporary science-based knowledge bases in Austria must be considered at the edge of semiperipheral/peripheral, ranking only in 22nd place with a SCI-share in mainstream journals of 0.65, well below comparable countries like Switzerland (1.64), the Netherlands (2.28) or Sweden (1.84) and in close vicinity to areas like countries like Brazil (0.65), Ukraine (0.58) or South Korea (0.55) (Gibbs 1995:76). It is interesting to note that this semiperipheral/peripheral position finds its proper reflection in the two volumes of the present review. In other words, one of the distinctive *merits* of the volume on *Science as Culture* lies in its *adequate* positioning of Austria's scientific landscapes and their cognitive shapes.

First, Friedrich Stadler's edition of *Science as Culture* makes it painfully evident that current knowledge production in Austria does proceed only marginally *via* innovative *local*, though globally embedded research traditions and programs. In sharp contrast to the 1920s and 1930s, neither in philosophy, nor in the social sciences or in the natural sciences for that matter, can one clearly identify asymmetrically embedded "local research traditions". Ironically, Heinz von Foerster, originator and "spiritus rector" of a "constructivist" program in the United States, made his contribution not on the status and the diffusion of the constructivist paradigm, but on the Vienna Circle. Likewise, Paul Watzlawick restricted his article to a comparatively minor reflection on goals and goal attainments far away from his constructivist contributions to communication and interaction patterns.

Second, the current Austrian knowledge bases exhibit, as becomes very clear from Stadler's volume, a typical pattern of semiperipheral/peripheral production, having, on the one hand, a small number of internationally well-recognized and well-established persons within an *international* research agenda like "transdisciplinarity" or "complex, dynamic systems", and replicating, on the other hand, the "state of the art" topics and "best practice" methods which are put forward and diffused *elsewhere*.

Third, the heterogeneity between different regional knowledge bases in Austria has been significantly reduced. Compared to the knowledge bases of the First Republic with their marked distinctions between Vienna and the remaining parts of Austria, the knowledge bases have become by now far more homogeneous, exhibiting mainly a diversity in topics and research fields. Thus, the differences between, say the "Salzburg knowledge base", the "Linz knowledge base" or the "Vienna knowledge base" result simply from a different "mix" in university institutes and research institutes while, at the same time, the *common* pattern of reception and reproduction prevails in all these different locations.

Fourth, the status of one of the few research programs with an asymmetrical international reception during the 1970s and early 1980s, namely the program of "evolutionary epistemology", comprising domains like ethology, cognitive science and philosophy and centered around Konrad Lorenz in Altenberg near Vienna, has entered a new phase, to borrow a term by Imre Lakatos, of "empirically *degenerating* problem-shifts" (Lakatos 1984). As the article by Erhard Oeser *implicitly* demonstrates, the current state of the art-*production* with respect to "consciousness" and "cognitive mechanisms" has clearly moved *away* from the current Viennese contributions.

Fifth, the identification of important *development patterns* in culture and in the arts has returned apparently, despite a long history of promising research strategies and research contributions in Austria, to its infant stages. Reading the articles on music and architecture, one cannot suppress the feeling that problems of the interlinkages of "art and society" have been investigated and analyzed within more theoretically guided frameworks already in previous decades.

Sixth, Stadler's volume does contain many "white spots" since major fields like economics, history, psychoanalysis, art history or psychology, let alone the natural sciences, are largely missing in this volume. Judged on a preliminary assessment, the inclusion of these areas would not alter the emerging overall picture. In these cognitive domains of discourse too, the diagnosed pattern of semiperipheral/peripheral knowledge production is clearly dominating.

Seventh, the volume on Felix Kaufmann reflects another feature of the current Austrian knowledge base, namely its non-innovative and replicative discussion with past research traditions and programs. Very seldom is the rich scientific heritage used in a way to *recombine* current state-of-the-art discussions in the field of methodology of science or within scientific disciplines with heuristics and research strategies of the past. On the contrary, the "interpretative game", far from being restricted to Austria, assumes an isolated discourse of its own, recombining *marginal* elements from current scientific programs with highly selective bits and pieces of the historical materials. While the introductory articles on Felix Kaufmann exhibit a different attitude, the remaining parts of the book reflect the above evaluation to a high degree.

In this manner, an interesting comparison can be established between two very different types of "knowledge bases" within a small region. The challenges and implications, following from the above cognitive features, will be dealt with briefly within the concluding section.

VI. CONCLUSIONS

The above reflections are clearly of high relevance for the shaping of current science and innovation policies within a regional or local basis. As one can easily see from the preceding discussions, one of the most challenging tasks here must be directed toward the establishment of new "centres of cognitive competence" with an *asymmetrical* reception pattern. The "self-organization pattern" of "competence centres" during the First Republic demonstrates both the possibility and the "reachability" of such achievements despite a *very* sub-optimal socio-economic environment as well as the improbability and unlikeliness of such occurrences in the *absence* of an interlinked network of *local* variations to *global* themes and approaches. Moreover, the trajectories, leading from peripheral status to core-status, are situated, as becomes clear from the contributions on complex systems in Stadler's volume on *Science as Culture*, within a multi-peaked fitness landscape with many local maxima, "lock-ins" and the like. Looking at the emergence of *local* "competence centres" at the fin de siécle as well as during the 1920s and early 1930s, it remains to be seen whether a sufficient "critical mass" of "genii loci" will emerge who can act in a similar fashion to two persons in Stadler's two volumes, namely to Felix Kaufmann as a typical representative of the "Vienna knowledge base" of the First Republic or to Heinz von Foerster, who has turned out to be a highly successful provider within the "Urbana knowledge base" especially through the founding of his "Biological Computer Laboratory" in 1957. Taking, finally, the contributions on complex and dynamical systems within Stadler's second volume seriously, even *small* variations in initial conditions might have strong and powerful consequences in the long run...

NOTES

1. In recent years, conceptions of "National Innovation Systems" (Lundvall 1992, Nelson 1993, 1996), or, alternatively, of "knowledge based economies" (Drucker 1993) have become a popular "description device" for a new phase transition which can be characterized in a very elementary manner as "knowledge matters". In the words of Lester Thurow:

   > *Today knowledge and skills now stand alone as the only source of comparative advantage. They have become the key ingredient in the late twentieth century's location of economic activity.* (Thurow 1996:68)

   The subsequent characterizations of "knowledge bases" follow a new research program for "knowledge based" processes in science, economy and society (Müller 1996a,b, 1997a,b,c) which has been developed within the OECD-context of "National Innovation Systems".

REFERENCES

F.I. Dretske, *Knowledge and the Flow of Information*. Cambridge, MA: The MIT Press 1981.

P.F. Drucker, *Die postkapitalistische Gesellschaft*. Düsseldorf: Econ-Verlag 1993.

M. Gibbons, et al., *The New Production of Knowledge. The Dynamics of Science and Research in Contemporary Societies*. London: Sage Publications 1994.

W.W. Gibbs, "Lost Science in the Third World", in: *Scientific American* 8, 1995, pp.76-83.

I. Lakatos, *The Methodology of Scientific Research Programmes. Philosophical Papers*, vol. 1. Cambridge: Cambridge University Press [3]1984.

B.A. Lundvall (ed.), *National Systems of Innovation – Towards a Theory of Innovation and Interactive Learning*. London: Pinter 1992.

M. Minsky, *Mentopolis*. Stuttgart: Klett-Cotta 1990.

K.H. Müller, *The Austrian Innovation System*, 7 vol. Research Report for the OECD. Vienna: IHS 1996a.

K.H. Müller, *Basic Architectures for Knowledge and Information Societies. Theory, History, Methodology, Data, Complex Modeling, Policy*. Habilitationsschrift: University of Koblenz 1996b.

K.H. Müller, *Lebensformen und multiple Risikogruppen. Neue Schichtungstypen für Wissens- und Informationsgesellschaften*. Sociological Series Nr. 14, Vienna: IHS 1997a.

K.H. Müller, *Selbstsichten, "implizites Wissen" und Gesellschaftsbilder. Ein kognitionstheoretischer Streifzug durch soziale Wahrnehmungsfelder*. Sociological Series Nr. 15, Vienna: IHS 1997b.

K.H. Müller, *Markt-Entfaltung und Wissens-Integration. Doppelbewegungen in der Moderne*. Frankfurt: Campus-Verlag (in preparation) 1997c.

R.R. Nelson (ed.), *National Innovation Systems: A Comparative Analysis*. New York: Oxford University Press 1993.

R.R. Nelson, *The Sources of Economic Growth*. Cambridge: Harvard University Press 1996.

H. Nowotny, *The Dynamics of Innovation. On the Multiplicity of the New*. Budapest: Collegium Budapest 1995.

L.C. Thurow, *The Future of Capitalism. How Today's Economic Forces Shape Tomorrow's World*. New York: William Morrow and Company 1996.

I. Wallerstein, *The Modern World System I. Capitalist Agriculture and the Origins of the European World Economy in the Sixteenth Century*. New York: Academic Press 1974.

I. Wallerstein, *The Capitalist World-Economy*. Cambridge: Cambridge University Press 1979.

I. Wallerstein, *Unthinking Social Science. The Limits of Nineteenth-Century Paradigms*. Cambridge: Polity Press 1991.

Institute for Advanced Studies
Stumpergasse 56
A-1060 Wien
Austria

EDITORIAL INFORMATION

With the fifth issue of the *Vienna Circle Institute Yearbook* the review section has grown. Thus it seems appropriate to once again state its objectives. Apart from addressing the topics of the *Yearbook*'s single volumes the review section should provide a regular international forum dedicated to works belonging to the following fields:

- Historical and philosophical studies on Logical Empiricism and related traditions
- Austrian philosophy
- Epistemology, history and philosophy of science
- Foundational problems in the natural and social sciences
- Foundations of logic and mathematics
- Analytic philosophy if related to the above topics
- Austrian and Central European intellectual emigration, cultural transfer

We encourage authors, editors and publishers to send in review copies or notify the editors about forthcoming works. Suggestions for books to be reviewed are also welcome, but they should be confirmed by the editors. Suggestions should arrive by the end of the year so that they can be included in the next yearbook.

Friedrich Stadler
Series Editor

Michael Stöltzner
Review Editor

HANS HAHN, *Gesammelte Abhandlungen/Collected Works,* Band 2/Volume 2. Edited by Leopold Schmetterer and Karl Sigmund. Wien, New York: Springer 1996.

HANS HAHN, *Gesammelte Abhandlungen, Collected Works,* Band 3/Volume 3. Edited by Leopold Schmetterer and Karl Sigmund. Wien, New York: Springer 1997.

The volumes under review contain Hahn's contributions on calculus of variations, real analysis, hydrodynamics, measure theory, Fourier analysis, complex analysis, and philosophy. A short review cannot deal with all of these topics. Hence, unavoidably, the following remarks on Hahn's contributions in some of these fields reflect the reviewer's preferences. It may well be the case that the reader finds other topics in these volumes at least as interesting.

Editing "Collected Works" of outstanding mathematicians of our century recently has become rather fashionable, perhaps because historians of mathematics usually shy away from twentieth century mathematics: even Bourbaki's "Abregé d'histoire des mathématiques" only covers the subject till the end of the last century arguing that it may be too early for a final verdict on the fruitfulness of the more recent lines of research. This may be true, but perhaps a final verdict isn't exactly what we should look for. "Collected Works" do not allow a final verdict, they offer an opportunity to get a glimpse of what has been achieved by our recent precursors. As Kahane, the commentator of Hahn's works on harmonic analysis, pointed out, the past of mathematics may turn out to be a gold mine. Often, promising new tendencies are rooted deeply in the past, and ignoring the past may hamper the future. Hence, "Collected works" are to be welcomed as a sort of substitute for the unwritten history of the discipline. They are useful for drawing the mathematician's attention to the historical dimension of their discipline.

Mathematicians are used to seeing their discipline from a Whiggish perspective. Most often, only the winner's approach is remembered, and more recent results and methods quickly supersede the ones of yesterday. Although the names of famous mathematicians are immortalized as epithets of important theorems, e.g., the Hahn-Banach theorem, little remains known about the work of those who lent their names. As Kahane points out, a typical example of this fact is provided by Hahn's works on Fourier series and related topics. Although his contributions are to be considered as "highly original, deep and far reaching", presently they are not particular well-known among contemporary mathematicians. Kahane sees a reason for this in the fact that the rival research programme of Wiener's based on *convergence in the mean* instead of *ordinary convergence* (as Hahn's) has turned out to be more successful, at least for the time being. But this may not be the last word.

While mathematics may be said to suffer from the neglect of the past, philosophy of mathematics, and, to a lesser degree, history of mathematics may be said to suffer from the complementary neglect of the present. In the case of philosophy this leads to the vice of elementarism according to which the philosophical understanding of mathematics can be achieved on the basis of the most elementary knowledge of arithmetics and geometry. This quite narrow-minded attitude completely misses the plasticity and flexibility of mathematical conceptualizations. Instead, it renders mathematics an ahistorical affair dealing with fixed "Platonic" ideas. Hahn's work on measure theory, for instance, evidences that nothing could be further from the truth. In this field he published extensively for almost twenty years from 1914 to 1933. One of the most important achievements of early measure theory was the creation of a powerful general notion of the integral. In common usage, one speaks of "the" integral as though this notion had a uniquely determined meaning once and for all. This is, of course, wrong. There was a bunch of candidates for the "true" concept of the integral, for example the concepts of Riemann, Stieltjes, Borel, and Lebesgue. Lebesgue's concept has turned out to be the most versatile and fruitful one clearly outscoring the more traditional Riemann integral. As Bauer explicates in his comments on Hahn's contributions to measure theory, this outcome of the competition wasn't that clear in the first quarter of the century when Hahn was working in this area. Several attempts were made to modify Riemann's definition of the integral so that it would accommodate unbounded integrands and encompass conditionally convergent integrals. One of them was Borel's notion which Hahn dealt with in his paper "Über eine Verallgemeinerung der Riemannschen Integraldefinition". As Bauer pointed out, the relation between the Borel and Lebesgue integrals was in dispute even among the protagonists Borel and Lebesgue for some time since they didn't know Hahn's results.

Next I'd like to consider somewhat more closely Hahn's contributions to the field presently known as real analysis. The latter aptly introduced by the comments of Preiss which may be considered as a paragon of a useful introductory comment. Real analysis, at least its more elementary parts, deals with objects such as continuous and differentiable real-valued functions. As the results of Hahn show the relations between these seemingly simple concepts are far more complex than any "intuition" would ever have imagined. For instance, there are continuous nowhere differentiable functions. Physically, these entities may be interpreted as continuously moving particles lacking velocities. Hence, pure mathematics may be said to have pointed at conceptual possibilities realized only much later in the empirical context of quantum mechanics. Or, to take another example which convincingly shows that intuition is a poor guide even in a quite simple-looking situation, let us mention one of Hahn's results concerning one of the fundamental results of the calculus. As everybody knows, continuous functions having the same derivative differ only by an additive constant. This is, as Preiss points out in his comment, actually not a knowable mathematical truth, but rather a problem, namely, to find out under which circumstances it is true, or, equally important, under which circumstances it is false. As is shown by Hahn in his paper "Über den Fundamentalsatz der Integralrechnung" the allegedly well-known truth becomes false if derivatives are allowed to have infinite values.

Hahn's proof is based on an ingenious argument invoking the famous Cantor set. As Preiss reported, till today Hahn's idea is the best approach to this type of problems, i.e., even if Hahn's results are superseded by more modern achievements his methods are still being used.

Summing up I agree with the reviewer of the first volume (in *Vienna Circle Institute Yearbook IV*) according to whom the editors have done a fine job with respect to the edition of Hahn's mathematical papers. This edition can be considered as an extremely useful reference for (historically minded) mathematicians and historians of mathematics alike. In particular I consider the three-tiered approach chosen by the editors as a particularly felicitous choice: introductory comments by expert contemporary mathematicians, Hahn's original contributions, and further bibliographical references for more in-depth studies of the mathematical and historical ramifications. To be sure, some of the comments could have been, for my taste, more ample in order to facilitate access for the non-expert. A minor point of criticism concerns the fact that the table of contents, the bibliography of Hahn's writings, and the way the commentators refer to Hahn's papers could have been arranged in better way thereby avoiding some unnecessary leafing back and forth. This has been solved in a more satisfying way in some other "Collected Works" released by the same publisher. Moreover, Kahane's comments contain too many typographical errors.

Since the *Yearbook* is a philosophical publication, I'd like to spend the rest of this review on Hahn's philosophical papers even if the amount of space dedicated to philosophy may appear somewhat unbalanced if the size of Hahn's philosophical output is compared with that of his mathematical writings. All philosophical papers included in the *Gesammelte Abhandlungen* have already been published elsewhere, either in the German edition of Hahn's philosophical works published in 1988, or in the English edition of 1980. The only exception is the publication of the short paper "Mengentheoretische Geometrie" and Hahn's part of the Hahn-Gmeiner exchange. Hence, for philosophers the philosophical writings of the *Gesammelte Abhandlungen* don't offer anything essentially new. Nevertheless, even for philosophers it might be interesting to get a glimpse of the other - mathematical - aspect of Hahn's scientific personality.

Empirical science, in particular physics, on the one hand and logic on the other, have long been considered as natural allies of the logical empiricist movement. Mathematics is a different case. It is not too obvious how mathematics may be related to the logical empiricist's philosophical world view. Hahn was one of the very few examples of a mathematician *and* a radical logical empiricist. One may well ask what was the specific contribution of Hahn the mathematician to the advancement of the logical empiricist movement. One answer may be provided by looking once again at Hahn's contributions in the mathematical field of real analysis. It may be used as a convenient link for connecting Hahn's mathematical and philosophical thought. Actually, it served Hahn himself as a source of inspiration for a penetrating and original attack on one of the backbones of traditional philosophy, to wit, the Kantian doctrine of the synthetic apriori character of geometry. The attack is put forward in his masterful paper "Die Krise der Anschauung" (1933). As is shown convincingly by the findings of Cantor, Peano and Hahn's own contributions in real analysis, geometric intuition is unreliable. Hence, it cannot claim to lead to apodictically certain knowledge.

As is well-known, the Vienna Circle cultivated a thorough-going anti-Kantian attitude. It was one of the deep-rooted convictions of the Circle that all allegedly synthetic apriori truths belonged to an obsolete metaphysics. As it seems, Hahn played a sort of an expert witness against Kantian philosophy of mathematics and science. As Hahn convincingly shows the thesis of the synthetic apriori character of geometry may not only be challenged by the well-known developments of non-Euclidean geometries and their applications in physics, it may also be attacked by the results of modern mathematics discovered by Cantor, Weierstraß, Peano and Hahn, for instance, the above-mentioned "un-intuitive" relation between the concepts of continuity and differentiability, to take another example, the phenomenon of space-filling curves.

According to Thiel's introductory comments on Hahn's philosophical writings "[t]hanks to their directness of and perspicuity of style, the texts are not in need of any commentary on their purpose or contents." I'm not quite sure whether this is the case. Logical Positivism used to present itself simply as nothing but rigorous good sense, no wiles, nothing more or less than the best of science presented in a format supplied by modern logic. This, however, is deceiving. Logical Empiricism, in particular, the radical version Hahn subscribed to, was not simple common sense. For instance, Hahn heavily relied on logical empiricist notions such as "Physikalismus", "Einheitswissenschaft", which are not common coins among the average philosophical reader. The same holds for Hahn's usage of Carnap's distinction between "inhaltliche vs. formale Redeweise".

As recent research on the history and development of the Vienna Circle has made amply clear, this group of philosophers, scientists and mathematicians was not one of those academic circles interested only in some abstruse epistemic and logical doctrines. Rather, for them the *connection* of philosophical and scientific work on the one hand and political activities on the other was essential for a general scientifically minded world view directed towards a just and enlightened society. This particularly held for the Circle's left corner Hahn belonged to as one of the most prominent members. He was a "homo politicus" and last but not least he played a crucial role in the foundation of the Vienna Circle. According to Karl Menger's testimony Hahn was a convinced socialist and active in the development of the social democratic movement. Regrettably, none of these facets of his personality show up in the *Gesammelte Abhandlungen*. As Friedrich Stadler reports in his recently published *Studien zum Wiener Kreis* Hahn wrote a lot of articles in newspapers and political journals taking side in the fight between social-democratic and liberal forces defending the democracy of the first Austrian republic, and the antisemitic clerical-conservative right attacking academic and political freedom. As it seems, the editors have somehow felt the need for embedding Hahn's work in a wider political and social context. Hence a collection of "Lebenszeugnisse" ("Testimonials") is added. Regrettably, none of them deals with Hahn's political and social activities. As a result, in the *Gesammelte Abhandlungen* the socio-political dimension of Hahn's work is lacking completely. Nothing is mentioned on Hahn's role as a leading member or even a founder of the Vienna Circle. As has been repeatedly reported Hahn played a key role in the struggle for Schlick to obtain the chair in the philosophical institute of the Vienna university. All this must have left some written traces which would have made interesting "Lebenszeugnisse". The *Gesammelte Abhandlungen* do not shed any new light on the politico-philosophical aspects of his personality.

Summing up one may say that the edition of Hahn's *Gesammelte Abhandlungen* is to be unconditionally welcomed as a happy event for mathematicians and historians of mathematics. For the first time, they make accessible the complete mathematical works of this important Austrian mathematician of the first third of the century. For philosophers, the *Gesammelte Abhandlungen* are of limited use. The already existing, much cheaper editions serve their purposes at least as well. Moreover, a certain editorial sloppiness must be noted: Due to inadvertence, the titles of the papers *Überflüssige Wesenheiten* and *Logik, Mathematik und Naturerkennen* are missing. Taking into account the fact that these volumes intend to be the final authoritative edition of Hahn's works, this is unacceptable. The bibliography of the *Gesammelte Abhandlungen* is based, as it seems, on the one that has been compiled shortly after Hahn's death in 1934. At that time, for obvious political reasons it might have been considered inopportune to emphasize Hahn's political writings. But some sixty years later it would not have been inappropriate also to take notice of this aspect of Hahn's work, at least partially. For people interested in the political and social context Hahn lived and worked the *Gesammelte Abhandlungen* are useless. This is a real pity. After all, to borrow a phrase recently used to characterize his comrade-in-arms Otto Neurath, Hans Hahn was a man between science and politics. This fact could and should have been reflected in the edition of his *Collected Works*.

Thomas Mormann

KURT GÖDEL, *Collected Works, volume 3: Unpublished Essays and Lectures*, edited by Solomon Feferman et al. Oxford University Press, 1995.

This is the third volume of a comprehensive and critical edition of the works of Kurt Gödel. The first two volumes comprised all his published papers. The present volume contains unpublished material, notes for lectures and drafts for papers. They are all preceded by detailed and informative introductions regarding their historical context, background and the origin of the particular manuscript.

The volume contains essays and lectures on Mathematics, Mathematical Logic, Philosophical Logic, Axiomatic Set Theory, Physics and Philosophy of Science. Altogether there are 17 manuscripts of which 15 appear here for the first time in print. Two papers are presented here in two different versions.

All these essays and lectures are most carefully written and remarkably rich. They all give considerable insight into Gödel's own achievements in Logic, Set Theory and Physics and also into his philosophical views.

In the sequel I shall give a brief glimpse to the various papers in the present volume.

Completeness and Incompleteness

No. 1 is Gödel's Königsberg lecture 1930 on his completeness theorem for first-order predicate logic (in the present manuscript Gödel only considers pure quantification theory without equality). It is a clear and enlightening exposition of the main ideas of the proof. He also mentions his 'recent' incompleteness theorem and concludes by a kind of Vaughtian test that second-order logic has no complete axiomatization. In this lecture Gödel seems to be rather cautious in saying anything about the philosophical significance ot these results.

No. 2 is a lecture, presumably his lecture at the DMV-Meeting in Bad Elster 1931 on undecidable statements in Arithmetic. Again, the mathematical side of his incompleteness theorem is clearly set out. A final remark (coronidis loco) concerns consistency proofs: "The consistency of a formal system (of the kind characterized above) can never be established by methods of proof more stringent than (or the same as) those formalized in the system in question; rather, for that one always needs some methods of proof that transcend the system".

No. 7 is Gödel's lecture on "Undecidable diophantine propositions", where various thoughts centered around Hilbert's tenth problem. The lecture was addressed to an audience of non-specialists. All relevant ideas had been carefully introduced. It must have been pure pleasure to listen to such a beautifully presented lecture!

Philosophy of Mathematics and Mathematical Logic

No. 3 is Gödel's invited lecture at the 1930 AMS-meeting in Cambridge/Mass on *"The present situation in the foundations of Mathematics"*,

No. 5 is Gödel's *"Vortrag bei Zilsel"*, 1938

No. 9 is Gödel's lecture at Yale, 1941, *"In what sense is intuitionistic logic constructive?"*,
No. 12 is Gödel's *Gibbs lecture* 1951,
No. 13 is an essay on the philosophy of Rudolf Carnap: *"Is mathematics syntax of language?"* in two different versions, and
No. 14 is a draft on *'the modern development of the foundations of mathematics in the light of philosophy'*.

A central point in Gödel's philosophy of mathematics is his conviction that mathematics has contents ("Inhalte") which precede any formalization and moreover that these contents cannot be fully captured by formalization. The sources for these convictions are on the one hand Gödel's platonistic standpoint and on the other hand his incompleteness theorems. In his 'Vortrag bei Zilsel' (No. 5) Gödel argues against Hilbert who viewed mathematics as a 'voraussetzungslose Wissenschaft' (presuppositionless science). In his Gibbs lecture (No. 12) Gödel seems to argue against statements like Dedekind's "Wir sind göttlichen Geschlechts und besitzen ohne jeden Zweifel schöpferische Kraft, nicht blos in materiellen Dingen, sondern ganz besonders in geistigen Dingen" ("We are of god-like nature and possess, without any doubt, creative power, not only in material things but especially in mental things."; cf. Dedekind's *Werke*, vol.3 (1932), p.489). Gödel, however, claims that 'mathematics is not our own creation' and he adduces a number of considerations favoring his standpoint called realism or Platonism. He thinks that sets and concepts 'form an objective reality of their own, which we cannot create or change, but only perceive and describe'.
The position Gödel takes in his early paper No. 3 comes close to Moritz Pasch ("Neuere Geometrie", Leipzig, 1882). Here logic only provides a certain standard of rigor while on the other hand all axioms have to be justified by empirical facts. The principle aim, however, is – as in his later Gibbs lecture – the incompletability of mathematics and to describe possible ways to enrich arithmetic in a convincing way. It is argued that assuming the existence of larger and larger infinite cardinals would be a possible way towards a stepwise completion of arithmetic.
No. 15 is the sketch of Gödel's *"Ontological proof"* (1941-1970) concerning the existence of an "ens perfectissimum". Gödel does not analyze Anselm's argument, who argued that from the notion of a *supreme being* (id quo nihil maius cogitari possit) it follows that its existence in the intellect implies its actual existence (in intellectu et in re). A variant of this argument was given by Descartes who replaced the 'supreme being' by a 'perfect being' (ens perfectissimum). Kant, however, found Descartes' argument not conclusive. (Kant's argument in modern model-theoretic terminology is the following: the existence of an object g is not proved by showing that the predicate $\exists x$: $x = g$ belongs to the complete type of g, but by showing that the type is 'realized'). However, what Gödel does in his note, is a formalization of the properties of an

ens perfectissimum in modal logic and then proving from certain axioms that the possible existence of a perfect being implies its necessary existence.

Manuscripts on Set theory

No. 6 and *No. 8* are lectures (Göttingen 1939, and Brown University 1940) on the relative consistency of the generalized continuum hypothesis GCH with the axioms of set theory, ZSF (Zermelo-Skolem-Fraenkel). In both papers he describes the hierarchy of constructible sets L_α ($\alpha \in \Omega$) and explains in detail why L is an inner model of set theory including the axiom of choice and the GCH. The letter "L" stands for "law". In his Göttingen lecture Gödel acknowledges his dependence on a few ideas of Poincaré and Russell (the ramified type theory). It is remarkable, however, that a crude form of the hierarchy of constructible sets already appears in Poincaré's "Sechs Vorträge ... gehalten zu Göttingen 1909" (Teubner Verlag, Leipzig 1910, pp.46-47). The essential difference is that Gödel had to use a so-called "gödelized" language while Poincaré used the ordinary meta-language.
No. 16 and *No. 17* are drafts of papers on the size of the continuum. The first paper attempted to prove from some axioms about scales of functions that the size of the continuum is $\aleph_2$. In the second paper from 1970 Gödel discussed some "highly plausible axioms about orders of growth" and indicated that this axiom "gives much more likelihood to the truth of Cantor's continuum hypothesis than any counterargument set up to now gave to its falsehood". The arguments in both papers are incomplete. But still, both papers are of some value since they are a stimulus for further research in the area.

Philosophy of space and time

No. 10 is Gödel's draft on "the relationship between theory of relativity and Kantian philosophy" (in two different versions) and *No. 11* is Gödel's 1949 lecture "On rotationg universes".

We do not know if Gödel would have agreed in publishing this selection of "finished and unfinished papers". We hesitate to give an answer. Gödel was an extreme perfectionist and did not want to publish work which was not in final form. However many of the lectures and drafts of this volume were considered by Gödel as "was ich publizieren könnte", and we should say that we are most happy that they are published now. Actually, this volume was a desideratum for a long time.

We also hope very strongly that volume 3 is not the last volume and that other volumes (comprising his letters and notes?) will appear soon.

Ulrich Felgner

JOHN W. DAWSON, JR. *Logical Dilemmas. The Life and Work of Kurt Gödel.* Wellesley, Mass.: A.K. Peters 1997.

John W. Dawson, Jr. is probably the person best acquainted with Kurt Gödel's *Nachlaß*; he has undertaken the cataloguing of the papers ("a scholarly midden", as he puts it, of some sixty boxes) and is co-editor of the *Collected Works* of Kurt Gödel.

The book which grew out of those experiences reflects his expertise, and gives a diligent account of many aspects of Gödel's life. Yet the reader should be aware that by far the most exciting of those aspects is the works and manuscripts of Gödel, while from the outside his life was short of adventurous incidents of broader interest. If the noblest aim of the biographer is to stimulate the readers' interest to study the original works by Gödel, then Dawson has succeeded as an author.

The book contains fourteen chapters, three appendices, and 16 figures. Chapters I and II are devoted to Gödel's ancestry, his childhood and student years. The comparatively few facts known about Gödel's early years are compensated for by descriptions borrowed from cultural studies by Zweig and Johnston. Written for a predominately American audience, Austrian history is sketched in some detail, while literary treatments of this kind are missing for Gödel's Princeton years.

Chapter III (*Excursus. A Capsule History of the Development of Logic to 1928*) is a crash course in the history of logic to motivate Gödel's results for readers without a background in logic.

The main part of the book is formed by Chapters IV to XII. Gödel's life and work from 1929 to his death in 1978 are documented in mostly chronological order. Based mainly on the *Nachlaß*, correspondence, Karl Menger's recollections, and Oskar Morgenstern's diaries, Gödel's personal and intellectual biography is developed without gaps.

Chapter XIII (*Aftermath*) summarises the events till Adele Gödel's death in 1981, and the story of Gödel's *Nachlaß*. Finally, Chapter XIV (*Reflections on Kurt Gödel's Life and Legacy*) concludes the text with a cautious attempt to sketch the features of Gödel's personality and epistemological convictions which made him such an outstanding figure in the field. Still, the causality implied by Dawson between Gödel's introspective rationality and his paranoia remains questionable. Illness is not necessarily an outcome of intellectual endeavour – a logical dilemma, so to say.

Appendix A contains a chronology listing the main dates in Gödel's life. Appendix B comprises the genealogies of the Gödel family, with the paternal line dating back to 1799, the birth of Carl Gödel. In Appendix C, Dawson included concise biographical sketches of famous late mathematicians who surrounded or influenced Gödel.

The book is directed to a mathematically mature audience of US American background. Unlike Hodges' *Enigma*, *Logical Dilemmas* is not a private book, yet this can be ascribed partially to Gödel's reclusive personality. However, other people, too, like Gödel's close friend Morgenstern or his brother Rudolf remain quite remote figures. The only notable exception is Gödel's wife Adele who is described in quite some detail. Despite mostly unfavourable recollections of Adele, Dawson stresses her utmost important role as a friend and protectress.

Gödel's research over fifty years is followed very closely. The ideas in his major papers and their reception in the logic community are documented and motivated in great detail, and attempts are made to convey Gödel's philosophical thoughts to the reader.

I recommend also reading the *Collected Works* if one wants to fully appreciate the biography. In fact, Dawson states that he designed the book as a complement to the *Collected Works*, and very often the reader is tempted to consult them.

Another important merit of the book is that it sheds light on scientific myths connected with Gödel, like the history of Gödel's alleged independence proof for the Continuum Hypothesis, Gödel's results in cosmology, and his letter to von Neumann in which he brings up the $P \neq NP$ problem.

A good deal of the book is devoted to Gödel's health condition. Having been subject to so much speculation, this topic certainly has to be addressed by the biographer. Anecdotal and voyeuristic episodes are inevitable here and have been exposed tactfully. Many facts of the Gödels' everyday life and their bureaucratic affairs have been collected, and contribute to a detailed picture.

Several working logicians of the younger generation are hardly mentioned in the text, although they did have intense scientific contact and discussions with Kurt Gödel while staying at the Institute for Advanced Study. One of them is not mentioned at all, another one quite infrequently, a third one whose eccentricity compares to Gödel's only, is mentioned just for his technical merits. I refer the reader to the recollections in Odifreddi's *Kreiseliana*[1] for details.

In Chapter XI, Dawson reports on how fastidious Gödel was with respect to van Heijenoort's translations of his papers, for "Gödel was concerned that van Heijenoort was not a native speaker of English ...". Ironically, the translations in *Logical Dilemmas* appear to be quite hasty in the few cases where the original text is cited. Grammatical mistakes in the original German texts are met frequently, the wrong endings and articles indicating that Gödel's shorthand is hard to read. Where the original texts of Oskar Morgenstern's diaries are included, the translation often misses the point. For example, the translation of Morgenstern's diary entry about Adele Gödel ("Wiener Wäschermädel") misses the ironic, yet affectionate, Nestroy-like tone of the original. Similarly, Gödel's grumbling about the quality of coffee in Nazi-occupied Vienna could hardly "take aback" anybody by Central European standards, but rather reveals a certain *façon de parler*.

Gödel surrounded himself almost exclusively with European emigrants – a fact not commented upon in the book.

One more word about linguistic questions: Dawson does not give evidence to support his severe statement that the (German) *Dialectica* paper "betrayed some loss of facility in his native language."

Unlike most logic books, biographies are bought, too, for the pictures. Dawson reuses only a few photographs known from the *Collected Works*, and thus the selection encountered may appear poor to the naive reader. There are few portraits it included, yet the photograph used twice – on the back cover and on page ii – shows Gödel at Amherst in 1967, when he was obviously in very bad physical condition. The only facsimile handwriting to be found among the illustrations is a page of Gödel's primary school arithmetic workbook.

In conclusion, *Logical Dilemmas* comprises a wealth of well-documented facts about Gödel's life. For the researcher, as well as for the generally interested mathematician or philosopher, it provides a readable introduction to the *Collected Works* and to the *Nachlaß* in general. No future biographical work about Gödel can circumvent Dawson's authoritative book.

NOTES

1. Piergiorgio Odifreddi (ed.), *Kreiseliana: about and around Georg Kreisel*, Wellesley: A.K. Peters 1996.

Helmut Veith

BRUNO DE FINETTI, *Filosofia della probabilità*, ed. by Alberto Mura, Milano: Il Saggiatore, 1995.

In 1979 Bruno de Finetti gave a course at the "National Institute for Advanced Mathematics" in Rome. Alberto Mura took part in that course and agreed with de Finetti to record it on tape. What appears in this volume is Mura's transcription of the 19 course sessions, taking the form of a seminar. The result is a very interesting book, containing many insightful remarks, which throw new light on the thought of the most outstanding representative of the subjectivist approach to probability.

In his preface Mura cites the text of a short note written by de Finetti to announce the course. It says: "The course, bearing a deliberately vague title [*On probability*], will deal with conceptual questions and controversies concerning probability: questions that it is necessary to settle in some way or other, in order to avoid the process by which reasonings are made being reduced either to a mere formalistic play using mathematical expressions, or to empty and oversimplified pseudophilosophical or allegedly practical statements". In his usual colourful and polemic style, de Finetti emphasizes from the outset the crucial role ascribed to foundational questions in dealing with probability. The objects of his polemics are specified in the introductory lecture, where de Finetti declares his aversion to the axiomatic approach to probability, on the one hand, and to the classical and frequentist interpretations of probability, on the other.

As is well known, de Finetti is, together with F.P. Ramsey, the founder of modern subjective probability. Working independently at about the same time, both suggested a view of probability taken as degree of belief, to be conveniently (though not necessarily) measured by means of fair betting quotients obeying the requirement of coherence. To the definition of probability as degree of belief, de Finetti was able to add the notion of exchangeability. Combined with Bayes' rule, exchangeability gives rise to the inferential method which is at the root of neo-Bayesianism.

This view of probability, that de Finetti claims to have developed between 1926 and 1928 (de Finetti 1976, p.283), was largely inspired by a genuinely empiricist philosophical attitude of the kind that permeated logical positivism. As a matter of fact, among the thinkers who influenced de Finetti's philosophy of probability, we find, by his own admission (See de Finetti 1931), E. Mach and H. Poincaré, whose names also appear in *Wissenschaftliche Weltauffassung: Der Wiener Kreis* among those of the inspirers of that scientific world conception which is the kernel of logical positivism. From Mach and Poincaré de Finetti derived two major ingredients of his "probabilism", namely sensationalism and relativism. These features, combined with pragmatism and operationalism, in de Finetti's hands give rise to an original philosophy of probability which is opposed not only to metaphysics, but to any form of realism (See my 1989). De Finetti remained faithful to this philosophy until his death in 1985.

Let me highlight de Finetti's position with the help of some passages taken from the book under consideration. First of all, according to de Finetti the only meaning to be ascribed to probability is subjective. Probability is the degree of belief entertained by someone relative to the occurrence of an event, on the basis of the information available to him. "It is meaningless to speak of the probability of an event, if not relative to a person's knowledge" (64). Probability is always relative to the "state of information" of those who make probability judgments, where the "state of information" includes not only the whole of past experiences, but also all sorts of personal attitudes and context dependent elements. Under the influence of Mach's positivism, de Finetti takes probability as an expression of an individual's feelings. As a consequence of this position, for de Finetti it is perfectly possible that two people, on the basis of the same information, make different probability evaluations. An interesting remark in this respect is contained in a footnote in chapter 2 where, commenting on the possibility envisaged above he says: "This claim marks the main difference between subjectivism and logicism: according to the latter, probability values, taken as rational degrees of credence, depend solely on the event under consideration and the state of information H_0. It follows that, their state of information being equal, two rational people should make the same probability evaluations" (89).

Subjective probability needs an operational definition to gain both applicability and admissibility. To this end, in the present volume de Finetti adopts the method of proper scoring rules, which avoids some of the problems pertaining to the method of bets. For de Finetti the operational definition of probability as degree of belief is at the same time the best possible foundation for the notion of subjective probability and the only one needed. Inductive inference, on the other hand, is based on Bayes' theorem, which for de Finetti represents the only acceptable way of addressing probabilistic inference.

At the beginning of chapter 5 we read that de Finetti came to the conviction that probability is subjective when he was a student and happened to read Czuber's book *Wahrscheinlichkeitsrechnung*, containing a short survey of the various notions of probability. On that occasion, he was convinced that there cannot be any "objective" notion of probability. "One should not mythicize probability and think that probability values exist in some sort of empyrean, like the Platonic world of ideas" (87).

Consequently, he regards the classical and frequentist notions of probability as equally untenable. The classical interpretation, based on the notion of "equally probable events" is actually rooted in a misunderstanding, because "both equiprobability and independence are subjective" (87). Things are even worse with frequentism, which according to de Finetti originates from the contradiction of *assuming* a link between probability and frequency, that should instead be taken as a matter for demonstration. "From the frequentist standpoint – he states – frequency converges to probability by definition. But unfortunately there is no guarantee for it. Frequentism originates from the mystification which pretends that probabilistic laws are certain" (132).

De Finetti sharply criticizes the basic notions of frequentism, such as those characterizing von Mises' sequences, namely randomness and infiniteness. So, "the idea of an infinite sequence of experiments" is seen as simply "meaningless" (130) for "even supposing that the universe will last billions of times longer than one thinks it will last, it cannot but have a finite duration" (124). Similarly, "the notion of an irregular sequence is a '*pseudonotion*'. And pseudonotions are always a great obstacle in dealing with whatever issue. One should avoid such conceptual outgrowths, originating from superstitions like that according to which the long awaited numbers of a lottery are more likely to be drawn, and other suchlike cabalistic beliefs" (134). The idea that frequentism is based on superstition inspires the title of Chapter 7: "Superstition and frequentism".

De Finetti also opposes Reichenbach's idea that updating probability judgments can be seen as a self-correcting procedure. For him updating what one thinks in view of new evidence does not mean changing opinion: "If we reason according to Bayes' theorem we do *not* change opinion. We keep the same opinion and we update it to the new situation."

Having criticized the frequentist pretence to *define* probability in terms of frequency, de Finetti clarifies the role played by frequencies in the *evaluation* of probabilities. Their role is reflected by the representation theorem, which shows precisely how information about frequencies interacts with subjective elements in making probability assignments. At the same time, the representation theorem performs the reduction of objective to subjective probability, since the probability values it leads to are subjective, and so is the assumption of exchangeability on which it rests. In the light of the representation theorem, objective probability becomes a useless notion, of which there is no need in inductive inference.

As a consequence of his uncompromising subjectivism, de Finetti never saw any need to tackle notions like "probability in physics". Probability for him is just the expression of the feelings of the subject who evaluates it, in science as in everyday life. The drawback of such an intransigent position is that subjective probability has never been taken seriously by physicists. The volume under consideration contains some interesting remarks on this point. Referring to the probability distributions belonging to statistical mechanics, he claims that "they provide more solid grounds for subjective opinions" (117). This suggests that for de Finetti probability evaluations in science are largely determined by scientific theories. However, it has to be kept in mind that such evaluations depend on "our way of theorizing phenomena, which is always open to revision" (118). As to the interpretation of theories, de Finetti takes a pragmatist and even instrumentalist attitude: "If science is to be taken at all seriously, it is always to be taken as an instrument. What would it mean otherwise? It would amount to building castles in the air" (119). This position opens the door to a less intransigent form of subjectivism where one can make sense of some notion of "objective chance" in tune with a non-realistic approach to scientific theories (See my 1996).

REFERENCES

B. de Finetti, 1931, "Probabilismo", in: *Logos*, pp. 163-219, English translation: "Probabilism", in: *Erkenntnis*, 31, 1989, pp.169-223.

B. de Finetti, "Probability: Beware of Falsifications!", in: *Scientia*, 70, 1976, pp. 283-303.

M.C. Galavotti, "Anti-realism in the Philosophy of Probability: Bruno de Finetti's Subjectivism", in: *Erkenntnis*, 31, 1989, pp.239-261.

M.C. Galavotti, "Probabilism and Beyond", in: *Erkenntnis*, 45, 1996, pp.253-265.

Maria Carla Galavotti

FRIEDRICH STADLER, *Studien zum Wiener Kreis. Ursprung, Entwicklung und Wirkung des Logischen Empirismus im Kontext*, Frankfurt/M.: Suhrkamp 1997.

Friedrich Stadler's study marks a significant step ahead in our understanding of the history of the Vienna Circle. Persuasively written and admirably documented the book gives a comprehensive account of the evolution and eventual dissolution of the Circle. It is Stadler's careful and detailed contextualization of that history which distinguishes his study most clearly from much preceding work on the Vienna Circle. He writes in this connection, for instance, that only "a pragmatically situating interpretation of logical empiricism" can fully account for the contested antimetaphysics of the Vienna Circle. "For antimetaphysics was, indeed, nothing new in the history of philosophy. Verification as a criterion of meaning provided the scientifically and ideologically motivated members of the Vienna Circle, however, with a suitable tool... against contemporary forms of irrationalism and universalism (from Heidegger to Spann)." (p.29f.)

By delineating in this way the historical context of the Vienna Circle, by tracing its origins in the rise of a scientific philosophizing that began in Austria with Bolzano and extended through Brentano and Mach, by exploring the antecedents of the Vienna Circle in the first two decades of this century, by showing, in particular, how the Vienna Circle interacted with a number of other intellectual groups in Vienna (such as the mathematical colloquium conducted by Karl Menger and the circles around Heinrich Gomperz and Ludwig von Mises), how it was engaged in a continuous struggle with the largely conservative and reactionary academic environment, and how it concerned itself with popular education (*Volksbildung)*, Stadler succeeds in showing that the Vienna Circle was indeed a central element in the culture of its time. One certainly takes from his book a deepened grasp of the intellectual, social, and political environment in which the Vienna Circle operated. But Stadler's account also makes clear why the Circle eventually dissolved. For the constant battering by ideological and political enemies (which accused the Circle of being destructive, liberal, anti-catholic, Jewish, non-Austrian, etc.) eventually made it impossible for its members to survive at the University and in Austria as a whole. Even before the "Anschluss" in 1938 the hostile climate had forced some members of the Circle to emigrate. The result, ratified, by Hitler's march to Vienna was the complete dispersion but, at the same time, also the internationalization of the Vienna Circle.

One of the most notable features of Stadler's work is his ability to describe with utter clarity and economy a large number of related but still distinctly different individuals, movements, and conceptions. His account of the contributions made to the development of logical empiricism by Bolzano, Brentano and his school, by Mach, Boltzmann, and Einstein are compelling and to the point. Stadler also succeeds admirably in describing the constitution of "the first Vienna Circle" initiated in 1918 by Hans Hahn, who may for that reason be considered the real founder of the group, its institutionalization in 1924 after Schlick's arrival in Vienna, and the Circle's public phase which began with the appearance of its manifesto in 1929, which also led to rapid international recognition, to the arrival of an increasing number of foreign visitors, and to contacts with sympathetic groups and individuals in Germany, England, Scandinavia, Poland, the United States, and other countries. In his treatment of this large historical panorama Stadler is anything but a reductionist intent on seeking to bring this rich variety down to a single formula. His theme is, rather, throughout the variability and variation of the ideas entertained within the Vienna Circle and in the wider domain in which it is situated.

One of Stadler's main conclusions is, indeed, "how multi-colored and heterogeneous the Vienna Circle was in respect to its mentality and its conception of science - despite its implicitly shared cognitive frame of identification." (p.17) He illustrates this heterogeneity throughout his book, but most specifically in Chapter 11 where he contrasts Schlick's and Neurath's contributions to the ideology of the Circle. He concludes that as a result of such investigations "the picture of the Vienna Circle appears much richer and more differentiated than the one we find, on the one hand, in [its own] programmatic statement or, on the other hand, in the relevant secondary literature - with a few very recent exceptions." (p.54) In his discussion of the Circle's well-known programmatic statement, *Wissenschaftliche Weltauffassung. Der Wiener Kreis*, Stadler emphasizes therefore that this text was largely written by Neurath and Carnap, representatives of the "radical" wing of the movement, and that it represented, in fact, predominantly their views from around 1929, whereas Schlick, one of the "moderates", distanced himself from both the content and the language of the text as propagandistic in style and dogmatic in formulation (p.375). While the unity of science was certainly an important concern for the members of the Vienna Circle, Stadler stresses rightly that the members of the Circle at no point agreed on the question what that unity would look like.

In describing the pluralistic character of the thought of the Vienna Circle, Stadler also draws attention to its dynamically changing character which was affected by both internal debates and by external influences. The shift from a predominantly phenomenalistic to a physicalistic conception of the base propositions was, as Stadler shows, brought about by internal discussions (motivated, perhaps, to some extent by Wittgenstein's *Tractatus)*, whereas the shift from a syntactic to a semantic conception of logic was due to the intrusion of Tarski's work into the thinking of the Circle. Stadler quotes approvingly Victor Kraft's statement, that "the Vienna Circle was so lively that it was constantly undergoing transformation to such an extent that one cannot pin it down to any particular doctrine." (p.255) What held the members of the Circle together was, so Stadler, "only a general commitment to a scientific philosophy" which refuses to take the subject as an "autonomous discipline sui generis, but postulates instead a mutual interdependence with the individual empirical sciences (natural or social sciences) - and/or with formal logic and mathematics." (p.42) As such, Stadler wants to understand scientific philosophy as an expression of the "spirit of modernity" (ibid.), and a late embodiment of the enlightenment tradition. The problem with such a characterization is, of course, that the concept of science is itself heavily contested (one need consider only that Hegel and Fichte also thought of their philosophizing as "scientific" and as continuous with the rest of science). The same holds for the notions of modernity and enlightenment. It is here, where one wishes Stadler had developed his characterization of the historical place of the Vienna Circle further, in particular, since he speaks of the significance of his own investigations as lying in their relevance to the current debate between modernism and postmodernism (p.19).

The question what has remained of the heritage of the Vienna Circle is, of course, not at all easy to answer. Some tendencies in contemporary Anglo-American "analytic philosophy" can be seen as continuations and reworkings of ideas first advanced by the Vienna Circle. But analytic philosophy is a much broader movement in which Aristotelian, Neo-Kantian, Wittgensteinian, phenomenological, and even decidedly metaphysical modes of thinking intertwine with assumptions inherited from logical empiricism. The question of the relation of philosophy to empirical science remains therefore contested; while some analytic philosophers would, no doubt, affirm the interdependence of the two which the Vienna Circle assumed, there is also a distinctive tendency to a separation, an insistence on distinctively philosophical ("transcendental") modes of reasoning, an inclination towards a disciplinary and "scholastic" treatment of philosophical questions. Characteristic of analytic philosophy is also that Wittgenstein (both early and late) has come to overshadow the Vienna Circle. The *Tractatus* and the *Philosophical Investigations* have become classical parts of the philosophical canon and are likely to remain there for some time while the writings of the Vienna Circle from Carnap's *Aufbau* down are considered dated, naive, and remain largely unread. Students of philosophy generally learn of Viennese thought, therefore, only through work and the eyes of Wittgenstein. Most surprisingly, contemporary analytic philosophy has begun to break down the boundaries which the Vienna Circle erected against its contemporary competitors. It is quite possible today for analytic philosophers to refer positively to and, in fact, to utilize ideas taken from Husserl and Heidegger, Sartre, Merleau-Ponty, and Foucault. Some interpreters have, in consequence, begun to speak of the rise of a new "post-analytic" paradigm of philosophy.

Given the fluid nature of the situation, we cannot blame Stadler for having failed to provide a definitive assessment of the Vienna Circle. History will settle that question. What Stadler has shown us decisively is the important role the Vienna Circle has played in the formation of philosophy in our century. His account is particularly compelling due to the large amount of documentation he has provided which includes a fascinating set of notes of what was actually said at meetings of the Circle (pp.275-359) and a previously unpublished interview with Karl Popper concerning his relations to the Vienna Circle (pp.525-545). The second half of Stadler's volume contains, moreover, a series of diagrammatic surveys of the Vienna Circle and a treasure trove of further documentary information, such as lists of contents of various publications of the Circle, short biographies of members of the Vienna Circle and associated figures (including extensive bibliographies) and a documentation concerning the assassination of Schlick. Stadler's own thoughtful presentation together with the documents thus provided mark a breakthrough in our understanding of the Vienna Circle to an entirely new level of scholarship. There is no doubt that this book will become an indispensable tool for anyone working in this area.

Hans Sluga

PAOLO PARRINI: *Conoscenza e Realtà. Saggio di filosofia positiva,* Roma & Bari: Laterza, 1995. An English translation appears under the title *Knowledge and Reality. An Essay in Positive Philosophy* within the Western Ontario Series in Philosophy of Science at Kluwer, Dordrecht.

During my first year of philosophy study at a German university I shared the (unjustified) prejudice of my colleagues that positivist texts are tedious if not boring, at least compared to Schopenhauer or Nietzsche, to Sartre or other French philosophers. Here is a book that shows the opposite and succeeds in keeping the esprit of an essay over 233 pages. Parrini's use of the word 'essay' alludes to Hume's *Essays*, more generally to the English tradition rather than to those contemporaries to whom philosophy is nothing but an essay. Unfortunately, Parrini pays an unnecessary high price for this form when the book's only footnote informs us that "Reasons of expressive form after all have led me not to burden the text with the usual references to authors and works used. I apologize to the reader."(3)[1] The expert will easier accept this excuse and, perhaps, enjoy the unencumbered reading, although, as a matter of fact, some footnote-like hints to other authors or even Parrini's own works are entirely useless without any reference. But, the omittance partly spoils one great merit of the book. As large parts of it are self-contained and Parrini succeeds in making most technical points quite transparent, the essay appears to me very readable for students. To him or her, hardly reading footnotes when going through it first, it becomes almost impossible to penetrate further into 'positive philosophy'. It would have been much more profitable to the freshmen to supplement the book by a two-page bibliography containing the basic references. The English translation to appear soon will contain references and bibliography in the standard fashion. It is a pity, that unlike the Italian original only few students will be able to afford this fascinating book due to its high price.

The main objective of *Conoscenza e Realtà* is "a rational reconstruction of the key concepts of epistemology: truth, objectivity, rationality"(226) from an empiricist viewpoint. A successful treatment of these issues requires a general frame of reference for detailed epistemological analyses. The general perspective intended by Parrini's positive philosophy stems from various traditions that all share an anti-absolutistic stance and the refutation of metaphysics, that had characterized Logical Empiricism. As its main weapon, the verifiability criterion of meaning, has not stood criticism, the positive philosopher courageously – 'Mut' is instilled into us already in the dedication – "is unhappily forced to go beyond the scope of pure and simple analysis of concepts and facts and to become a philosopher in the full sense against his own will"(9). But the philosophical enterprise is pragmatically relativized in various respects. Parrini expresses this by modifying Wittgenstein's famous metaphor: "The ladder which I will endeavor to build ... aims to achieve the maximal balance between the semantic and ontological component and the component connected with epistemology and philosophy of science. Its supporting axles have to be calibrated in order to achieve an equilibrium as close as possible to that which characterizes the ladder we confidently climb up in daily life."(43)

The first chapter "Old and new Positivisms" sketches the crisis of the standard conception of scientific theory which was mainly shaped by Carnap and Hempel. The desired rejection of any knowledge *synthetic a priori* was achieved by the verificationist criterion of scientific meaning and a strict distinction between analytic and synthetic statements. The criticisms raised against both which "to a large extent have hit the mark"(19) not only concerned the problem of metaphysics, but also "the relevance of experience in the validation and refutation of our hypotheses and theories."(19) This disastrous outcome was broadly conceived as a proof of the failure of any positivist approach and ended up in the all-pervasive relativism of the 'new philosophy of science'. Here Parrini emphatically disagrees. The failure of the verificationist doctrine only precludes the quick linguistic refutation of metaphysical statements, but it does not devaluate Mach's understanding of anti-metaphysics as an anti-absolutistic and critical attitude. The author emphasizes that Logical Empiricism had many facets and points – as many do nowadays in this respect – at Neurath's multifarious criticisms 'from within' the Vienna Circle. But, Parrini goes back even further until the roots of the Vienna Circle's synthesis of empiricism and rationalism, to Mach, Duhem and Kant. Historical comparison between the classical and modern positions is Parrini's strategy to develop his conception of positive philosophy that is characterized by "relativism, empiricism, antimetaphysical objectivism"(24).

Relativism is the first notion addressed. Logical empiricists did not actually disregard subjectivity as long as it was not directed at knowledge *synthetic a priori*. On the contrary, making explicit the arbitrary points of a scientific theory was conceived as the first clue to attain objectivity. In the same manner, Parrini aims at rescuing the objectivity of scientific knowledge by assenting to large parts of the relativists' criticisms, but by rejecting their 'absolutistic' anti-objective and anti-empiristic claims. "This effort of conceptual mediation will result in what I call 'positive philosophy', as distinct from the old and new positivisms of the past."(42) To Parrini's mind, the latter have not been sufficiently 'positive' because they were "never fully successful in formulating their directive ideas by correct extrapolation from the philosophical analysis of scientific knowledge".(42)

Parrini partly assents to Feyerabend's criticism of neoempiricist methodology that emphasizes that all observations are theory-dependent, that all our experience is already segmented by language, and that there is a pragmatical distinction between observation and theory, only. Against the last claim, however, Parrini sustains a theoretical status of analyticity that rests on "the existence of a distinction *in principle* between language and theory"(66). Such a distinction is precluded by accepting Quine's identification of both dogmas of empiricism. Parrini admits the failure of verificationist reduction, but argues in favor of a relativized distinction *in principle* between analytic and synthetic statements. "The boundary is certainly fluid. Theory change can provoke language change, and *sentences* which were considered analytic can become synthetic and vice versa."(66) Dropping verificationism only entails *theoretical* relativism, but not *linguistic* relativism. This was already stated clearly by Duhem insisting against Poincaré's linguistic conventionalism that one cannot construct the language of a physical theory without disposing of this theory itself. Accepting moreover Duhem's holism it becomes possible that assertions that are void of empirical content if considered alone are nevertheless testable indirectly in conjunction with other assertions of the theoretical system. In this way, Parrini obtains also a characterization of the classical dispute between Reichenbach and Schlick: Schlick's point on the conventionality of general relativity was a *global* one, while Reichenbach was referring to the dependence of *local* problems on theoretical presuppositions that cannot be reduced to mere conventions.[2]

The main challenge of relativism for the empiricist is that "epistemic relativity endangers the possibility of a genuine, non-circular empirical control of our cognitive claims"(77). Parrini judges Mary Hesse's *network model* as capable "to restore the possibility of empirical control without denying the full theory-dependence of observation"(91) by assuming a contextual and pragmatic difference *in principle* between observational and theoretical language. Hesse argues that given the infinite complexity of the empirical world the proper use of words in new situations requires criteria of gradual recognition of physical similarities. Although, as Wittgenstein observed, the latter criteria do not suffice for the functioning of words with extralinguistic reference, nonetheless some words have to be introduced directly. Already the physics and physiology of the situations provide some non-verbalized point of view that allows to recognize repetitions and differences. To my mind, this first building block of the network model bears striking similarities to Ernst Mach's foundation of science in instinctive experiences that are structured as economical theories in order to allow recognition and communication of similarities between given situations and already known facts. The network model, however, is anything but sensualistic. Also terms that are usually counted as theoretical might be introduced by direct association of empirical laws. "So, the network model rests upon the recognition of two salient aspects of descriptive expressions with extralinguistic reference: i) none of them can work exclusively by means of empirical associations; ii) even those introduced and learnt by means of linguistic associations are in principle susceptible to empirical applications,"(98) as there are no *a priori* limits to what we can measure. According to Parrini, the network model shows that in any given moment the are some statements which are the product of a correct application of observational terms to empirical situations. Hence, there is no *a priori* exclusion of empirical control.

In order to defend its non-metaphysical *objectivity*, positive philosophy uses key elements of Kantian philosophy. Nonetheless, Kant's 'formal idealism' as a general concept is rejected. In rejecting the absolutistic *synthetic a priori* without the verificationist criterion, Parrini looks at attempts to reconcile Kant's philosophy with modern science. Cassirer maintained Kant's notion of the objectivity of scientific knowledge by reinterpreting it as the ideal limit of the variable theories and conceiving the transcendental principles as ultimate invariants of thinking which give unity to the process of attaining knowledge. But, Cassirer's attempt fails due to the autonomy of observation pointed out by Reichenbach as a consequence of Duhem's holism. To every given system of presuppositions one can imagine experiences contradicting it. Preti's attempt to combine logical idealism and positivism has shed light on the very important connection between metaphysical realism and skepticism that results from the Kantian thinking. But, to Parrini's mind, Preti's version of transcendental criticism does not succeed in proving metaphysical realism as meaningless. Once again the ultimate dependence of the transcendental elements on experience leads into a dilemma unresolvable to Kantian philosophy between relativism and metaphysical realism. "If we acknowledge the modifiability of the cognitive presuppositions under the pressure of conceptual innovation and of the data of experience, we end up with unpleasant consequences with respect to Kant's original aspirations. On the one hand, it will no longer be possible to warrant cognitive objectivity by appealing to the universal and necessary character of the principles of knowledge. On the other hand, the tendency to interpret scientific change as the progressive approximation to knowledge of transcendent reality will gain momentum."(136)

According to Parrini, the main reason for the failure of Popper's fallibilism was precisely that it was implicitly based on metaphysical realism and the absolutistic use of the concepts of rigor and risk. Similarly as Popper's verisimilitude, 'inference to the best explanation' is based, too, on an erroneous connection between the success of a theory and its truth. Neither is this historically justified, nor does positive philosophy a priori fail in explaining the factual success of science. Accordingly, interpreted itself as a scientific theory, metaphysical realism performs quite badly. But, as Parrini retains a contextualized distinction between analytic and synthetic statements, metaphysical realism can be separated from scientific empirical realism by using Carnap's framework concept, although it cannot be deemed as meaningless. As science always acts within a historically mutable framework, absolutistic claims can never be scientific. The same is true for radical relativism. Not only does it contradict the factual ideals of scientists towards objectivity and truth, it is also self-contradictory as a non-contextualized principle. To Parrini the latter is no "philosophical quibble"(172), but decisive. As it is not without problems to ground a rational norm, such as objectivity, on a self-contradiction, one should emphasize that here unlike transcendental pragmatism, a large dose of relativism is kept by positive philosophy.

Parrini conceives truth and objectivity not as absolute rules, but "as unitary regulative ideals or values that transcend the particular presuppositional structures"(176) and call for a maximally organic and unitary description of empirical reality. In this perspective, the notions of objectivity and truth become merely formal categories in the Kantian sense that are specified in any concrete application. As they do not require any continuity of a particular theoretical framework, accepting objectivity and truth as empty ideals suffices to turn the philosophical problem of incommensurability into a logical or scientific one. Accordingly, to the empiricist the emptiness of the ideals is advantageous because philosophy should not head at specific contents if it wants to avoid metaphysics. Seen from the historical perspective, this well-contrived incorporation of Kant's regulative principles into positive philosophy appears to me as a key merit of Parrini's book. Having in mind that the anti-Kantian attitude of the Vienna Circle was directed against the absolutistic claims of a *synthetic a priori*, Parrini indicates a possible fruitful discussion between neokantianism and logical positivism on the pragmatic foundations of knowledge that never took place.

Taking positive science as the paradigmatic example for an empty ideal of objectivity, all non-scientific absolutistic claims – be it metaphysical realism or Kantian transcendentalism – cease to be even cognitive. As Parrini wants to avoid plain scientism, he concedes cognitive content also to conceptual syntheses of experience other than science, poetry for instance. Apart from the rejection of absolutism, the criteria for scientificity become part of science proper. "The main philosophical difficulty arises when ... we face more or less marked interpretative divergences about the meaning to attribute to the resulting [from the empirical and logical controls] data, due to further differences in the original theoretical, methodological, and axiological presuppositions. The solution of the contrast and the choice of the criteria that should guide us in the decision are thorny problems. They are what is commonly understood as the problem of scientific rationality."(201) This shares many features of objectivity. First of all, any absolutistic justification or delegitimation fails. Neither can rationality be reduced to formal criteria, nor is it codifiable in a set of certain explicit norms, as it always has a tendency to transcend them. On the general level scientific rationality appears as an empty category that is filled in the course of the scientific enterprise. Beyond this descriptive character, rationality contains a normative aspect, too, that consists – quite analogous to the good will in Kantian ethics – in the norm 'Be rational!'. Scientific rationality embraces certain forms of intuition, instinct, belief, as those can be argued about by referring to analogies, earlier applications of these rational criteria, etc. Accordingly, scientific discoveries are, at least partially, rational. They can even lead to modifications of the norms of scientific rationality.

In the end, Parrini admits that in his conception "the alternative to relativism and absolutism has to be paid for by emptying of all content the fundamental notions of truth, objectivity, and rationality."(232) Parrini's general scheme would be idle and useless if it were not intimately connected to the level of particular experiences and, after all, to a good dose of Duhemian scientific common sense. Apart from its negative stance against absolutism and radical relativism, the general scheme also intends to provide a framework for all our scientific regulative ideals. The scheme does not suffice, to my mind, to yield the best explanation or the most rational choice in all cases. Here I think it is time to recall Neurath's concept of bounded rationality, or the famous 'auxiliary motive'. Sometimes, if the boat is already leaking and we lack sufficient knowledge how to repair it, any decision is more rational than further reasoning about what the most rational solution might be. The 'auxiliary motive' appears to me precisely of that type of formal and empty rationality, Parrini has constructed. It may involve instincts, but also a use of the classical regulative ideas without justification by scientific content. But, unlike spiritism it is shaped in such a way never to preclude being filled by scientific content.

Alluding to Neurath, Kant, and Mach, I intended to emphasize one lesson Parrini's essay teaches. Discovering the complexity of the real (or only reconstructed) debates on Logical Empiricism offers still many perspectives to surmount certain agonies felt within present philosophy of science. The other lesson is that striving for philosophical generality is a necessary and still rewarding enterprise. Writing a compelling book does not require a verbose metaphysical outfit.

NOTES

1. The present review is based on the Italian original and does not comment on the English translation which is still in preliminary form. Nevertheless, for the sake of terminological consistence I have decided to take all quotations from there, apart from this footnote that does not appear there.
2. To my mind, Parrini's characterization matches very well Michael Friedman's recent analysis: "Geometry, Convention, and the Relativized A Priori: Reichenbach, Schlick, and Carnap", in: Wesley Salmon, Gereon Wolters (eds.): *Logic, Language, and the Structure of Scientific Theory,* Pittsburgh & Konstanz, 1994, pp. 21-34.

Michael Stöltzner

MICHAEL HEIDELBERGER, *Die innere Seite der Natur. Gustav Theodor Fechners Wissenschaftlich-philosophische Weltauffassung*, Frankfurt/Main: Klostermann, 1993.

Michael Heidelberger has presented a book that is convincing in many respects. On almost 400 pages written in an exceptionally fluid style, the author argues that Fechner's natural science of the "Tagesansicht" ("daylight view") is a quite significant achievement. Already the book's title contains an ambitious program. Heidelberger who aligns himself with the philosophical tradition of Logical

Empiricism, links Fechner's *Naturphilosophie* to scientific modernity even in those cases where Fechner's philosophizing concerns the 'inner side of nature'. This classical philosophical endeavor is rooted deeply in the tradition of German *Naturphilosophie* of the late Romanticism. But Heidelberger's interpretation departs from the traditional lines of reception that considered Fechner a postidealist whose thinking contained religious and mystic speculations as a central element. Instead, he shows how many perspectives, quite promising for contemporary philosophy of science, were for a long time buried in Fechner's work. To arrive at this new lucid perspective, Heidelberger had to unearth the multifarious aspects of Fechner's thinking and to sort out those relevant for posterity. The reconstruction of Fechner's way of reasoning through several stages is particularly appealing.

Fechner's very specific theory of cosmological evolution constitutes the core of his philosophy. In a very detailed analysis, Heidelberger develops its methodological foundations, constantly emphasizing Fechner's deep and far-reaching empiricism and the lasting importance of concepts shaped by him. This holds, in particular, for his notion of causality which – through Ernst Mach – became pioneering for modern epistemology. For Fechner, the law of causality consisted in relatively free, but nonetheless categorically valid, functional dependencies and relations between causes and effects. Heidelberger demonstrates that this important step originated in Fechner's metaphysically motivated opposition against contemporary materialism and the latter's strictly deterministic understanding of the law of causality. The author's detailed reconstruction shows how the conceptual basis of Fechner's all-pervasive indeterminism developed and resulted in a – highly modern – conception of cosmological evolution that comprised the inanimate nature as well.

From a historical perspective, the indeterminism issue is paradigmatic for the highly productive synthesis between the different aspects of Fechner, the *Naturphilosoph* of late Romanticism and the devoted natural scientist. According to Heidelberger, this combination of logical analysis of scientific concepts and a consequent empiristic attitude paved the way for the Vienna Circle and Logical Empiricism. Unfortunately, he is rather sketchy about the connections to the Berlin group of scientific philosophers in which the influence of both Mach and Fechner was mediated by Joseph Petzoldt. Here a footnote does not suffice. Moreover, Gestalt theory played a key role in the Berlin 'Society for scientific philosophy' which finds in Fechner's psychophysics – a topic studied in depth as such by the author – one of its ancestors, too.

The book makes Fechner's immense philosophical creativity tangible to the reader. Heidelberger convincingly argues that Fechner's non-reductive materialism as expressed in the 'Tagesansicht' is still a highly reputable scientific research program. Yet precisely in this reconstructive and analytic part of Fechner's thinking he locates the gist of Fechnerian philosophy. But one has to be aware of the fact that the metaphysic-free analysis also constitutes the basis of Fechner's constructive metaphysics. Heidelberger admits that Fechner's interest was primarily directed to his *Naturphilosophie* as the 'philosophical closure' of natural science – or as the metaphysics induced by the latter. Heidelberger carefully works out the first part because it is only Fechner's materialistic and empiristic world conception that can be uncovered as a significant ancestor of modern philosophy of science. Heidelberger's method of interpretation from today's perspective exhibits its didactical fruitfulness particularly with regard to Fechner's conception of the indeterministic process of nature that is guided by a general tendency toward stability. Hence, it appears that the paradigm of self-organization accompanies, from the very beginning, indeterministic physics. Moreover, Fechner is convincingly shown to be the first physicist to consistently apply probabilistic methods.

Heidelberger translates parts of Fechner's speculative metaphysics into a modern conceptual framework. In the introduction, the author emphasizes that it surprized him to discover these metaphysical roots of contemporary philosophy of science. A paradigmatic case are the forgotten connections between Fechner's *Kollektivmasslehre* and Richard von Mises's mathematical foundation of modern probability theory and statistical physics. Fechner's revolutionary book originated in his attempt to devise a rigorous formulation of his indeterministic conception of cosmological evolution. Heidelberger finds in the *Kollektivmasslehre* already both ingredients of von Mises's notion of the collective (lawlessness of its members and the existence of the limit of relative frequencies) and, accordingly, his definition of probability as the limit of relative frequencies. This connection between the Fechnerian Mach and the Machian von Mises is illustrated by Heidelberger's excellent analysis of the relation between Fechner and Mach. The core of the latter's disagreement lies in their attitude towards atoms. Whereas a realistic stance fitted very well into Fechner's indeterministic 'inductive metaphysics', Mach's antimetaphysical move dispensed with both as far as possible.

One of the most interesting findings of the book under review consists in the lucidly elaborated insight that Fechner's thinking has exerted a multifarious and persisting influence on *Austrian philosophy* that concerned some of its key issues. This subtle and profound relation does not figure significantly – apart from brief observations on Fechner's influence on Mach – in the various historically-oriented works on the *Austrian philosophy* topos – whatever their approach to the Neurath-Haller thesis is. *Die innere Seite der Natur* is an indispensable supplement to the recent books of Haller, Smith, and Stadler which contributed so essentially to the revision of the 'received view' on the Austrian tradition.

With regard to the broader context of the Austrian tradition, Heidelberger has primarily focused on Mach and von Mises. The further development of Fechner's naturalistic and evolution-theoretic approach that encompasses not only Mach but also the philosophy of the interwar period, is merely mentioned in passing. Heidelberger's analysis provides a number of promising perspectives for subsequent work. One finding is Ludwig von Bertalanffy's Ph.D. thesis on Fechner. This suggests to trace back biological System Theory, at least in part, to a peculiar reading of Fechner's theory of self-organization. In a wider horizon, Evolutionary Epistemology might find facets it shares with Fechner's consequent naturalism, to wit, his psycho-physical parallelism.

The few lacunae diagnosed so far are largely compensated for by the many pioneering perspectives this fascinating and comprehensive book opens up. Perhaps, only the curious reviewer feels such shortcomings at all. However, for the historian some critical remarks might be in order concerning Heidelberger's methodology. He quite successfully brings to the fore the empiricist, that was hidden in many previous interpretations. Nonetheless, the conclusiveness of Heidelberger's argumentation, on the other hand, slighly obscures the coherence of Fechner's actual thinking. Let me sketch two closely related points. Firstly, the philosophical implications of Fechner's deep religiosity – studied succinctly in the first part of the book – are not pursued any further. This enables a splendid analysis of the intimate connection between the naturphilosophical-metaphysical and the analytical-empirical thrusts of Fechner's philosophy. Its aesthetical and ethical aspects, however, are encapsulated. In Fechner's own view, the latter were part and parcel of his very comprehensive concept of nature. This deprives Heidelberger of the possibility to include Fechner's firm belief in the 'goodness of nature' in his philosophical reconstruction. Also the poetic images of nature, their subjective and optimistic traits, cannot be conceived in this way.

My second remark concerns the relation between Heidelberger, the philosopher of science, and Heidelberger the historian of philosophy of science. The philosopher ensures Fechner's fertility for modernity by elaborating on the antimetaphysical aspects of his philosophy. Here, however, the historian falls behind in that he neglects the 'turn of philosophy' proclaimed by the Vienna Circle and subsequently perpetuated by much of modern philosophy of science. Not taking into *historical* account the good reasons for the vehement rejection of metaphysics and *Naturphilosophie* can only partially be compensated for by the convincing interpretation of Mach's antimetaphysical stance as a reaction against Fechner's metaphysical realism. Hence, Heidelberger cannot explain the connection between the refutation of metaphysical speculation about the 'inner side of nature' and the rather broad success of the empirical-analytic research program that characterized modern philosophy of science.

These methodological inclinations are the reason that, at least sometimes, Heidelberger's generally convincing story of rediscovery and rehabilitation smacks a bit Whiggish. Nevertheless, the book represents an almost mandatory reading as it provides many new historical insights that are quite significant for contemporary philosophy of science.

Veronika Hofer

MARTIN CARRIER: *The Completeness of Scientific Theories* (The University of Western Ontario Series in Philosophy of Science, Vol. 53), Dordrecht: Kluwer Academic Publishers 1993.

Philosophy of modern physics seemed not to be among the strongholds of post-war German philosophy. As this book testifies, Martin Carrier is among the few exceptions. He makes valuable and original contributions to some important problems in philosophy of science: the theory-ladenness of observation, the possibility of universally valid theories, and the status of physical geometry. I will report first on his main theses and then point to some open problems.

Einstein-Feigl completeness. Physical theories can serve two purposes. They may be intended to describe some physical system under investigation (in which case we can call them *object theory*) or they may be used for establishing a relation between pointer readings of measurement instruments and some object theory (in which case Carrier calls them *observation theory*). An object theory is called Einstein-Feigl complete, EF-complete, if it operates as its own observation theory. Such a theory can establish a link between its own concepts and empirical procedures – just by relying on itself.

Is EF-completeness a desirable feature of a theory or not? It seems that yes: physical theories aim at laws generally valid in a domain of application. The wider the domain of application the better the theory. If a theory does not only apply to some object but also to the measurement instruments used for observing the object, it is a better theory than one applying just to the object. In particular we would expect the ultimate goal of physics, universally valid theories, to be EF-complete.

Theory ladenness of observation. But EF-completeness has its nasty sides. For example, it may give rise to a particular awful form of theory-ladenness of observation. In the tradition of Hanson and Kuhn observation is regarded as theory-laden because theoretical proclivities of the experimenter influence perception. A somewhat weaker claim is that theories influence the linguistic expression of observations. Even seemingly innocent observation statements like "The temperature outside is 15 centigrade" or "The position of the pointer is x" involve terms like temperature or position which are theoretical. On different theories of thermodynamics or space these observation statements will have a different meaning. This second kind of theory ladenness emerges not only on a context account of meaning (as Carrier seems to suggest) but also on a double language account and probably on any other account of meaning.

A third kind of theory ladenness was pointed out by Duhem, Reichenbach, and Lakatos. When designing an experiment to measure some physical quantity, the reliability of the experiment is established on grounds of some physical laws. These laws are part of the observation theory and influence observation as well. If the observation theory is flawed the observations will display systematic errors. The nasty side of EF-completeness now is this: an EF-complete theory is involved in establishing the reliability of the experimental tests to which it is itself submitted. It may happen that flaws of an EF-complete theory lead to systematic observation errors rendering the experimental test of the theory unreliable. A flawed EF-complete theory may pass the unreliable test it sets itself.

Threats to experimental accessibility and ways of escape. A paricularly strong form of EF-completeness, called strong EF-completeness on the hypothesis level, is present if a law under investigation necessarily enters into the theoretical analysis of all experimental procedures measuring one of the quantities in this very law. Carrier presents two types of situations in which this kind of difficulty threatens to seriously impair testability of EF-complete theories.

A theory suffers from a *Sneed bottleneck* if one of the quantities in some law of the theory can be evaluated only by recourse to that same law. In this case every experiment must yield values for that quantity which are in accordance with this law. No counterevidence can possibly emerge. The situation is even worse if a *Reichenbach loop* is present: two or more quantities occuring in some law can be evaluated only by recurrence to that law. In this case it is even possible to choose a value for one of these quantities arbitrarily and to adjust the other so that the law in question is satisfied. Again, any experimental counterevidence seems to be precluded. In both situations we encounter what Carrier calls *non-Duhemian test problems*: not only is it impossible – as argued by Duhem – to definitely refute by experiment some hypothesis because some other hypothesis in the theory-net can be blamed for counterevidence, but the emergence of counterevidence is outrightly excluded.

So how is it possible to seriously test a hypothesis if all observation is theory-laden? The ususal answer to this is that the hypothesis under investigation must not be part of the observation theory. Giving up EF-completeness saves testability by avoiding Sneed-bottlenecks and Reichenbach-loops. Carrier suggests a different answer: EF-complete theories can be seriously tested even in the presence of Sneed-bottlenecks and Reichenbach-loops by strategies which he calls non-instantiative testing and external fixing.

A paradigm example of a Sneed-bottleneck – the term derives from this being the pet example for Sneed's problem of theoretical terms – is the test of momentum conservation assuming that only collision experiments can be performed. In order to refute momentum conservation it seems necessary to measure velocities of two bodies before and after the collision as well as their masses and obtain values violating $m_1 \Delta v_1 = m_2 \Delta v_2$. Since by assumption masses can only be determined in collision experiments, namely from some observed velocity values using $m_1 / m_2 = \Delta v_1 / \Delta v_2$, it is impossible in a collision to measure velocities and masses violating momentum conservation. Carrier elegantly exorcises the structuralist's demon of Sneedian test problems by pointing to what a sober physicist would do: repeat the experiment. If the ratios $\Delta v_1 / \Delta v_2$ measured in subsequent experiments with the same two bodies vary, momentum conservation is refuted. This is an instance of a *non-instantiative testing strategy* allowing for counterevidence even in the case of Sneek-bottlenecks. Alternatively, the test problems associated with Sneed bottlenecks can be overcome by *externally fixing* some quantity by appeal to an additional hypothesis.

EF-completeness of general relativity and the conventionality of physical geometry. A Reichenbach-loop prominently figures in general relativity. In his 1928 argument for the conventionality of physical geometry Reichenbach points to the possibility of universal forces deforming rods in some way independent of the material. Such forces cannot be detected because two rods which are of the same length at some place and time will be so everywhere and always. In the presence of universal forces length measurements will give results suggesting a distorted geometry g^*_{ab} instead of the true geometry $g_{ab} = g^*_{ab} + U_{ab}$. Both the undistorted metric g and the correction term U are only defined by this equation, and g can be inferred only from U which in turn can only be inferred from g. This circularity prevents the inference of g from the uncorrected g^*. To aggravate the situation a similar circularity arises from distortions by differential forces depending on the material of the rod: corrections of differential forces in length measurements require laws describing the distortions of rods, which in turn use the concept of length. These test problems are due to EF-completeness and prevent experimental data from uniquely determining the geometry. Thus different geometries may be empiricially equivalent and the choice of a geometry becomes a matter of convention.

Carrier argues that these threatening test problems can be overcome and that at the same time general relativity is EF-complete. The circularity of differential corrections can be remedied for example by what he calls the *Grünbaum helix*. Assuming some geometry g^1 entering the distortion law one arrives at corrected length measurements suggesting a geometry g^2 usually different from g^1. Feeding g^2 into the distortion laws one arrives again at corrections which suggest a geometry g^3. If the sequence of geometries converges – and Grünbaum argues it does – the limit geometry applied to the deformation laws yields as corrected length measurements this very geometry. The method of the Grünbaum helix is often useful but not generally applicable. Alternatively the geometry could be determined by identifying geodesic paths, which are not subject to differential distortions. In any case the testability of general relativity remains unimpaired by its EF-completeness; the threatening restrictions to testability do not materialise.

Reichenbach's argument for the conventionality of geometry was based on a claim that as a matter of principle it is impossible to experimentally determine the true geometry. Carrier finally dismisses the arguments of Reichenbach and modern improvements thereon; he claims that it is possible to attach empirical indicators to the key concepts of the theory and to infer the theoretically relevant quantities from undistorted as well as distorted evidence – just by resorting to general relativity. But still he adheres to the claim that physical geometry is in an interesting way conventional. How is this possible? Simply by pointing to an example. In the Thirring-Deser account gravity is represented by a field acting in flat Minkowski space. It affects the length of rods and the operation of clocks to the effect that measuring instruments display a curved metric instead of the true flat one. The Thirring-Deser account of gravitation and general relativity are thus empirically equivalent and satisfy some addition conditions required for conventional alternatives. Whether physical geometry is curved (as implied by general relativity) or flat (as implied by the Thirring-Deser account) is a matter of convention. Both theories are EF-complete in that they can themselves supply empirical indicators of their key concepts, but they are empirically equivalent because the same pointer readings will serve to determine different quantities. Reichenbach's argument suggests that test problems of EF-complete theories imply conventionality. Carrier, however, argues that that EF-complete theories need not have test problems and still may allow for conventional alternatives.

Open problems. One of the main merits of the book is that it does not close a field of investigation but rather opens one. I will now point to some open problems which seem promising for further investigation.

Carrier does not say much about how EF-completeness is related to other forms of completeness figuring in the literature, as for example universal validity of a theory, semantic completeness, and completeness of formal system. It seems that universal validity of a theory is neither necessary nor sufficient for EF-completeness. A theory is EF-complete if it describes in addition to the object system also the measurement device and its interaction with the object system. But this of course does not imply that it applies to any physical system. On the other hand, if a theory is universally valid and therefore applies to any system, it admittedly applies to measurement devices as well. But the concepts relevant for a description of the measurement device need not be part of a universally valid theory. A measurement device operating on the chemical level need not be fully described by some form of high energy physics even if the laws of high energy physics apply to it; chemical concepts simply may fail to be translatable into high energy physics. But perhaps under some additional assumption of reductionism does universal validity imply EF-completeness? If this can indeed be confirmed, Carrier's results on the testability of EF-complete theories bear importantly on the testability of universally valid theories.

The relation between EF-completeness and semantic completeness of formal systems is another matter of interest. Tarski calls the language in which a formal system is formulated *semantically complete* if the system contains (1) names for expressions of the language and (2) semantical concepts like truth or meaning. An EF-complete theory can establish a relation between pointer readings and expressions of the object theory. In a loose sense the pointer readings refer to states of the object system and thereby to expressions of the object theory. Furthermore, the semantics of a theory worth being called empirical relies at least partially on experiments. Measurements thus seem to have a semantic dimension which in case of an EF-complete theory is also part of the object theory. There is much work left in spelling out precisely the relation between EF-completeness and semantic completeness.

Carrier does not say anything about the relation of EF-completeness to the completeness of formal systems. (A formal system is called *complete* if every true statement is provable.) But the fact that he denies for EF-complete theories any restrictions of principle to measurability and testability suggests that there is no analogue in EF-complete theories to Gödel-like incompleteness.

Carrier nicely elaborates that there is no significant relation between EF-completeness and EPR-completeness. A theory is EPR-complete if its pure states say all there is to say about a system. Contrary to the suggestions of Einstein, Podolsky and Rosen, it seems that quantum mechanics is indeed complete in this sense (unless we admit the completion by hidden variables which would have to be non-local in the sense that joint expectation values of space-likely separated events need not factorise). What Carrier suggests is that quantum mechanics in the Copenhagen interpretation is not complete because it does not apply to measurement devices; and in von Neumann's interpretation it is not complete because it cannot offer a satisfactory description of quantum measurement processes. According to Carrier quantum mechanics can be EF-complete only in the many-worlds interpretation. He claims that the many-worlds account satisfactorily solves the quantum measurement problem. But even in the many-worlds account the wave function determines the splitting of the universe into different worlds only after some basis of the Hilbert space of the measurement device has been specified. Carrier is well aware of this problem but does not seem to regard it a serious objection to the many-worlds interpretation. This is where I cannot agree. I think there is no fully satisfactory solution of the measurement problem in sight which does not modify the equations of motion in one way or another. As long as unmodified quantum mechanics cannot account for the possiblity of measurements it cannot be regarded EF-complete. I wonder whether the difficulties of quantum mechanics to account for measurements are due to quantum mechanical popularities or reflect some yet undiscovered test problems for EF-complete or universally valid theories.

Another point of disagreement concerns Carrier's trust in the Grünbaum helix as an instrument to solve self-reference problems arising in connection with EF-complete theories. According to this approach, if there are two hypotheses relating two quantities to one another and if the relation between the two quantities admits only one solution, then this solution will be approximated by a series of values for these quantities obtained by starting from some arbitrary value of one quantity, calculate the value of the other, calculate a modified value of the first, and so on. But obviously this cannot be the whole story. Assume that some differentiable function f describes the modified value of one quantity as function of the original value of this quantitity. (In Grünbaum's original example f would associate to some initially assumed geometry g^1 the modified geometry g^2 as resulting from measurements corrected by calculations assuming g^1.) Assume additionally that f has only one fixed point x_0. (So in the example there is only one geometry whose application in the correction process yields in turn the same geometry as the result of measurements.) Still, if $f'(x_0) > 1$ one can start from any x other than x_0 and $f \circ f \circ f \circ f(x)$ will not converge. So by itself the existence of a unique solution of the two equations involved in a Grünbaum helix will not guarantee convergence to the solution.

Thomas Breuer

JAN FAYE, *Niels Bohr: His Heritage and Legacy*, Dordrecht: Kluwer 1991.

JAN FAYE/HENRY J. FOLSE (eds.), *Niels Bohr and Contemporary Philosophy* (Boston Studies in the Philosophy of Science, Vol.153), Dordrecht: Kluwer Academic Publishers 1994.

Our centuries saw two deep revolutions of physics with far ranging consequences for the way we view the world. The revolution brought about by the Theory of Relativity, or rather the two Theories of Relativity, the Special and the General one, implies that the notions of space and of time are not absolute. Rather, they depend on the relation of the observer to the observed phenomena. This is deeply rooted in Mach's principle: "Physics should only make statements about observable quantities". The clear analysis as provided by Albert Einstein of what it really means to actually measure distances in space and differences of time implies that neither space nor time can be absolute.

The consequent mathematical formulation by Einstein has found numerous experimental confirmations, not the least being the day-to-day application of relativity theory in the operation of the global positioning system (GPS). Without relativistic corrections positions determined by that system would be significantly wrong. Yet it is important to realize that relativity theory does not at all touch the question of whether the phenomena we observe enjoy existence independent of observation.

Much more radical questioning of our everyday view of the world has been provided by the second revolution – Quantum Physics. That theory, born out of the necessity to describe atomic phenomena, to date also has found such widespread experimental confirmation that, to say the least, it is very unlikely to turn out to be incorrect some day. Yet it has been realized from the early days of Quantum Mechanics that this theory leads to very deep conceptual issues as evidenced most remarkably by the discussions between Niels Bohr and Albert Einstein in the late 20s and early 30s of our century. In these discussions and in many analyses of the implications of quantum physics, gedankenexperiments have played a prominent role. It is interesting to note that in the last decade or two, because of significant technological progress, we have seen a number of novel experiments which not only provide realization of many of these famous gedankenexperiments but which even go far beyond what the forefathers dared to dream. That experimental development has brought forth more clearly the conceptual problems in the understanding of quantum mechanics and this again is leading to a renewed debate about its interpretation and about its meaning for our "Weltanschauung" (worldview).

An interesting phenomenon is that, while the Copenhagen interpretation of Quantum Mechanics is considered to be the most widely accepted one, the writings of Niels Bohr, who provided the deepest formulation of that interpretation, are rarely read by practicing physicists. Furthermore, many of those who actually read Bohr's papers perceive his writings as complicated as obscure. In fact, it appears that the working interpretation of most physicists is a rather naive one which does not do justice to the depths of the problems raised by Quantum Mechanics.

In part I of *Niels Bohr: His Heritage and Legacy*, Jan Faye, who is a associate professor of philosophy at the University of Copenhagen, provides a careful analysis of the relation of Niels Bohr's thinking to the work of the Danish philosopher Harald Høffding (1843–1931). Apparently, a small circle of scientists including Niels Bohr's father Christian Bohr, who was a physiologist, met regularly for discussions in each other's homes. Niels Bohr appears to have participated in these discussions. Evidently it is very difficult to judge correctly the specific influence Høffding had on Niels Bohr. But on the occasion of Høffding's 85th birthday, Niels Bohr writes: "... the philosophers... have furthered a widespread understanding of the relativity and complementarity of all human concepts...". And about the same time Høffding writes in a letter: "Bohr declares that he has found in my books ideas which have helped scientists in the 'understanding' of their work...". Thus, in part I of his book entitled "Høffding as Mentor" Faye argues convincingly that Bohr benefited very significantly from Høffding's thought. It might even be that some of the concepts introduced by Bohr into the interpretation of Quantum Mechanics, like complementarity, might actually have originated with Høffding, though this is very hard to prove, because notions in philosophy and in physics sometimes imply different concepts.

Part II of the book is entitled "Bohr and the Atomic Description of Nature". Here Faye gives a number of quotes by Bohr related to questions like the nature of phenomena, his analysis of the role of language and communication, etc. This analysis is quite interesting and deep. It is particularly important that Faye carefully compares Bohr's position with various general philosophical lines of thought like realism, idealism, etc. Yet I have problems when Faye tries to categorize Bohr, in that case as an "objective anti-realist". While it seems to be clear that Bohr's concept of objectivity implied an essential role of prediction, and therefore an essential role of the limitations of the description of an apparatus in terms of classical language, it also seems to follow that Bohr believed in the existence of a world prior to and independent of our observations. Besides this shortcoming of trying to categorize Bohr in terms of some –ism, the book is an excellent overview of his thinking with many careful analyses and quotations.

Niels Bohr and Contemporary Philosophy is a collection put together by Jan Faye and Henry J. Folse, who is Professor in the Department of Philosophy at Loyola University. It contains contributions by many of the leading philosophers of science. Just to give some examples, in her contribution Catherine Chevalley presents a careful analysis of the role of "Anschauung" and of symbol in Bohr's interpretation of Quantum Mechanics, including again a reference to Høffding. It is interesting to realise that to Bohr Quantum Mechanics essentially implied a failure of "anschauliche" pictures. Apparently to him "Anschaulichkeit" was good for children. This already carries an essential point of the spirit in which Bohr should be read today. That point is that Bohr's contribution is not so much a defense of the interpretability of Quantum Mechanics but rather it opens a vista onto a new philosophy and epistemology.

That this lesson is not appreciated by all is clearly demonstrated in James T. Cushing's contribution "A Bohmian Response to Bohr's Complementarity". His contribution even contains objectively wrong sentences, like: "However, anyone who has actually tempted to read much of Bohr's writing on causality and complementarity finds it, at best, obscure." This statement certainly is not correct for the author of the present book review. Cushing's contribution also demonstrates very nicely to what lengths adherents of pre-quantum worldviews have to go in order to save their concepts from the attack of Quantum Mechanics. It is not surprising that even Einstein, who himself was an adherent of pre-quantum objectivity, did not consider the Bohmian approach to be satisfactory.

Further in the book it is interesting to read in the contribution of David Vavrholdt entitled "Niels Bohr and Realism" that Niels Bohr rejected a "God's Eye View" of the world. He writes: "we are 'in' the world and therefore we cannot see it from 'without', nay, we cannot even ascribe any sense to this world". Bohr also seems to have had a healthy suspicion of transcendental views in general. It might be justified to speculate that this suspicion might even have originated through Ernst Mach's influence on Høffding.

There are many other excellent contributions to the Bohr discussion in the book. Some of them suffer again from the problem of trying to categorize Bohr's thinking into some kind of -ism, discussions which do not seem to bear significant fruit. It might be wise to follow the advice of Don Howard, who in his contribution ("What makes a Classical Concept Classical?") declares that "we need to return to Bohr's own words, filtered through no pre-conceived philosophical dogmas". It is to be hoped that heeding his advise will lead to a new appreciation of Niels Bohr, who might very well turn out to be one of the most significant philosophers of the 20th century. The books reviewed here are an important contribution to that debate and they are a must for anyone interested in the interpretation and the philosophy of Quantum Mechanics.

Anton Zeilinger

CHRISTIAN THIEL, *Philosophie und Mathematik*, Darmstadt: Wissenschaftliche Buchgesellschaft 1995.

Philosophy of mathematics deals with a number of issues as, for instance, the question as to the essence of the mathematics and its objects (the so-called ontological issue). Then there is also the epistemological question of whether mathematical knowledge is possible in the classical sense, and the question as to what the validity of mathematical theorems and theories is based on. Yet another issue is the question as to the interaction between mathematics and physics on the one hand and between mathematics and philosophy on the other.

Whereas generally the problem of infinity is ascribed to the first group of issues, the question of the extent to which mathematics has also been a paradigm and propaedeutic for philosophical theories must be ascribed to the latter group of issues.

Also belonging to the philosophy of mathematics is the question as to the typical activities of the mathematician, the typical "features" of mathematics.

Finally, the role of logic must be addresssed as well as the question of its essence and on what its validity is grounded. (Here one already notices that also historical questions as well as mathematics as part of cultural history must be dealt with in this context.) Mathematics, above all, never ceases to develop, and its 'focus' and 'fashions' constantly change. This, of course, has repercussions on philosophy and the status of mathematics among other disciplines. As is known, mathematics shows two different faces: humanistic and scientific. How and according to what aspects the emphasis is to be placed is ultimately also an issue to be studied in the philosophy of mathematics.

What role does intuition play in mathematics? Why is it admitted that our intuition is "violated" by rules and laws, while at the same time rules and laws are admitted that contradict our logical capabilities. And why (or to what extent?) do we have powers that can be attributed to intuition and those that can be attributed to logical, i.e., inferential powers? (In the course of over two millenia history, philosophy and mathematics have shown many, in part contradictory responses: beginning with Platonism, via empiricism, rationalism, all the way to positivism, this century's analytic philosophy and finally today's evolutionary epistemology.)

The role of aesthetics, even that of religion, should be studied in connection with the philosophical problems of mathematics.

An equally important theme is the study of the question why and to what extent the rules, concepts and laws of mathematics that are situated in a sort of transcendent, socalled "third world" (K. Popper) find application in the "real world", in real problems. (A question that is most plausibly answered in connection with evolutionary epistemology. Unfortunately evolutionary epistemology in this sense is not addressed in this book. Y. Rav could have been considered in this context to the extent that he deals with the philosophy of mathematics from the perspective of evolutionary epistemology.)

The question of the essence of mathematical metaphors, models and of how mathematicians deal with them today – especially in regard to the computer and its uses – is of course related to the applications of mathematics. The "philosophy" of applying computers to mathematics and the question of whether mathematics as such is subject to changes is not dealt with in the book. Also the question as to an "experimental mathematics", or the role of (computer) experiments and the role of pictorial representations that have become possible today in such diverse forms (e.g., in the topology or the theory of dynamic systems) for mathematics is hardly addressed. Can pictorial representations (e.g., computer-generated ones) have "cognitive value", and what role do they still play today for the discipline of mathematics? How can they contribute to new insights and to the formation of concepts? And do they not already do precisely this?

Generally speaking, the problem of concept formation (at least in mathematics), of course, belongs to the themes already addressed and is dealt with again and again, e.g., under the aspect of mathematical "objects".

From the outlined perspective, it is not surprising that there cannot be one single book that can deal with the philosophy of mathematics in the way described – as a "totality" as it were. This book, too, thus only addresses one segment of the philosophy of mathematics, even if a very important, comprehensive one, and does so in a profound and professional manner. In particular the various perspectives of the 'objects' of mathematics and the related possible points of view form a central focus of this book. Yet the 'activities' of the mathematician – as a something typical – do not come off badly; both points of view are dealt with again and again – often in an almost ideological contrast. In the process – and this may in some quarters be seen as a slight restriction – in many chapters the author basically follows the approach and theories of Paul Lorenzen; this never takes place in a concealed way. Rather, many ideas of the well-known logician and foundational theoretician can be understood better than before! And this is something I see as positive about this book.

It is also essential to state clearly that this book does not so much deal with the philosophy of mathematics per se. Instead, it seeks to provide an introduction to considerations that mathematicians and philosophers of this century have dedicated to the relation of mathematics and philosophy. And this is something the author generally succeeds well in doing in the way intended.

I will now list the main themes of this book, since this allows me to describe the book well. (Some of the themes can be dealt with without reciprocal dependencies, so that they often stand as a self-contained theme).

1. The object of mathematics, 2. mathematics and reality: the problem of application, 3. the mathematical way of thinking, 4. Mathematics, logics, metamathematics, 5. counting and the concept of numbers, 6. construction and abstraction, 7. propositions on infinite domains, 8. diagonalization, the continuum, transfinite numbers, 9. axiomatization and formalization, 10. limits to the axiomatic method, 11. finitary and constructive procedures, 12. the concept of structure in mathematics, 13. geometry as the theory of forms, 14. is there unique a fundamental discipline for mathematics?, 15. antinomies and paradoxes, 16. the foundational crisis and the foundational debate.

It is not the job of a reviewer to write a symposion-style commentary on, or to deal with parts of the book from a different angle, even if he or she wanted to. Rather, he or she has the task of judging whether the themes of the book are dealt with exhaustively and whether they provide stringent answers. This is certainly the case in this book. In order to fully recognize this one would have to read the entire book, it is not enough to select individual themes. The reader would then see some things inaccurately or one-sidedly. It is, however, very important to point this out since the themes of this book could entail an eclectic reading and further pursuit of individual areas of interest. All of these remarks, however, are certainly not meant to cast a bad light. They do not lessen the quality of the book in any way!

The book focuses largely on the foundational issues of the 20th century to which this book can still contribute simply because it – as already said – often (and among other things) also addresses themes that Paul Lorenzen worked on. Such work has hardly been addressed in secondary literature to date and certainly not in such lucid form.That the interactions between mathematics and philosophy seem to be dealt with from only one perspective, namely from mathematics to philosophy, cannot be explained by the author's approach. Rather it is motivated by the essence of these two disciplines: "If mathematics serves as model for philosophy, it then has most likely influenced its development through the model (or countermodel) of its way of thinking and less through individual discipline-specific results." (p.51) Of course, one could also contradict this somewhat: recall, e.g., the influence exerted by the "discovery" of the coordinate method and analytic geometry (as well as the algebraization of geometry, the old "queen of mathematics") on the thinking of, say, R. Descartes, as well as in a further sense on "rationalism" or on its "coming into being". Something similar could be said, for instance, of Hilbert's axiomatic perspective of geometry as well – somewhat later – of Wittgenstein's Tractatus philosophy. The author was essentially right, especially when he, by contrast, stated: "If philosophy had exerted any influence on mathematical thinking, then mostly on general issues and their treatment." (p.51)

In summary this book gives a "valid", comprehensive and interesting account of important aspects of the relations between mathematics and philosphy (certainly an enrichment of already existing literature). However, certain perspectivees and areas could still be dealt with in greater detail and clarity (for instance, Lakatos' role for 20th century philosophy and the understanding of mathematics as a "quasi-empirical discipline" or experimental (and computer) mathematics, modern algorithmics, semantics and systematics of formal languages, or the meaning of complexity theory. Put differently: The meaning of non-classical logics for the philosophy of mathematics, e.g., G. Chaitin's algorithmic information theory, artificial intelligence, etc. Yet let me stress once again that I do not see this as a shortcoming of the book but only as the truism already alluded to that the very general topic of the book can always only be addressed from local viewpoints. As a mathematician I am of course reviewing this book from a very special angle. However, I do think that mathematicians, in particular, will greatly profit from this book, since large portions of it are actually geared to the expert. On the other hand, the 'philosophical character' of the book make the mathematician all the more aware of the philosophical implications of his doing and the 'objects' he is working with. The professional philosopher, in turn, can gain an understanding of the importance of mathematics within his discipline. And precisely this was the author's intention!

This book pursues similar interests as those pursued by Gerhard Frey thirty years ago in his *Einführung in die philosophischen Grundlagen der Mathematik* (Introduction to the Philosophical Foundations of Mathematics). Christian Thiel's approach and his account are an excellent complement - albeit in a different terminology and from a somewhat different angle - to G. Frey's interests.

Hans-Christian Reichel

JESÚS PADILLA GÁLVEZ (ED.), *El Círculo de Viena, Reconsiderado* (special issue of *Arbor*, Madrid), December 1996

The reconsideration of the Vienna Circle is not just limited to the German- or English-speaking countries. In December of 1996 the Spanish journal *Arbor*, a publication of the *Consejo Superior de Investigaciones Científicas* (CSIC), put out a special issue on the Vienna Circle. The contributions by four Austrian and four Spanish authors, represent the proceedings of a meeting held in Madrid. In the introduction, the editor of the issue, Jesús Padilla Gálvez lists some of the desiderata of the reception of the legacy of logical empiricism in the Spanish-speaking world. To be sure, the dogmatic view of Vienna Circle thought was not limited to Spain. Padilla Gálvez, however, attributes some of the difficulties of giving an objective historiographical account to the problem of language, more specifically, to the fact that the only texts accessible to the Spanish reader have been in English, which, in many instances, was not the language in which the most important texts of the Circle were written.

All of the authors who contributed to this issue (re)consider aspects of the philosophy of science and of the philosophy of logic developed by the Vienna Circle from a historigraphical and systematic perspective. The first article by Rudolf Haller critically examines the reasons for this new interest in a movement of the past and the resistance offered by critical rationalism, history of science and ordinary language philosophy, which, ironically enough, were influenced by the Vienna Circle. Haller uses Neurath's work to focus on one specific reconsideration of the Vienna Circle, since it clearly reflects the transformation of the interpretation of logical empiricism. He also shows how the first Vienna Circle discovered the historico-systematic relativity informing any theoretical progress. Several articles in the issue deal with Carnap – Carnap as a precursor of the representationalist view of scientific theories (Juan Bautista Bengoetxea and Xabier Eizagirre), Carnap's rejection of metaphysics (Werner Sauer) and Carnap's descriptive account of metalogic vs. Gödel's arithmetic account of metalogic (Jesús Padilla Gálvez). The first article provides a structuralist reading of Carnap's representational model, developed in his "Aufbau", and relates this to recent studies in the philosophy of science. In his discussion of Carnap's rejection of metaphysics, Sauer argues that the formal criteria of analysis employed by Carnap changed between "Logischer Aufbau der Welt" und "Logische Syntax der Sprache". Padilla-Gálvez reconstructs the controversy on Kurt Gödel's proposal of an arithmetic metalogic of 1931 so as to analyze Carnap's metalogical proposal. After demonstrating how intimately related Carnap's proposal is to Alfred Tarski's, the author makes some general remarks on the recent reception of Carnap's ideas.

The Austrian historian of science Friedrich Stadler unearths a lesser known aspect of the Vienna Circle: Karl Menger's circle, the Mathematical Colloquium of 1928-1936. This colloquium, a parallel initiative to the Schlick Circle, was a group of mathematicians and logicians uniting figures such as Karl Menger, Kurt Gödel, Abraham Weld, among others, and attracting such distinguished guests as John von Neumann, Alfred Tarski, Antonio Flores de Lemus. As Stadler shows with a wealth of archival material, the Mathematical Colloquium figured significantly in the innovation and development of modern mathematics and logic as well as in philosophy of science.

The last two articles in the journal deal with issues of probability theory. Andrés Rivadulla demonstrates that Popper was ultimately unable to offer an argument that would objectively prove the impossibility of inductive probability. As the author, concludes, this does not, however, automatically establish the possibility of inductive probability. The failure of Popper's proposal is compared with problems of the Bayesian theory of induction. Schramm deals with the notions of probability discussed by the Vienna Circle, showing how their interpretation was linked with that of the limit value and that of logical interpretation in a strict sense. He shows how the discussions of Mises and Reichenbach clarified some of the difficulties in the understanding of probability.

This special issue provides worthwhile discussions that should facilitate the rediscovery of the Vienna Circle in the Spanish-speaking world. Some prior knowledge on the reader's part is required, however.

Camilla R. Nielsen

KURT BLAUKOPF, *Pioniere empiristischer Musikforschung: Österreich und Böhmen als Wiege der modernen Kunstsoziologie* (Schriftenreihe Wissenschaftliche Weltauffassung und Kunst, vol.1), Wien: Hölder-Pichler-Tempsky, 1995.

In a splendid addition to scholarship, Kurt Blaukopf extends our knowledge about early sociology by introducing us to authors not typically associated with the founding of the discipline. Normally we trace the beginning of sociology to Auguste Comte; and when we look to sociological pioneers in the sociology of aesthetics, there also is a list of "classical founders." Yet Blaukopf successfully wets our appetite to expand this list with additions presented in the auspicious initial volume of the new *Schriftenreihe* of the *Institut Wiener Kreis* by employing stories connected with Mozart leaving the employ of the Salzburg Court and establishing himself as free-lance musician in Vienna in 1781. He recalls for us the delightful episode with the critical remark of Emperor Josef to Mozart after the 1782 premiere of *Die Entführung aus dem Serail*: *"Gewaltig viel Noten lieber Mozart!"* (rendered as "Too Many Notes, my Dear Mozart" in *Amadeus*) to which Mozart replied "*Gerade so viel, Eure Majestät, als nötig ist*" ("just as many as are needed"). And with scholarly erudition, Blaukopf cites Mozart's early biographer Niemetschek on this episode who reported Mozart's enlightened reaction to this exchange as follows: "Mozart did not regard this as [the emperor's] independent judgment but as imitative chatter (*nachgesagt*)."

Kurt Blaukopf, an eminent Austrian sociologist, in *Pioniere empiristischer Musikforschung* incorporates these and other biographical details of Mozart's life and of his surviving sons and wife into an analysis of the political and social reforms introduced during the rules of Emperor Josef II (1780-90) and Leopold II (1790-92). These permitted the rise of empirical social study. For the history of the sociological perspective in music in Austria, Blaukopf traces the intellectual developments of these short years of enlightenment. They resulted in the appointment of non-Catholics (August Gottlieb Meißner) and Jews (Ignaz Jeitteles) to university chairs in aesthetics and contrary to Catholic tradition, honored men of science and liberal thought (Mozart the Free Mason included) in public settings. Blaukopf reminds us that conservative retrenchment occurred soon again in Austria. But important seeds of empirical approaches to aesthetics had been planted.

Early in the book Blaukopf turns to sociological pioneers such as Johann Friedrich Herbart (1776-1841) for whom the setting of normative "standards" was to be subject to "the controls of experience" – whether in science or aesthetics. Specifically, for a work of art to be regarded as such required that there was consensus that the work was of such stature. What these writers had in mind was that *absolute aesthetic* judgments were constrained in their relevance, if they had any at all besides for the person propounding them.

The timeliness of expanding our horizon of sociological pioneers backwards in time becomes obvious from Blaukopf's presentation of the stories about Mozart and how he connects them in this new book with concepts and material familiar to us today. In doing so he invites us to inquire into present day applications of these protosociological views to studies of audiences, of symphonic repertoires, or the roles of government in arts and music.

In the case of Mozart we are reminded of the situation so familiar to practicing musicians that there is social and economic interdepence between themselves and audiences. Without sponsors, subscription audiences, the manufacture, distribution and sale of printed music the sounds of Mozart's creations would not have been realized in performance so that they could be responded to, aesthetically, economically, social-psychologically.

When Blaukopf goes on to tell the unfamiliar history of the sociological pioneers of the Habsburg Empire we recognize how difficult, and even *subversive* of the political nature of university education and hence the established aesthetic *canon,* it was for these scholars to point to the interdependence between the political atmosphere, those holding political power, and the intellectual climate as explanations of the creation and dissemination of ideas about music. The civil servants monitoring university appointments would remove scholars from teaching posts for violating accepted views.

Times and political settings have changed – since Imperial Austria. Yet when we contrast contemporary public attitudes about the responsibilities of government for musical life in Austria with those in the United States, Europeans accord far more unchecked authority to politicians over *controlling* the arts and musical culture than is practiced in the US. While debates about financing the arts from tax revenue reverberate in Europe, their focus tends to be about the *level* and the targeted audiences and not on the very obligation of government to support the arts. That government has an obligation to promote (and regulate) the arts and culture appears to remain axiomatic to European thought.

In twenty-three chapters Blaukopf traces the Austrian heritage in the sociology of aesthetics from Mozart's contemporaries to the present. In this journey additional vignettes and biographic detail such as the forced early retirement of Professor Niemetschek, the introduction of a realistic aesthetic by Robert Zimmermann, Ernst Mach's influence towards interdisciplinary cooperation, the roles of Eduard Hanslick, Carl Stumpf, and Guido Adler in the context of the emergence of a Vienna school of the sociology of music and the pioneering foresight of the establishment of the *Phonogrammarchiv,* along with a detailed bibliography, provide fascinating reading. The book rounds out our knowledge of the intellectual background of European sociology and offers credence to Blaukopf's approving citation of Otto Neurath's claim of 1936 that "it is most appropriate to regard developments in Vienna and the other centers of the former Austro-Hungarian monarchy as a chapter in the intellectual history of Europe" (19).

This review cannot attempt to survey nor to summarize the wealth of detailed information brought to us by Blaukopf. In the opinion of this reviewer what makes this book so special is that the sociological perspective brought to bear on historical detail makes history come alive in illuminating important issues of the day. There are some books which one regrets to lay down after having read them, because one wishes that they would go on and on. Blaukopf's *Pioneers of Empiristic Musical Scholarship* is one of these.

K. Peter Etzkorn

KURT BLAUKOPF (ED.), *Philosophie, Literatur und Musik im Orchester der Wissenschaften: Wissenschaftliche Weltanschauung und Kunst,* (Schriftenreihe Wissenschaftliche Weltauffassung und Kunst, vol.2), Wien, Hölder-Pichler-Tempsky, 1996.

Some collections with discrete studies by scholars who represent a variety of disciplines induce readers to search out those chapters that till familiar soil. In such cases it is often unrewarding to guess why such congeries of contributions are bound together into one book. This cannot be said of Volume II of the Series *Wissenschaftliche Weltauffassung und Kunst* of the Institut Wiener Kreis on *Philosophie, Literatur und Musik im Orchester der Wissenschaften,* edited by Kurt Blaukopf. The editor obviously inspired his authors to contribute texts that support one another as would the members of an orchestra whose individual sounds continue to reverberate after the last downbeat of the conductor.

There are five major contributions including one by the editor Kurt Blaukopf, Wendelin Schmidt-Dengler, Herta Blaukopf, Gabriele Johanna Eder, and Kurt K. Neumann. They focus on various aesthetic, artistic, educational, philosophical and political themes which are shown to be centrally related so that readers gain a fuller understanding of the intellectual life of the late 19th and beginning 20th centuries in the Central Europe of the pre-World War I Austro-Hungarian empire. Individually and jointly they stimulate the asking of parallel questions about the future of multi-culturalism and cultural diversity in the United States or other multi-ethnic "nation" states such as the Commonwealth of Independent States (Russia) at the end of this century. Might debates about the fates of political correctness or the disputes about "the" literary canon in US universities show similarities to the history of the rise and decline of *Herbartismus* in Prague's universities and political life when no less a composer than Leoš

Janáček is shown to have been influenced by Durdík's discourse on Herbart's aesthetics?

Kurt Blaukopf's lead article "Die Orchestrierung der Wissenschaften. Otto Neuraths Konzept und die Kunstforschung" composes the framework for harmonizing the current set of interdisciplinary studies. The intellectual foresight of Bernard Bolzano adumbrated the influx of a formalized sociological perspective into scholarship dealing with the various arts *(Kunstwissenschaften)* in Vienna. Kurt Blaukopf illustrates this process by tracing it in literature, various art media, and particularly in music. Guido Adler's founding of modern musicology, Richard Wallaschek's building of bridges to ethnology serve to show how encounters with positive philosophy when extrapolated to Karl Popper's methodology bear fruit by not missing the beauty of the integration of the forest because of detailed fascination with individual trees. Referring to Friedrich Jodl and Robert Zimmermann who establish connections between art and science, Kurt Blaukopf hails the impact of the Vienna Circle in advancing linkages between the arts and sciences when he notes that members of the Vienna Circle never were hostile to artistic expressions but directed their criticism against *selbstzweckhafte Begriffsgymnastik* (hollow conceptual elaborations).

Wendelin Schmidt-Dengler "*Παιδιᾶσ χάριν*: Zu Literatur, Philosophie und Philologie im Wien der Jahrhunderwende" informs about educational objectives and conditions in the writings of father Theodor and son Heinrich Gomperz and their many followers in the Vienna Circle. They were members of what the author describes as the last generation whose intellectual culture was deeply shaped by having mastered access to antiquity through the taxing labor of the study of classical languages. The authors of "Jung Wien" still shared this tradition of humanist education. When the older Gomperz applies his thorough knowledge of the classics by reflecting on conditions concerning educational opportunities for women in the Vienna at the turn of the [last] century, he finds no support in their writing that would allow one to extrapolate from the repressed expressions of female accomplishments as depicted during the classical era to the realistic potentials of women in the contemporary world. On the contrary, Gomperz avers that empirical insights about the present were not to be repressed but would be promoted by the intensive study of the classics; for, indeed, the dialogue about the present was enriched by the rigorous encounter with minds schooled in the interdisciplinary, or pre-disciplinary, modes of classic and humanistic inquiry.

Herta Blaukopf's "Positivismus und Ideologie in der Germanistik: Aus den Anfängen der österreichischen Sprach- und Literaturforschung" connects with the previous section by following up with detail and providing further context as would a movement in a musical suite. She draws attention to the influential role of professorial appointments in modernizing scholarship in Austria by breaking up the circulus vitiosus between secondary schools and higher education. By appointing experts from outside Austria to university positions scholars who had not graduated from the local gymnasia began teaching university students. When these graduated to teach in secondary schools they now encountered professors with different educational backgrounds than their own. Herta Blaukopf, by following the career of Hermann Bonitz, documents how this former headmaster of gymnasia in Stettin and Berlin became a prominent force in reforming and establishing a specific variety of Austrian germanistic scholarship. Characteristic for this innovation is its empirical, positivistic methodology that leans on Auguste Comte and draws on Herbart's philosophy.

Through its historical depths and allusions to the present, Herta Blaukopf's chapter provides a quasi *cantus firmus* among the individual voices in this volume. Her analysis shows how the establishment of Austrian germanistics pioneered practices that remind one of strategies employed by builders of contemporary academic fiefdoms in our universities. For example, Bonitz early established a journal with a broad based readership among Austrian high school teachers (*Zeitschrift für die österreichischen Gymnasien*) that served both as a pulpit for propagation and as an organ for the defense of educational reforms. Further, an interdisciplinary association of congenial scholars was founded (*Historisches Kränzchen*) with such influential members as the future minister of education Wilhelm von Hartel, Ottokar Lorenz, and the elder Theodor Gomperz. Finally, arranging support for the official appointment of Bonitz's students Wilhelm Scherer and Richard Heinzel to other university chairs, also helped to consolidate the reforms. Scherer eventually left Austria for Strasbourg. Blaukopf engagingly presents the journeys of Bonitz, his students, the introduction of natural science methods into linguistics, and various intrigues in the academy with solid scholarship. Rare insight is offered on linkages between secondary and higher education and on Austrian cultural politics with their eventual culmination in the *Wiener Kreis*.

Gabriele Johanna Eder in "Guido Adler (1855-1941) and Alexius Meinong (1853-1920): Interdisziplinarität als Methode zur Lösung wissenschaftlicher Probleme" extracts from the give and take in the extensive private correspondence between Adler, the musicologist, and Meinong, the philosopher, how their views concerning their respective fields of scholarship changed. Eder regards the philosopher's influence on the emerging field of musicology to have been more significant than Adler's on Meinong. Musicology, for example, benefitted from Adler's examination of several of Meinong's aesthetic concepts for their applicability to music. Musicological analyses were to be distinct in their emphases on descriptive methods for capturing the specifics of musical activities which could incorporate as separate dimensions the historical setting in which listeners reacted aesthetically. The special significance of Eder's contribution to this volume could easily be overlooked since she modestly leaves to the very end the observation that neither Adler nor Meinong acknowledge in their autobiographies the scholarly interaction that she documents from their private letters. Her welcome chapter supports the admonition that scholarship on questions of intellectual relationships benefits from placing written products into their social contexts. The intriguing question, nevertheless, remains why did these scholars not acknowledge their mutual debts to each other more publicly?

The final study in the collection by Kurt K. Neumann, "Die 'Allgemeine Ästhetik' von Josef Durdík: Ihr Einfluß auf die Genese einer positivistischen 'Schönheitswissenschaft' auf die moderne Kunst," also invites comparative reflections with the present. Neumann reminds us that professors of philosophy Leopold Rembold and Bernard Bolzano were deprived of their academic freedoms as university teachers. The *venia legendi* was removed from them because of their views. Neumann provides succinct introductions to the influence of Durdík in the politicized academic climate and intellectual life of Prague. This wide-ranging chapter spans the times from his studies at Prague University, his years there as professor, his support for the appointment as professor of Thomas Masaryk (1882) and inquiry into the sources for subsequent differences of opinion with him and the eventual official disregard of his aesthetic through Stalinist party ideologues. Notwithstanding these changing fates, Neumann avers that Durdík's influence through his students and as precursor of structural linguistics qualifies him to be ranked as a prominent representative of Czechoslovakian intellectual history. Moreover, Neumann concludes that Durdík, because of his intellectual connections to Bernard Bolzano and Robert Zimmermann, and the subsequent exploration of their joint heritage by such representatives of logical positivism as Philipp Frank and Rudolf Carnap, should be included in the pre-history of the *Wiener Kreis* and, perhaps, be considered a potential founding force of a future *Prager Kreis* of philosophers analogous to the Circle of Vienna Philosophers.

The editor and the individual authors must be thanked and the publisher congratulated for having documented this phase of interdisciplinary cooperation from the perspective of the epistemology of science. We are the richer for their writings bring to the foreground aspects of the intellectual setting in which the arts and music of the *fin de siecle* were created.

K. Peter Etzkorn

HANS SLUGA/DAVID G. STERN (EDS.), *The Cambridge Companion to Wittgenstein*, Cambridge: Cambridge University Press 1996.

Ludwig Wittgenstein can probably be regarded as the most enigmatic philosopher of this century. His works have now been taken on by the renowned *Cambridge Companions* series. It is no simple task which editors Hans Sluga and David Stern have set themselves in their attempt to fulfil the goals defined in the cover text – particularly when the subject is Wittgenstein: "to use contributions from an international team of scholars to illuminate the character of the whole body of the work so that students and non-specialists can gain access to him". This in itself implies the existence of a consensus – with regard at least to the basics – on precisely this character of Wittgenstein's thinking. The existing multiplicity of Wittgenstein interpretations offers eloquent proof that such is still not the case and in fact a more modest and welcome goal of such a "companion" would be to dispose of all the repeated attempts to claim Wittgenstein for whichever interpretation is in fashion.

Both Sluga in his clear introduction and Stern in his informative contribution to "Availability of Wittgenstein's Philosophy" stress this as the main thought of their editorial activity and of their own interpretative approach. A conclusive view of Wittgenstein's work cannot be an issue here, nor indeed a presentation of defintite statements. However it is imperative to deal with the need – which is particularly marked in the case of Wittgenstein – to "rediscover one's own thoughts in his". The philosophy of the Vienna Circle and the so-called "ordinary language school", functionalism and Zen Buddhism are only a few of those who have claimed to be influenced by Wittgenstein – often without offering evidence or even a good reason.

As Sluga stresses, it cannot be denied that Wittgenstein himself tended to express himself in a deliberately ambiguous way. Misunderstandings are not inevitable given an approach which treats all philosophical questions as if they were an illness and sets as a goal the recognition that all philosophical theories are meaningless – and which at the same time is unmistakably an explanation and also wishes to be one? What philosophical relevance should be ascribed to a "therapy" which does not wish to present a solution to the philosophical problems – and still considers them to be located just beyond the written works? At first sight that must appear to be a contradiction and it is possible that Wittgenstein himself failed to achieve "both".

The authors of the *Cambridge Companion* seem to have tacitly agreed to treat these difficulties in a way which mainly involves characterizing Wittgenstein's thinking as negative; in comparison to other philosophers and philosophies and as "antithetical" to all common positions. Thus Wittgenstein's philosophy is supposed to be principally "anti-dualistic", "anti-materialistic", "anti-behaviouristic", "anti-referential", "anti-objectivistic", "anti-empirical", "anti-transcendental", etc. An approach which, in view of the Wittgenstein misunderstandings, can be extremely productive – but which can also often lead astray.

Wittgenstein's critical aim is emphasized in Robert Fogelin's article which subsumes under "referentialism", many attempts to philosophically explain intellect, reason, number, generality, against which Wittgenstein's "anti-referentialism" is supposed to have been directed. Fogelin's approach concentrates on Wittgenstein's later philosophy. However, the fact the *Tractatus* had already dealt with the idea of the rejection of the certainty of our reference to the "factual" can be understood as that which gives meaning to characters and content to thoughts, is presented in Thomas Rickett's article "Pictures, logic and the limits of sense in Wittgenstein's *Tractatus*". It principally deals with the comparison of Russell's conception of logical relations and Wittgenstein's criticism of it. This line of thought is also pursued in the thoroughly worthwhile article by Donna Summerfield about "Fitting versus tracking: Wittgenstein on Representation". In virtue of what do thoughts have content, if that what thoughts are directed to, can fail to exist?" "How is it possible for signs to point to what is not there?" These are the guiding questions on the basis of which Summerfield displays the picture-theory of thc *Tractatus* as a solution to problems which derive from what she calls "tracking" and "fitting" theories in their explanations of intentionality. This article is of interest not only because of the illuminating interpretation it provides of the picture-theory of the *Tractatus*, but also because the way in which Summerfield shows the failing of theories which the *Tractatus* is opposed to is relevant for the discussion of the problems plaguing current naturalistic and mentalistic accounts of intentionality.

"Fitting theories" which regard resemblance between thoughts and what they represent as a necessary condition for thoughts to have content threaten to get into an infinite regress if one seeks to determine a unique way in which some features of the representation, "mental images", "mental structures", would "fit" onto what they point to, so that the meaning of signs could be fixed in a distinctive way. And nothing – according to Summerfield's view – can stop that regress, which derives from the explanation "fitting-theories" offer, because about every feature of a representation, as long as it belongs to "the side of representation", one can once again raise the question of how that something "fits" onto the world.

To stop the regress without declaring something like a "brute intentionality" to be the condition of its own possibilty and without getting into the problems of "tracking theories" that presuppose the existence of the designated entities, was – according to Summerfield – the goal of the picture-theory of the *Tractatus*. And the trick she ascribes to Wittgenstein consists in distibuting both functions of signs, reference and meaning, to two sorts of signs. Propositions have meaning because they share the form of representation with what is represented. And it is just this form that cannot be interpreted further, because it is determined by a non-accidential combination of atomic objects, which names refer to. It was only by this radical distinction between names which "track" objects and don't mean them, on the one hand, and propositions with no reference but sense as a means of fitting facts, on the other hand, that Wittgenstein in his *Tractatus* was able to evade the menacing regress and to provide an answer to how propositions can be meaningful but false.

Wittgenstein himself abandoned his *Tractatus*-solution later, because he regarded the talk of absolutely simple objects as senseless and thought of rules according to which names are supposed to combine as not radically different from pictures and mental images. But he maintained his prime concern – to draw limits to what cannot be said – from his early to his later work. This is something emphasized by all authors, concerned with various questions of that later work. Three articles, all of which deal in a general way with the significance of

Wittgenstein's later philosophizing, show how Wittgenstein himself viewed his "second philosophy" as a realization of this intent.

In his essay "Philosophy as Grammar" Newton Garver argues that the basic idea of Wittgenstein's later thinking is no longer the structure of language being determined by the structure of reality but the determination of our view of reality by language. Garver calls this approach a "descriptive way" of philosophizing. He sees it as the completion of Kant's endeavors whereas Kant had been necessarily unsuccessful given the fact that he chose the wrong point of view. Unless self-critique of reason is undertaken by reflection on the means and structure of language it will not be able to avoid just the criticism it launches. Kant ultimately failed because he chose scientifically "objective" facts as his starting point. Kant's and Wittgenstein's mutual intents consist, according to Garver, in setting limits to our thinking in order to give way to morality and faith. However, by placing the emphasis on reducing Wittgenstein's earlier and later work and Kant's *Critique of Reason* to a common denominator he doesn't leave much space for differences. Yet there are distinctions made – but only in comparison to other thinkers, philosophers, linguists, and psychologists, from Aristotle to Derrida, whose writings unfortunately always were "less descriptive" than Wittgenstein's.

Bloor and Scheman also emphasize that our definitions and concepts, our language and our diction determine our view of the world, so that it is – logically – due to our agreement to speak of something as some thing or other. In his discussion of G.E.M. Anscombe's criticism of Wittgenstein's "linguistic idealism" David Bloor concludes that Wittgenstein's position as well as Hume's can only to be made accessible in a sociological way. Since it is not just any "facts of matter" determining the acceptance of what we talk about, Wittgenstein, according to Bloor, treats facts as a matter of principle in the same way sociology treats "social facts" such as money, institutions etc.: as exclusively depending on our mutual relations. Self-relatedness means construction which is to be found likewise in institutions and performative actions.

Naomi Scheman goes one step further by integrating an explanation of meaning into a conception of human life, and looking upon this as the question as to when we would be ready to feel at home with our words. She laments "our inability to recognize home when we are in it"(p. 385) and reports on different "forms of life". The author considers it to be an insight of Wittgenstein's philosophy that a certain "set of practices" can only be justified at the price of disqualifying the others, unless we could succeed in detecting commonalities among differences and in achieving mutual recognition of different ways of living.

You may ask why the author has chosen Wittgenstein and not any other philosopher. It is surely not a fact but obviously a consensus that Wittgenstein is better qualified for showmanship – Scheman also reveals to the sympathetic reader her personal motivation for her dealing with Wittgenstein: Derek Jarman's movie provided to her the importance of being able to move on ice...

An approach to Wittgenstein, that doesn't use his "negativity" as a license for supporting one's own aims, but reflects on it in comparison with one's own explanatory interests is represented by Michael Kober's remarks upon "On Certainty" und Hans Sluga's on the concept of the Self. Kober's article "Certainties of a world-picture" tries to show among other things that the "collective external idealism" Wittgenstein is said to have pursued cannot, compared to our certainties, withstand a detached skepticism.

Hans Sluga in his stimulating article "Whose house is that? Wittgenstein on the Self" presents his critical reflections upon Wittgenstein's, in his opinion, rather unsatisfactory approach on the foundation of ourselves being conscious of being subjects. It had been Wittgenstein's "anti-materialism, -cartesianism, -freudianism, and -behaviorism" that hindered him in giving a "positive account" of awareness covering one's own individuality in its development and in its distinction from others. Michel Foucault's interpretation is described as more fruitful insofar as it shares Wittgenstein's criticism of the interpretation of the self as an object that we refer to in the mind, but exceeds it with the idea of "self-formation". In Sluga's compelling account of Wittgenstein's criticism of dualism, seeing body and mind as an exclusive possibility of the reference of "I" is presented as the consequence of a thinking that takes everything that needs to be explained as a piece of physical objectivity. Thus the question how selfconsciousness ought to be explained becomes the question of how we can become conscious of, know or have a self, no matter if it is material or immaterial.

This "relation" can obviously not be clarified, if overcoming the seperation from our self is the condition for any knowledge or our identity. The farmer on a painting cound not enter the house he owns – as Sluga, quoting Wittgenstein from the *Philosophical Investigations*, begins his paper. That the self, in this sense, cannot be owned, thus remained Wittgenstein's position in the *Philosophical Investigations*. Giving full details about Wittgenstein's *Tractatus*-position of the self that nobody can have because of its forming – like the "I" that could have it – the boundaries of the world, Sluga dedicates less consideration to the context of the reflections on the use of "I" in the *Philosophical Investigations*. The use of "I" as being expressively specified by Wittgenstein and the functioning of the so-called "use as an object" of "I" was obviously meant for identifying oneself with certain statements thus show the connections between these different usages – a connection which is not made clear by Wittgenstein and whose conditions, for Sluga can only be revealed by the assessment or consciousness of ourselves.

Various contributions are concerned with Wittgenstein's comments on questions concerning the logical necessity, normativity and facticity topics which have attracted Wittgenstein all his life.

Hans Joachim Glock's essay "Necessity and Normativity" deals with Wittgenstein's account of logical necessity. The priority of this article is set up rather on the opinion that is confronted with Wittgenstein's view than on

Wittgenstein's own statements and comments on logical necessity. Glock follows W.V.O. Quine's arguments for declining necessary propositions as a special kind of propositions, to point out on the one hand that Quine's theory about the function of linguistic communication and of evolving a scientific theory is based on differentiating logical and empirical propositions. On the other hand it is shown that a critique of conventionalism – here's where Wittgenstein's and Quine's concerns meet – requires a normative interpretation of language, so as to not run the risk of hitting upon a kind of conventionalism – and therewith an inexplicability of the correct use of words. Exactly this normativity ought to be found, says Wittgenstein according to Glock, in the importance of specific propositions, not in the abstract nature of their supposed reference. In is therefore both "accidental" as subjected to changes, which propositions play the role of standards for a self-evident descripton of nature. However, the functioning of distinct propositions as norms of representation, that cannot itself be put to the test for the justification of their normative character, makes up their necessary character – differing from empirical propositions. Glock also points out that the concept "logical necessity" makes sense to Wittgenstein solely in the context of the normativity of language.

Steve Gerrard's and Cora Diamond's contributions deal with Wittgenstein's thinking on logical necessity as well. The line of argumentation of Gerrard's article is already expressed in its title "A philosophy of mathematics between two camps": Wittgenstein's views of logical necessity, of the concepts of numbers amd mathematical propositions are, – rather inconsistently – characterized as situated "between" those two poles they disapprove of: the psychological concept of mathematics on the one hand and a meaphysical one on the other: Diamond's article "Wittgenstein, mathematics, and ethics. Resisting the attraction of realism" unfortunately deals with a rather unrepresentative reading of Wittgenstein's opinion of ethics as moral realism by Sabina Lovibond.

The *Cambridge Companion* also contains contributions from Stanley Cavell and Barry Stroud, two authors who can certainly be seen as some of the most important and influential interpreters of Wittgenstein. In his article "Notes and afterthoughts on the opening of Wittgenstein's *Investigations*", which is also reproduced in Cavell's *Philosophical Passages: Wittgenstein, Emerson, Austin*, Cavell describes his own reception of Wittgenstein. Old lecture preparations, extracts from *The Claim of Reason* and *Conditions Handsome and Unhandsome* and *Wittgenstein* quotations and recent commentaries are strung together in various formats. This form is certainly intended to demonstrate the point of the article, which is also emphasized in the choice of content: "How difficult it is to make a start in philosophy!" Unfortunately this mixture of formats is thoroughly detrimental to the reading of Cavell's sensitive commentaries and observations.

The main theme of Stroud's exceptionally lucid article "Mind, meaning and practice", is a discussion of Wittgenstein's "use theory" of meaning, which is developed over considerable lengths as a response to Saul Kripke's famous interpretation. According to Stroud, the "skeptical" interpretation of

Wittgenstein's theory of meaning, as represented by Kripke and many others, rests on a basically false opposition of "use" and "facts" as a possible "measure" of understanding – which was also repeatedly denied by Wittgenstein. As Stroud points out, it is precisely this standpoint of possible "guidance" which is mistaken, since it leads by necessity to a regress: Everything which could "guide" us must itself first be understood in order to be usable as such a "signpost".

The incorporated rejection of the explanation of meaning through reduction to non-intentional facts – the regress of the interpretation makes all possible facts compatible in some way or another with the understanding of the meaning of a sign and therefore indicates exactly none – does not break down, since then the use of a specific word would be associated with no truth conditions whatever. However precisely this position gives rise to the viewpoint that the conditions for correct use are determined solely by the society game of defining correctness. For the establishment of the conformity between the individual's use of words and the usage "agreed" by society, cannot explain the actual phenomenon of meaning since it cannot provide a testable content for the alleged agreement.

Thus to interpret Wittgenstein's reference to "use" as that which explains the meaning as the opposite to there being something which can be used correctly or incorrectly, would mean that we could no longer explain the facts of our daily, practical, understanding manner of using words. How someone intends or understands an expression, or which rules exist in a society for the correct use of an expression are facts – which, however, cannot generally be adequately specified without using the same concepts with which the speaker is familiar. Here too, it seems that a regress exists, related to the one above, which – as Stroud writes – appears to be "philosophically dissatisfying or disappointing". However whatever remains to be done lies outside the realm of philosophy.

Gabriele Mras

ACTIVITIES OF THE INSTITUTE VIENNA CIRCLE

SURVEY 1997

Conferences/Lectures

International Symposium:
Alfred Tarski and the Vienna Circle – Austro-Polish Connections in Logical Empiricism
Polish Center for Sciences, Vienna, June 12 to 14, 1997

5th Vienna Circle Lecture:
Walter Thirring (Vienna)
University of Vienna, December 1, 1997

in connection with:

Lecture Series:
Results and Problems of the Exact Sciences (2nd Course)
Further speakers: Helga Nowotny, Peter Schuster
University of Vienna, December 1 to 5, 1997

Lectures:
Ernst Tugendhat (Free University, Berlin)
Rationalität und intellektuelle Redlichkeit
University of Vienna, April 16, 1997

Hubert Schwyzer (University of California, Santa Barbara)
Subjectivity in Descartes and Kant
Institute Vienna Circle, April 24, 1997

Michael Heidelberger (Humboldt-University, Berlin)
Zum Leib-Seele-Problem seit Fechner und Mach
University of Vienna, October 9, 1997

Martin Carrier (University of Heidelberg, Germany)
Empirische Hypothesenforschung ohne Felsengrund
University of Vienna, October 27, 1997

John Michael Kroiß (Humboldt-University, Berlin)
Ernst Cassirer und der Wiener Kreis
University of Vienna, November 1997

Research Projects/Workshops

Scientific World Conception and the Arts – Art, Art Theory and Studies in Art in the Scientific Debate (2nd stage)
International Workshop:
Languages of Arts – Language of Science
Institute Vienna Circle, February 2, 1997

Bilateral Project Austria-Hungary:
Philosophy of Natural Sciences
Institute Vienna Circle and Eötvös University of Budapest
One Workshop in Vienna and Budapest in Spring respectively Autumn 1997

Workshop:
The Philosophy of Imre Lakatos – Its Roots, Content and Limitations
University of Vienna, September 12 to 14, 1997

Publications 1997

Konstruktivismus und Kognitionswissenschaft. Kulturelle Wurzeln und Ergebnisse. Zu Ehren Heinz von Foersters. Ed. by Karl. H. Müller, Albert Müller and Friedrich Stadler. Wien-New York: Springer 1997 (= Publications of the Vienna Circle Institute)

Institute Vienna Circle Yearbook 5/97:
Game Theory, Experience, Rationality. Foundations of Social Sciences, Economics and Ethics. In Honor of John C. Harsanyi. Ed. by Werner Leinfellner and Eckehart Köhler. Dordrecht-Boston-London: Kluwer 1997

Wendelin Schmidt-Dengler, *Wissenschaft und Literatur der Moderne*. Wien: Hölder-Pichler-Tempsky 1997 (= Publication Series *Scientific World Conception and the Arts*, Vol. 3)

Peter Weibel, *Wissenschaft als Kunst – Kunst als Wissenschaft*. Wien: Hölder-Pichler-Tempsky 1997 (= Publication Series *Scientific World Conception and the Arts*, Vol. 4)

Library and Documentation

Acquisition of sources and research literature about the Vienna Circle and its influence, acquisition of the library of Kurt R. Fischer and of various archival materials in Austria and abroad, acquisition of ten original Isotype-Boards by Otto Neurath's son Paul Neurath (Vienna/New York).

PREVIEW 1998

International Conference:
Logical Empiricism in North America
Together with HOPOS (History and Philosophy of Science Working Group) and ZiiS (Center for International and Interdisciplinary Studies, University of Vienna)
Harvard University, Cambridge MA, USA, May 7 to 9, 1998

International Conference:
On the Nature of the Political - Hans Kelsen, Carl Schmitt, Eric Voegelin
University of Vienna, June 11, 1998

International Symposion:
Epistemological and Experimental Perspectives of Quantum Physics
Erwin-Schrödinger Institute, Vienna, September 3 to 6, 1998

Pilot Project:
International Summer University: "Scientific World Conception"

Research Projects:
Liberalism and Logical Empirism
Together with ZiiS,

Index of Names

Not included are: Notes, Figures, Illustrations, References

Vienna Circle Institute Yearbook

1. F. Stadler (ed.), Scientific Philosophy: Origins and Developments. 1993 ISBN 0-7923-2526-5
2. H. Pauer–Studer (ed.), *Norms, Values, and Society*. 1994 ISBN 0-7923-3071-4
3. W. DePauli–Schimanovich, E. Köhler and F. Stadler (eds.), *The Foundational Debate*. 1995 ISBN 0-7923-3737-9
4. E. Nemeth and F. Stadler (eds.), *Encyclopedia and Utopia*. The Life and Work of Otto Neurath (1882–1945). With the first publication of Otto's Neurath's full manuscript on "Visual Education" and the documentation of the *Otto Neurath Nachlass* (Haarlem, The Netherlands). 1996 ISBN 0-7923-4161-9
5. W. Leinfellner and E. Köhler (eds.): *Game Theory, Experience, Rationality*. Foundations of Social Sciences, Economics and Ethics. In Honor of John C. Harsanyi. 1998 ISBN 0-7923-4943-1

KLUWER ACADEMIC PUBLISHERS – DORDRECHT / BOSTON / LONDON